Guidelines for Molecular Analysis
in Archive Tissues

2.2 Preanalytical Time Interval (PATI)

The PATI can be divided into Interval I and Interval II.

2.2.1 Interval I: From the Body to the Surgical Table (Temperature 37°C or More)

This interval is also called "warm ischemia time." It depends on:

- Type of operation
- Modality of intervention
- Ability of the surgeon

Time between ligation of arteries and removal:
Time negligible for:

- Brain
- Breast
- Liver
- Lymph nodes
- Skin

Time from a minimum of 30 min to 1 h for:

- Stomach
- Colon
- Lung
- Pancreas
- Thyroid

Effects:

- Tissues are kept in hypoxic conditions at 37°C for variable time.
- Remarkable loss of RNA and antigen degradation if the time is longer than a few minutes [12, 13].
- If elimination of blood from ablated tissue is not immediate and complete, the hemoglobin coming from rapid tissue hemolysis can inhibit the downstream RT-PCR procedure.

Additional Caveats:

- Areas near heath cutting
- Treatment with Lugol solution (used for detecting the presence of starch in the cells)

2.2.2 Interval II: From the Surgical Table to the Pathology Lab

The options can be:

1. Tissues left fresh
2. Tissues immersed in formalin
3. Tissues preserved under vacuum

2.2.2.1 Tissues Left Fresh

Temperature: Room Temperature (about 20°C).

In some realities, tissues (free in a vessel or in a bag) are transferred to the grossing room. After grossing, tissue specimens can be either stored at −80°C for frozen-section histology or fixed in a fixation solution (see Sect. 2.4 and Chap. 3).

- Time interval between the surgical table and the pathology lab: relatively short, but depends on the internal organization of the hospital.
- Time in pathology lab before grossing: variable, from a few minutes up to several hours. Optimal: 30 min.
- Up to 4 h in most Pathology labs, according to Grizzle [14].

Dangers:
Effect of delay on:

- Structure
- Proteins (antigens)
- Nucleic acids

Merits:

No Fixation (material available for fresh banking)

Drawbacks:

- Drying of tissues (Even those left in the refrigerator)
- Loss of antigens and RNA related to the time spent at room temperature before grossing

Fresh Tissues Immersed in Stabilization Solutions

A possible option is the use of new tissue stabilization solutions in order to specifically recover intact RNA from fresh frozen tissue specimens. One of the most commonly used reagents of this type is RNA*later* (Applied Biosystems). It rapidly permeates tissues and inactivates RNases by precipitation, which eliminates the need to immediately freeze or formalin fix tissues after their removal. It is a practical solution because, after resection, the sample can be left in RNA*later* solution for at least 1 week at room temperature, 8 weeks at 4°C, or indefinitely at −20°C, theoretically without compromising RNA integrity. Tissue in RNA*later* can be subsequently processed as a fresh frozen sample for downstream molecular analysis. RNA purification can be performed, for example, by the standard TRIzol-based protocol (see product details at http://www.ambion.com/techlib/prot/bp_7020.pdf).

However, contrasting opinions have risen about its effective conservative properties on tissue architecture and macromolecule integrity. Some studies indicate that tissue samples stored in RNA*later* and conventionally processed for histology have excellent morphology and immunohistochemical stain [15]. Others observed uneven results and decreased reactivity using different antibodies [16]. Some studies reported that results of RNA analysis in RNAlater-treated samples are comparable to those obtained in fresh frozen tissues [15, 17]. Others suggest that freezing should still be preferred over RNA*later* treatment [18]. Recently, RNA*later* has been compared to standard tissue fixatives [16]. RNA*later* has also been tested as a pretreatment solution before alcohol-based tissue fixation, but at present the introduction of this prefixation step affects RNA quality [19].

2.2.2.2 Tissues Immersed in Formalin

Variables:

- Temperature: room temperature (generally).
- Time interval: variable from a few minutes up to days but it depends on the internal organization of the hospital.
- Formalin: penetration is initially fast (1 mm/h), but then becomes much slower (1 cm/24 h). This is followed by fixation (slow) reaching subtotal binding plateau at 24 h [20, 21].
- Fixation time: should be at least 6–8 h for 3 mm thick specimens [8, 22]. This is also applicable to fixation before and after grossing.

Merits:

In small blocks, formaldehyde rapidly affects structure, antigens, and nucleic acids (preservation/denaturation).

Drawbacks (*in large specimens*):

- Degradation continues in deep areas (not reached by the fixative)
- Frozen tissue banking is hampered
- Formalin containing vessels are heavy to carry
- Spilling of formalin may occur
- Fumes are dispersed while grossing
- Nurses refuse to handle this "carcinogen" in the surgical theater (without hoods)
- Tissue is forgotten by the surgeon because it is "already safe in formalin"

2.2.2.3 Tissues Preserved Under Vacuum

Temperature: 0–4°C.
Time interval between the surgical table and the pathology lab: up to days.

Preservation under vacuum (U.V.) is easy to handle (inside the surgical theater). Tissues are immersed (packaged) in a plastic bag (with identification label), then into the U.V. machine (relatively small, semiprofessional). In a matter of seconds, the tissue is U.V.

The bag is left in the fridge, then transferred to the pathology lab inside a plastic container (with icing devices) [11]. After grossing, the tissue specimens can be easily processed as fresh tissues or immersed in fixation solutions (see Sect. 2.4 and Chap. 3).

Merits:

- No more formalin in the surgical theater (except for small specimens, where prefilled tubes are employed)
- No spilling
- No fumes
- No drying of tissues
- Colors preserved
- Lack of insulating air around tissues allows fast cooling
- Tissues (bags) light and easy to carry
- Structure (DNA, RNA, Antigens) preserved up to days
- Banking (selective) allowed
- Demonstration of operated tissues is convincing for students and surgeons

2.2.3 Consequences of PATI on Gene Expression Levels

PATI variably affects the expression profile both at the mRNA and the protein level [2]. These expression alterations represent a biological response to the detrimental effects of PATI rather than to the pathological condition.

Evaluation of ischemia effects on differential gene expression is currently focused on the detection of genes that are up-regulated. This strategy is pursued because gene down-regulation could be incorrect due to an artifact of RNA degradation (caused, for example, by a prolonged warm ischemia time) rather than to the cell response to tissue injury.

Several studies have reported that overexpression changes are time and tissue dependent (Table 2.1). Significant changes have been found not only in hypoxia-related genes (HIF-1α, c-fos, HO-1), but also in cytoskeletal genes (i.e., CK20) and in tumor-associated antigens (i.e., CEA).

Therefore specific precautions should be taken when gene expression studies are performed on target genes that could be affected by a physiological reaction of the cell to a decrease in oxygen availability.

Table 2.1 List of genes that are significantly overexpressed in different tissues injured by PATI

Gene	Tissue	Time threshold for overexpression detection
HIF1 alpha [2]	Colon	8–10 min
c-fos [2]	Colon	8–10 min
HO-1 [2]	Colon	No increase within 30 min
CK20 [2]	Colon	25–30 min
CEA [2]	Colon	10 min
EGR1 [23, 24]	Prostate	Within 1 h
Jun-B [23, 24]	Prostate	Within 1 h
Jun-D [23]	Prostate	Within 1 h
ATF3 [23]	Prostate	Within 1 h
PIM-1 [24]	Prostate	Within 1 h
p21 [24]	Prostate	Within 1 h
Krt-17 [24]	Prostate	Within 1 h
DUSP [24]	Prostate	Within 1 h
S100P [24]	Prostate	Within 1 h
TNFRSF [24]	Prostate	Within 1 h
WFDC2 [24]	Prostate	Within 1 h
TRIM29 [24]	Prostate	Within 1 h

2.3 Fixation

Fixation is required for the preservation of the tissue specimens that will be submitted for histopathological examination. Currently, formalin is the most commonly used fixative in tissue processing because it ensures an optimal preservation of tissue morphology, although other fixatives are specifically used in some pathology departments in Europe (i.e., Bouin's solution [25]).

The quality of nucleic acids obtained from formalin-fixed and paraffin-embedded (FFPE) tissues varies enormously among laboratories. In some cases, but not in all, it is possible to extract well-preserved nucleic acids even from very old paraffin blocks [26]. Several fixation-related variables can affect the recovery of macromolecules. The failure of their optimization may result in underfixation or overfixation of the tissue, with a consequent high variability of the downstream results.

Fixation effects on macromolecule recovery depend on several elements:

- The *time of fixation* is not standardized. During the week, the histopathology lab procedures are quite standardized, but any variation can double the time of fixation and this fact is not usually reported anywhere. This can happen with the samples collected just before the weekend, or any holiday.
- The speed of fixative penetration is affected by the *type and thickness* of tissue. There is large diversity in the size of surgical specimens. Needle biopsy specimens are small (about 1.5 mm in diameter and 20 mm in length) and fix rapidly compared to larger tissue samples, while excisional specimens are 1 cm in diameter or more [7].
- The *volume of fixative* can vary widely during the handling of surgical specimens. Since the minimum formalin:tissue ratio is 10:1, the volume of fixative represents a problem for large surgical specimens, especially those weighing more than 100 g [7]. Optimally, small sections should be obtained from the fresh specimen as soon as possible and allowed to fix after dissection.
- The effect of the *type of fixative* solutions on tissues is well known, especially for formalin [27] and Bouin's solution [25]. Concerning RNA, for example, variable levels of mRNA degradation can be obtained depending on the fixative. Similarly, selective RNA component (mRNA or rRNA) degradation can be observed [28].

Several molecular methods can be used to estimate the levels of quality and quantity of macromolecules extracted from fixed tissue specimens. They are described in detail in Chap. 17.

Several alternative fixation procedures have been proposed, with the aim of improving the preservation of nucleic acids and proteins, as compared to formalin fixation [29, 30]. In general, alcohol-based fixatives such as methacarn [31] or a combination of alcoholic fixatives and microwave treatment [28, 32] reach the important goal of an improved preservation of nucleic acid integrity, face to a good compromise of morphological preservation.

2.4 Tissue Processing

Tissue processing is the stepwise replacement of fixative with alcohol (dehydration) followed by the clearing step, which replaces the alcohol with an organic solvent, usually xylene. This process is fundamental for paraffin embedding and is usually performed automatically. Parameters affecting this process include time, temperature, and the presence of vacuum [6]. It has already been reported that tissue processing can affect the recovery of nucleic acids from FFPE; longer tissue processing times seem to result in better quality RNA [6]. In the absence of an exhaustive dehydration process, residual water could be trapped in the tissue with the subsequent RNA hydrolysis [6].

References

1. Huang J, Qi R, Quackenbush J, Dauway E, Lazaridis E, Yeatman T (2001) Effects of ischemia on gene expression. J Surg Res 99(2):222–227
2. Spruessel A, Steimann G, Jung M, Lee SA, Carr T, Fentz AK, Spangenberg J, Zornig C, Juhl HH, David KA (2004) Tissue ischemia time affects gene and protein expression patterns within minutes following surgical tumor excision. Biotechniques 36(6):1030–1037
3. Bray SE, Paulin FE, Fong SC, Baker L, Carey FA, Levison DA, Steele RJ, Kernohan NM (2010) Gene expression in colorectal neoplasia: modifications induced by tissue ischaemic time and tissue handling protocol. Histopathology 56(2):240–250
4. Masuda N, Ohnishi T, Kawamoto S, Monden M, Okubo K (1999) Analysis of chemical modification of RNA from formalin-fixed samples and optimization of molecular biology applications for such samples. Nucleic Acids Res 27(22):4436–4443
5. van Maldegem F, de Wit M, Morsink F, Musler A, Weegenaar J, van Noesel CJ (2008) Effects of processing delay, formalin fixation, and immunohistochemistry on RNA recovery from formalin-fixed paraffin-embedded tissue sections. Diagn Mol Pathol 17(1):51–58
6. Chung JY, Braunschweig T, Williams R, Guerrero N, Hoffmann KM, Kwon M, Song YK, Libutti SK, Hewitt SM (2008) Factors in tissue handling and processing that impact RNA obtained from formalin-fixed, paraffin-embedded tissue. J Histochem Cytochem 56(11):1033–1042
7. Hewitt SM, Lewis FA, Cao Y, Conrad RC, Cronin M, Danenberg KD, Goralski TJ, Langmore JP, Raja RG, Williams PM, Palma JF, Warrington JA (2008) Tissue handling and specimen preparation in surgical pathology: issues concerning the recovery of nucleic acids from formalin-fixed, paraffin-embedded tissue. Arch Pathol Lab Med 132(12):1929–1935
8. Goldstein NS, Ferkowicz M, Odish E, Mani A, Hastah F (2003) Minimum formalin fixation time for consistent estrogen receptor immunohistochemical staining of invasive breast carcinoma. Am J Clin Pathol 120(1):86–92
9. Wolff AC, Hammond ME, Schwartz JN, Hagerty KL, Allred DC, Cote RJ, Dowsett M, Fitzgibbons PL, Hanna WM, Langer A, McShane LM, Paik S, Pegram MD, Perez EA, Press MF, Rhodes A, Sturgeon C, Taube SE, Tubbs R, Vance GH, van de Vijver M, Wheeler TM, Hayes DF (2007) American society of clinical oncology/college of American pathologists guideline recommendations for human epidermal growth factor receptor 2 testing in breast cancer. Arch Pathol Lab Med 131(1):18–43
10. IARC (2006) Formaldehyde, 2-butoxyethanol and 1-tert-butoxypropan-2-ol. IARC Monographs on the evaluation of carcinogenic risks to humans, vol 88
11. Bussolati G, Chiusa L, Cimino A, D'Armento G (2008) Tissue transfer to pathology labs: under vacuum is the safe alternative to formalin. Virchows Arch 452(2):229–231
12. Ohashi Y, Creek KE, Pirisi L, Kalus R, Young SR (2004) RNA degradation in human breast tissue after surgical removal: a time-course study. Exp Mol Pathol 77(2):98–103
13. Stanta G, Schneider C (1991) RNA extracted from paraffin-embedded human tissues is amenable to analysis by pcr amplification. Biotechniques 11(3):304, 306, 308
14. Grizzle WE (2001) The effect of tissue processing variables other than fixation on histochemical staining and immunohistochemical detection of antigens. J Histotechnol 24(3):213–219
15. Florell SR, Coffin CM, Holden JA, Zimmermann JW, Gerwels JW, Summers BK, Jones DA, Leachman SA (2001) Preservation of RNA for functional genomic studies: a multidisciplinary tumor bank protocol. Mod Pathol 14(2):116–128
16. Paska C, Bogi K, Szilak L, Tokes A, Szabo E, Sziller I, Rigo J Jr, Sobel G, Szabo I, Kaposi-Novak P, Kiss A, Schaff Z (2004) Effect of formalin, acetone, and RNAlater fixatives on tissue preservation and different size amplicons by real-time PCR from paraffin-embedded tissue. Diagn Mol Pathol 13(4):234–240
17. Chowdary D, Lathrop J, Skelton J, Curtin K, Briggs T, Zhang Y, Yu J, Wang Y, Mazumder A (2006) Prognostic gene expression signatures can be measured in tissues collected in RNAlater preservative. J Mol Diagn 8(1):31–39
18. Wang SS, Sherman ME, Rader JS, Carreon J, Schiffman M, Baker CC (2006) Cervical tissue collection methods for RNA preservation: comparison of snap-frozen, ethanol-fixed, and RNAlater-fixation. Diagn Mol Pathol 15(3):144–148

tissues can be considered a good possibility to standardize the tissue treatment.

Recent studies give contradictory results about the relevance of temperature and time in the preanalytical treatment of tissues and these conditions are far from being clearly determined. The major problem is related to the hypoxic conditions of the tissues, which could occur in surgical specimens when the blood vessels are closed during the surgical treatment, even before tissue ablation. Conventionally, thirty minutes are considered the limit for a conservative treatment of the tissues [1], before fixation or freezing. However, it is difficult to accept that this time is not related to gene expression changes. Small modifications of gene expression have already been reported 5 min following the excision of the tissues [2], as a result of hypoxic conditions, but the alterations are not similar for all genes [3]. So early fixation is suggested, as it happens in small biopsies for which the clinical procedures are almost optimal. In larger surgical specimens, other variables, such as the hypoxia effect during the surgical intervention, the size of the tissues, the speed of fixative penetration, and the time for reduction of tissues to small fragments for the histological examination, have to be considered. This is still an open question that needs further studies and standardization.

The policy planned and experienced in our laboratory is that Phosphate-buffered Formalin (PBF) should be employed on tissue specimens after grossing and for a limited time (from 5 h, for relatively small specimens, up to 24 h, for large specimens). The reason for this policy is linked to observations of the uneven results of nucleic acid extraction and protein analysis of paraffin-embedded tissues (PET) fixed for long times in formalin [4, 5].The length of fixation in PBF influences the quality of RNA, which is found to be better in tissues fixed for 12–24 h, while longer fixation progressively decreases the quality of biomolecules [6, 7]. In addition, an optimal preservation of determinants predicting drug responsiveness in breast cancer is now regarded as mandatory and requires a formalin fixation not exceeding 48 h [8, 9].

Formalin fixation often starts in the surgical theater, where large specimens are immersed in PBF for transfer to the Pathology Laboratory. This procedure has drawbacks, since the degree of fixation of superficial and internal areas will be different. In addition, surgical nurses dislike handling this toxic, potentially cancerogenic fluid [10].

We consider that in order to achieve the goal of standardizing fixation times, grossing has to be done on fresh tissues and that, in order to assure a uniform fixation, this has to be done on 3–4 mm thick tissue slices in cassettes and for strictly definite times: for a minimum of 5 h for small specimens and for an average of 24 h (maximum 48 h) for large specimens.

Among the preanalytical variables associated with tissue harvesting and processing, the most critical one is the transfer of tissues from the surgical theater to the Pathology Laboratory. In our opinion, time and conditions of transfer need to be controlled, as a prerequisite to standardize fixation.

In our hospital (a large regional and University Hospital), tissue specimens to be transferred from the surgical theaters to the grossing room in the Pathology department are not immersed in formalin any more, but instead processed under vacuum (U.V.) [11]. This alternative procedure has further advantages, as tissue banking, cell cultures, and gene expression profiling are still feasible in U.V.-preserved tissues.

Upon the specific request of the surgical theatre personnel, the U.V. policy has been adopted in the whole hospital, which is presently considered as formalin-free as small specimens are treated in prefilled vessels, while large blocks are transferred U.V. in plastic bags (formalin is used in the Pathology Laboratory, but under proper hoods).

No detrimental effect, either on the morphological patterns or on the immunohistochemical features or gene expression profiling, was observed for tissues kept U.V. at 4°C over the weekend (up to 72 h).

In conclusion, it is practicable and feasible to start standardizing the length of fixation in PBF, a first step for a complete standardization of preanalytical variables. As a result, we shall achieve a more reliable evaluation of antigenic and genetic parameters, which nowadays represents a mandatory requirement in histopathological diagnoses.

In the following section, the detailed description of the technical problems connected with the preanalytical time interval (PATI) and with the formalin fixation step is presented. In the context of fixation, the use of fixatives alternative to formalin is reported.

Preanalytical Time Interval (PATI) and Fixation

2

Lorenzo Daniele, Giuseppe D'Armento, and Gianni Bussolati

Contents

L. Daniele (✉), G. D'Armento, and G. Bussolati
Department of Biomedical Sciences and Human Oncology,
University of Turin, Turin, Italy

2.1 Introduction and Purpose

Histopathological analysis (i.e., microscopical examination of tissue sections leading to related interpretation and diagnosis) on formalin-fixed and paraffin-embedded (FFPE) specimens is the goal of routine diagnostic histopathology. Owing to the recent advances in molecular approaches, however, this clinical material has also become a precious source of macromolecules for genomics and proteomics studies.

Before being analyzed, human tissues are submitted to a series of essential treatments that can directly affect their use in downstream molecular applications. All such procedures fall under the designation of "preanalytical treatment." Basically, we should consider the following steps:

- *Surgical removal of tissue*
- *Grossing (in the pathology lab)*
- *Fixation, to be followed by paraffin embedding, or*
- *Freezing, to be followed by tissue banking*

While grossing, paraffin embedding, freezing, and tissue banking are well defined in the literature, our aim is to reach a standardized definition of the time interval between surgical removal and tissue grossing and fixation, keeping in mind that the ultimate goal is to obtain an optimal preservation of morphological structure, nucleic acids, and proteins.

Preanalytical treatment of tissues is one of the most variable and debated questions in molecular analysis. Information about the time between ablation of tissues and tissue fixation or the time and the temperature of fixation procedures is hardly ever available. On the other hand, early fixation of the

G. Stanta (ed.), *Guidelines for Molecular Analysis in Archive Tissues*,
DOI: 10.1007/978-3-642-17890-0_2, © Springer-Verlag Berlin Heidelberg 2011

have found their way into routine clinical practice. The main reason for this slow adoption of RNA as an analyte for routine testing is that many RNA markers have been identified for fresh-frozen tumour material, which harbours good-quality RNA. However, this is not the material routinely obtained in hospitals. We can also consider that fixation can be a sort of standardization of tissue treatment, easily and highly improvable in the hospitals as reported in the next chapters. Advantages and disadvantages of FFPE tissues compared with fresh-frozen tissues are reported in Table 1.1.

One of the major reasons to perform molecular analysis in archive tissues is that in clinical routine, all human tissues are fixed and paraffin embedded. It could be a hard task to change the routine in hospitals, because this routine also means standardization of the clinical procedures. A histological diagnosis is necessary to define the lesion and to be able to perform an efficient microdissection. Furthermore, for any clinical molecular determination, histological confirmation is absolutely indispensable.

Another reason in favour of the use of FFPE tissues for molecular analysis is related to the fact that increasingly more often in clinical practice, the tissues taken from patients for diagnosis are very small, and they represent the only tissues available for molecular analysis.

Table 1.1 Comparison between FFPE and fresh-frozen tissues

	FFPE tissues	Fresh-frozen tissues
Availability	High number	Limited number
Clinical information	Sometimes retrospective long-term clinical information	Often short-time clinical information
Morphology	High quality	Low quality
Microdissection	Accurate	Less accurate
Variability range	Representative of large clinical variation	Low level of variability because of the limited number of cases
Follow-up	Very long	Often very short
DNA	Degraded, but can be analyzed	Well preserved
RNA	Degraded, but can be analyzed also at the quantitative level	Well preserved
Proteins	Cross-linked/new effective methods of analysis are available	Well preserved

Reference

1. Bevilacqua G, Bosman F, Dassesse T, Hofler H, Janin A, Langer R, Larsimont D, Morente MM, Riegman P, Schirmacher P, Stanta G, Zatloukal K, Caboux E, Hainaut P (2010) The role of the pathologist in tissue banking: European consensus expert group report. Virchows Arch 456(4):449–454

Archive Tissues

Giorgio Stanta

1

Content

G. Stanta
Department of Medical Sciences,
University of Trieste, Cattinara Hospital,
Strada di Fiume 447, Trieste, Italy

In clinical practice, a large number of tissue samples are removed from patients for several purposes. Biopsies are performed to better define a clinical diagnosis, while a surgical therapy is decided to treat the patient with the ablation of tissues. The tissue specimens are fixed as soon as possible to avoid autolysis and putrefaction, and then they are paraffin embedded to allow very thin sections to be cut for histological examination. These tissues are usually formalin fixed and paraffin embedded (FFPE). After a few sections are cut from the paraffin block, most of the tissues are stored in hospital pathology archives for decades and for this reason they are called "archive tissues" (AT). It is possible to estimate that every year in Europe 25–30 million cases of tissues taken for biopsy or surgery procedures are treated in hospitals, and in many cases these are multiple tissue samples. This huge number of tissues represents the largest available collection of human material. This material is also matched with the clinical records and represents a giant virtual bio-banking system in which even rare diseases with sometimes long follow-up periods are immediately available for translational research.

These are the only tissues suitable to tackle in a comprehensive way the vast heterogeneity of human diseases and to take clinics to targeted medicine. They can be used for the discovery phase of new biomarkers/therapy targets, for their validation, or for their implementation into clinical practice [1]. Normal tissues, either from the component surrounding the lesion or from negative biopsies, are also available. The use of archive tissues in clinical research can greatly accelerate the process of translation of the basic knowledge in molecular biology to molecular medicine. Although the use of expression profiling for identifying new molecular markers to help diagnosis and guide treatment of cancer has been an intensive field of research in recent years, only a few molecular markers

Part I

Archive Tissues

SD	Standard Deviation
SDS	Sodium Dodecyl Sulphate
Sec	Seconds
SELDI	Surface-Enhanced Laser Desorption and Ionization
SILAC	Stable Isotope Labelling With Amino Acids In Cell Culture
SNP	Single Nucleotide Polymorphism
SOP	Standard Operating Procedures
SSC	Saline-Sodium Citrate Buffer
TAE	Tris base, Acetic Acid, and EDTA
TBE	Tris-Borate-EDTA
TBS	Tris Buffered Saline
TBST	Tris Buffered Saline with Tween
TCA	TrichloroAcetic Acid
TE buffer	is derived from its components: "Tris", a common pH buffer, and "EDTA"
TEMED	N,N,N',N'-e-*Tetramethylethylenediamine*
TFA	Trifluoroacetic Acid
TMA	Tissue Microarray
TOC	Total Organic Carbon
TOF	Time of Flight
TRITC	Sheep Antidigoxigenin-Tetramethyl Rhodamine Isothiocynate
TSG	Tumor Suppressor Genes
U	Unit of enzyme activity
UDG	Uracil DNA Glycosylase
UNG	Uracil N-Glycosilase
U.V.	Under Vacuum
U.V.	Ultraviolet Light
w/w	weight in weight percentage

LOH	Loss of Heterozygosity
MALDI	Matrix Assisted Laser Desorption/Ionization
mCGH	metaphasic-Comparative Genomic Hybridization
mDNA	mitochondrial DNA
MEGA	N-Decanoyl-N-MethylGlucamine
MGB	Minor Groove Binder
min	Minutes
MINT	Methylated In Tumour, loci
miRNA	MicroRNA
MLPA	Multiplex Ligation-Dependent Probe Amplification
MMLV enzyme	Moloney Murine Leukemia Virus Reverse Transcriptase enzyme
MMR	Mismatch Repair
MOPS buffer	3-(N-morpholino)propanesulfonic acid buffer
MS	Mass Spectrometry
MS	Microsatellite Sequences
MSA	Microsatellite Analysis
MSI	Microsatellite Instability
MS-MLPA	Methylation Specific Multiplex Ligation-Dependent Probe Amplification
MSP	Methylation-Specific PCR
NAS	Non Absorbing Substrata
NCI	National Cancer Institute
NSCLC	Non Small Cell Lung Cancer
NT	Nick Translation
NTC	No Template Control
OD	Optical Density
OECD	Organization for Economic Cooperation and Development
OGP	n-Octyl b-D-GlucoPyranoside
on	overnight
PAGE	Polyacrylamide Gel Electrophoresis
PATI	Pre-analytical Time Interval
PBF	Phosphate-buffered Formalin
PBL-DNA	Peripheral blood leukocyte DNA
PBS	Phosphate Buffered Saline
PCR	Polymerase Chain Reaction
PET	Paraffin Embedded Tissue
PMR	Percentage of Methylated Reference
PVDF	Polyvinylidene Difluoride
QC	Quality Control
qPCR	Quantitative PCR
RCF	Relative Centrifugal Force
RIN	RNA Integrity Number
rpm	revolutions per minute
rt	Room Temperature
RT-MLPA	Reverse Transcriptase-Multiplex Ligation-Dependent Probe Amplification
RT-PCR	Reverse Transcription-Polymerase Chain Reaction
SAM	S-Adenosyl Methionine

dd water	double distillated water
DEPC	Diethylpyrocarbonate
DHPLC	Denaturing High Performance Liquid Chromatography
DIG	Digoxigenin
DLU	Digital Light Units
dNTP	deoxy Nucleotide Triphosphate
DTT	Dithiothreitol
dTTP	deoxy Thymidine Triphosphate
dUTP	deoxy Uridine Triphosphate
EDTA	Ethylenediaminetetraacetic Acid
ELISA	Enzyme-Linked Immunosorbent Assay
EMEA	European Medicines Agency
EMQN	European Molecular Genetics Quality Network
EPS	Emergency Power System
EQA	External Quality Assessment
ESI	Electrospray Ionization
EtBr	Ethidium Bromide
EtOH	Ethanol
EXB	Extraction Buffer
FA	Formamide
FBS	Fetal Bovine Serum
FDA	Food and Drug Administration
FFPE	Formalin Fixed Paraffin Embedded Tissue
FISH	Fluorescence In Situ Hybridization
FITC	Fluorochrome Conjugated Avidin-Flourescein Isothiocyanate Antibody
GCP	Good Clinical Practice
GCLP	Good Clinical Laboratory Practice
gDNA	Genomic DNA
GIST	Global Internal Standard Technology
GLP	Good Laboratory Practice
H&E	Hematoxilin and Eoxin
HIER	Heat Induced Epitope Retrieval
HM	Homozigous
HNPCC	Hereditary Non-Polyposis Colorectal Cancer (Lynch syndrome)
HPLC	High Performance Liquid Chromatography
HRM	High Resolution Melting
HRP	Horseradish Peroxidase
HT	Heterozygous
ICAT	Isotope Coded Affinity Tags
IHC	Immunohistochemistry
IPC	Internal Positive Control
IQC	Internal Quality Control
ISH	In Situ Hybridization
ISO	International Organization for Standardization
ITO	Indium-Tin-Oxide
Kb	Kilobases
LCM	Laser Capture Microdissection
LNA	Locked Nucleic Acid

Abbreviations

1-D SDS PAGE	one Dimensional Sodium Dodecyl Sulphate polyacrylamide gel electrophoresis
1-DE	one Dimensional Electrophoresis
Ab	Antibody
a-CGH	array-Comparative Genomic Hybridization
ACN	Acetonitrile
ACS	American Chemical Society
ALL	Acute Lymphoblastic Leukemia
AML	Acute Myeloid Leukemia
AMV enzyme	Avian Myeloblastosis Virus Reverse Transcriptase enzyme
APS	Ammonium Persulfate
AS	Absorbing Substrata
AT	Archive Tissue, fixed and paraffin embedded tissues
BARQA	British Association of Research Quality Assurance
BEC	Blanck Extraction Control
BM	Bisulfite Modification
bp	Base Pair
BSA	Bovine Serum Albumin
CAP	College of American Pathologists
cDDP	cis-diamminedichloroplatinum, Cis-platinum
CEP	Chromosome Enumeration Probes
CEPH	Centre d'Etude du Polymorphisme Humain
CGH	Comparative Genomic Hybridization
CHAPS	3-[(3-Cholamidopropyl)dimethylammonio]-1-ropanesulfonate
CHCA	α-Cyano-4-hydroxycinnamic acid
CIMP	CpG Island Methylator Phenotype
CIN	Chromosomal Instability
CISH	Chromogenic In Situ Hybridization
CLIA	Clinical Laboratory Improvement Amendments
CpG	Cytosine–phosphate–Guanine dinucleotide
CPM	Counts Per Minute
CRC	Colorectal Cancer
Ct	Cycle Threshold
CTRL	Control
Da	Dalton
DABCO	1,4-diazabicyclo2.2.2octane
DAPI staining	4,6-diamidino-2phenylinodole dihydrochloride

Contents

Also a chapter dedicated to forensic methods has been added. Since nucleic acid degradation is a common problem with FFPE tissues, useful suggestions can be obtained also from this issue together with specimen identification through DNA analysis.

Most of the methods here reported are more or less similar to those used in fresh cells and tissues, but the modifications made in the protocols are necessary to obtain a positive result in human FFPE tissues.

In the intention of the authors all the methods and protocols are described for a laboratory direct use, and the addition of explanatory notes should help even less experienced researchers, but a basic experience of laboratory research methods is required. If during the reading and practical use of the protocols errors or unclear and insufficient explanations are found, please contact directly the IMPACTS Group by e-mail (secretary@impactsnetwork.eu).

Bioethical norms are of pivotal importance in the use of human tissues for research, but we avoided ethical considerations because of the lack of uniformity in the directives on clinical residual tissue research in European countries. The only general suggestion is to contact the local ethical committees.

I would like to thank very much all the contributors of the IMPACTS group that collaborated not only in the preparation of the chapters but also in the multicentric project of methods validation, and took part in the extensive discussions held within the IMPACTS meetings.

I would like to thank Serena Bonin, who has worked with me for the past fifteen years, for her continuous interest and effort in developing molecular methods in archive tissues.

I would like to specially mention Isabella Dotti for the work done in the preparation of the DNA and RNA methods, and Valentina Faoro for the groundwork and the assembling of the proteomics chapters.

This book could have not been edited without the continuous effort of Valentina Melita, Danae Pracella, and Renzo Barbazza who dedicated a lot of time to reading, correcting, and improving the comprehension and the presentation of the protocols.

Trieste, January 2011   Giorgio Stanta

Preface

These guidelines are devoted to disseminate molecular methods that can be used to analyze DNA, RNA, and proteins in formalin-fixed and paraffin-embedded (FFPE) tissues. We have called them guidelines because this is the first method book that is entirely dedicated to archive tissues. They are addressed to pathologists, biologists, and biotechnologists who do research in FFPE tissues. The protocols presented in these guidelines derive from the experience of the laboratories participating in the European project "Archive tissues: improving molecular medicine research and clinical practice" – IMPACTS (www.impactsnetwork.eu). The 20 participants, from 11 European countries, are some of the most experienced groups in molecular analysis of fixed and paraffin-embedded human tissues. Among them are some of the groups that developed this type of molecular analysis for the first time.

These guidelines are mostly dedicated to the homemade protocols that were developed, tested, and even validated by multicentric analyses performed by the IMPACTS groups. Experience in basic molecular methods is necessary to develop molecular pathology activities in human tissues. This allows research without too many technical limitations. In this way the researcher is also capable to better evaluate commercial kits and validate them. Also commercial kits are reported in some protocols, especially when the authors have consolidated habits in kit utilization or when advanced techniques are used without being directly developed by the participants. Thus the reported commercial reagents and kits reflect in the same way the experience of the IMPACTS laboratories. However, in our experience standardization of molecular methods by commercial kits is not sufficient to guarantee reproducibility; good laboratory practice is absolutely necessary to obtain reliable results. Sometimes a slightly different "lab jargon" is maintained in the different chapters; this reflects the authors' consuetude but it does not compromise its clarity. Basic footnotes are repeated in many chapters to make it easy to access the single protocol.

We recognize that semi- or fully automated instruments for nucleic acids extraction from FFPE could be very useful in establishing standardization and method reproducibility. In the nearest future they will represent an important task for any molecular pathology laboratory devoted to routine diagnostic molecular analyses in FFPE. However, we did not mention them in these guidelines, because we did not have the chance to compare the performances of the different commercial options.

These guidelines are divided into different parts related to tissue processing and macromolecule preservation, molecular analysis of DNA, RNA, and proteomics, and a short chapter about internal quality control. Small chapters about the new developing technologies are also included, these are often confined to the major research centres, but, as past experience has showed us, they sometimes diffuse very quickly.

Editor
Prof. Dr. Giorgio Stanta
Department of Medical Sciences
University of Trieste
Cattinara Hospital
Strada di Fiume 447
Trieste
Italy
stanta@impactsnetwork.eu

ISBN 978-3-642-17889-4   e-ISBN 978-3-642-17890-0
DOI 10.1007/978-3-642-17890-0
Springer Heidelberg Dordrecht London New York

Library of Congress Control Number: 2011925943

Cover design: eStudioCalamar, Figueres/Berlin

Printed on acid-free paper

Springer is part of Springer Science+Business Media (www.springer.com)

Giorgio Stanta
(Editor)

Guidelines for Molecular Analysis in Archive Tissues

 Springer

19. Benchekroun M, DeGraw J, Gao J, Sun L, von Boguslawsky K, Leminen A, Andersson LC, Heiskala M (2004) Impact of fixative on recovery of mRNA from paraffin-embedded tissue. Diagn Mol Pathol 13(2):116–125

20. Fox CH, Johnson FB, Whiting J, Roller PP (1985) Formaldehyde fixation. J Histochem Cytochem 33(8):845–853

21. Helander KG (1994) Kinetic studies of formaldehyde binding in tissue. Biotech Histochem 69(3):177–179

22. Goldstein NS, Hewitt SM, Taylor CR, Yaziji H, Hicks DG (2007) Recommendations for improved standardization of immunohistochemistry. Appl Immunohistochem Mol Morphol 15(2):124–133

23. Dash A, Maine IP, Varambally S, Shen R, Chinnaiyan AM, Rubin MA (2002) Changes in differential gene expression because of warm ischemia time of radical prostatectomy specimens. Am J Pathol 161(5):1743–1748

24. Schlomm T, Nakel E, Lubke A, Buness A, Chun FK, Steuber T, Graefen M, Simon R, Sauter G, Poustka A, Huland H, Erbersdobler A, Sultmann H, Hellwinkel OJ (2008) Marked gene transcript level alterations occur early during radical prostatectomy. Eur Urol 53(2):333–344

25. Bonin S, Petrera F, Rosai J, Stanta G (2005) DNA and RNA obtained from bouin's fixed tissues. J Clin Pathol 58(3):313–316

26. Iesurum A, Balbi T, Vasapollo D, Cicognani A, Ghimenton C (2006) Microwave processing and ethanol-based fixation in forensic pathology. Am J Forensic Med Pathol 27(2):178–182

27. Srinivasan M, Sedmak D, Jewell S (2002) Effect of fixatives and tissue processing on the content and integrity of nucleic acids. Am J Pathol 161(6):1961–1971

28. Dotti I, Bonin S, Basili G, Nardon E, Balani A, Siracusano S, Zanconati F, Palmisano S, De Manzini N, Stanta G (2010) Effects of formalin, methacarn, and fineFIX fixatives on RNA preservation. Diagn Mol Pathol 19(2):112–122

29. Cox ML, Schray CL, Luster CN, Stewart ZS, Korytko PJ, MK KN, Paulauskis JD, Dunstan RW (2006) Assessment of fixatives, fixation, and tissue processing on morphology and RNA integrity. Exp Mol Pathol 80(2):183–191

30. Lykidis D, Van Noorden S, Armstrong A, Spencer-Dene B, Li J, Zhuang Z, Stamp GW (2007) Novel zinc-based fixative for high quality DNA, RNA and protein analysis. Nucleic Acids Res 35(12):e85

31. Puchtler H, Waldrop FS, Meloan SN, Terry MS, Conner HM (1970) Methacarn (methanol-carnoy) fixation. Practical and theoretical considerations. Histochemie 21(2):97–116

32. Stanta G, Mucelli SP, Petrera F, Bonin S, Bussolati G (2006) A novel fixative improves opportunities of nucleic acids and proteomic analysis in human archive's tissues. Diagn Mol Pathol 15(2):115–123

Formalin-Free Fixatives

3

Isabella Dotti, Serena Bonin, Giorgio Basili, and Valentina Faoro

Contents

3.1 Introduction and Purpose

Currently, formalin-fixed and paraffin-embedded (FFPE) archival specimens represent the most abundant resource of human tissues for clinical research. Formalin fixation is the standard procedure for routine histological examination of tissues because it rapidly and permanently preserves the tissue architecture by protein cross-linking [1]. Several staining and immunohistochemical protocols have already been validated in formalin-fixed tissues for diagnostic and prognostic purposes. In the last few years, formalin-fixed specimens have also become amenable for molecular studies at the DNA, RNA, and protein level, thanks to the development of specific procedures for extraction of nucleic acids [2–5] and, more recently, of proteins [6, 7] (see Part XI for more details). Well-established molecular approaches (FISH, endpoint and real-time amplification, western-blot…), combined with innovative microdissection procedures (laser capture microdissection, LCM), are now widely used for qualitative and quantitative detection of macromolecules in already existing archives of tissues [4, 8–13]. The use of conventional formalin for tissue fixation, however, presents some limitations that interfere with translational studies. Recovery of nucleic acids and proteins of high quality and in good quantity from FFPE tissues is still a challenge. Indeed, formalin cross-links nucleic acids and proteins, and modifies nucleic acids by the addition of monomethylol groups to the bases, and therefore impairs extraction efficiency and quality of macromolecules [14, 15]. Changes in formalin concentration, temperature, and pH can also contribute to the modification of macromolecules, reducing their accessibility to molecular studies [16, 17]. The time of fixation has not been definitively standardized.

I. Dotti (✉), S. Bonin, and G. Basili
Department of Medical, Surgical and Health Sciences,
University of Trieste, Cattinara Hospital, Strada di Fiume 447,
Trieste, Italy

V. Faoro
Department of Medical, Surgical and Health Sciences,
University of Trieste, Trieste, Italy

G. Stanta (ed.), *Guidelines for Molecular Analysis in Archive Tissues*,
DOI: 10.1007/978-3-642-17890-0_3, © Springer-Verlag Berlin Heidelberg 2011

For formalin fixation, this variable is particularly important as its speed changes during fixative penetration, and the number of cross-links among proteins increases over time. All these factors result in limitations in the reproducibility and reliability of the results, no matter what technical approaches are applied to formalin-fixed tissues. Importantly, formalin is also highly toxic and carcinogenic, and its use is allowed only in a fume hood.

Alternatiive fixatives have been proposed to overcome the limitations of the use of formalin. Among them there are several alcoholic and nonalcoholic solutions (e.g., ethanol [18], methanol [19], acetone [20]), mixtures of fixatives (Carnoy [21], methacarn [22], Bouin's solution [23], the zinc-based fixatives [24]), and commercial new fixatives (e.g., FineFIX [25], UMFIX [26], RCL-2 [27], HOPE [28]), Alcolin [Diapath, www.diapath.com] (see Table 3.1 for the characterization of some of them). In order to be considered good substitutes of formalin, these fixatives should provide an optimal compromise between good tissue histology for the morphological examination, good immunoreactivity for immunohistochemical analysis, and good macromolecule preservation for downstream molecular analyses. The advantages of preserving morphology and macromolecules are even better appreciated when microdissection (mechanical or laser-assisted) of specific cell populations is included in the analysis. The use of fixatives with these characteristics will increase the applicability of clinical samples in translational studies.

3.2 Morphologic and Immunohistochemical Analysis

Histological examination by microscopic observation of formalin-fixed tissue sections represents the standard procedure to which pathologists from any hospital are accustomed. For this reason, a fixative can currently be considered a potential substitute of formalin in this kind of analysis if it offers tissue morphology as similar as possible to that provided by the conventional aldehyde-based fixation.

Recent studies evidenced that tissues are adequate for histological examinations, giving results almost comparable to formalin fixation, when fixed with laboratory-prepared alcoholic fixatives such as 70% ethanol, Carnoy's, methacarn [27, 29–32] or, less consistently, with acetone [33].

New commercial fixatives (e.g., FineFIX, UMFIX and RCL2) have also been tested and their conservative properties in tissue histology have been reported [25–27]. The recent Alcolin fixative represents another valid alternative to formalin, since it has already shown promising results in some of the pathology laboratories involved in the IMPACTS project [30].

It has been observed that histology evaluation may be variably improved by microwave-assisted fixation and processing according to the fixative [25, 30, 34] and the tissue under investigation [35].

Alcohol-based fixatives also ensure immunohistochemical results comparable to those obtained with formalin fixation. This has been seen for both laboratory-made alcohol-based fixatives, such as 70% ethanol, Carnoy's solution and methacarn [27, 29, 33, 36], and patented new fixatives (FineFIX, UMFIX, RCL2 and Alcolin) [25, 27, 37].

3.3 Molecular Analysis

The recovery of intact molecules from fixed and embedded tissues is essential for the analyses at the DNA, RNA, and protein level as the quality of the starting material highly affects the reproducibility of the assay and the reliability of the results. Moreover, a good DNA, RNA, and protein yield is also desirable when analyses are performed on limited archival samples, such as small biopsies or microdissected specimens. Several studies have shown that protein-precipitating alcoholic reagents yield better results than aldehyde-based fixatives in terms of quality and quantity of macromolecules. This happens because alcoholic fixatives are precipitating agents that neither cross-link proteins nor induce chemical changes in nucleic acids.

In terms of DNA and RNA recovery, it has been shown that extraction efficiency and integrity of nucleic acids isolated using alcohol-based fixatives are in many cases comparable to those of unfixed samples. Thanks to their preservative properties, alcohol-based fixatives represent the ideal substitutes to formalin fixation in several molecular studies (methylation studies, SNPs

Table 3.1 Comparison between conventional formalin and some alcohol-based fixatives

	Formalin	Methacarn	FineFIX	UMFIX	RCL2	Alcolin
Fixation time	Routinely for 12–48 h at room temperature [45]	Variable condition reported for about 24 h at RT	Up to 50 min in microwave-assisted procedure [44]	For 1 h–24 h at RT depending on the thickness of the sample [37]	For 24 h either at RT or at 4°C	For 24–48 h at RT
Fixation mechanism	Methylene bridges formation [1]	Dehydration with protein precipitation/coagulation [14]	Dehydration (with protein coagulation)	Dehydration (with protein coagulation)	Dehydration (with protein coagulation)	Dehydration (with protein coagulation)
Toxicity	High	High	Not present	Not present	Not present	Not present
Morphology	Well-known morphology	Low level of shrinkage [27]	Morphology is preserved and is similar to formalin-fixed tissues [25]. Some coarctation can be present depending on the sample	Morphology is preserved and similar to formalin-fixed tissues [26, 37]	Morphology is preserved and similar to formalin-fixed tissues [27]. Some coarctation can be present depending on the sample	Morphology is preserved and similar to formalin-fixed tissues. Some coarctation can be present depending on the sample
IHC	Formation of methylene bridges can oppose antigen-antibody binding	It is reported that antigen retrieval is not required [15]	Depending on the antibody, sometimes higher intensity of signal	It is reported that the immunoreactivity for a number of antibodies is slightly stronger: the antigen retrieval is not necessary [26]	Immunoreactivity very similar to that seen in formalin-fixed samples. Optimization of the immunostaining procedures is required due to the high antigen preservation [27]	To be evaluated
DNA and RNA integrity	Chemical modification and degradation. DNA up to 400 bp: RNA up to 300 bases [4, 46]	Excellent fixative for preserving tissue RNA. DNA up to 1,900 bp [33]; RNA up to 700 bases [14]	Good quality. DNA up to 2,400 bp; RNA up to 600 bases [25]	DNA and RNA quality comparable with those extracted from fresh tissues [26]	Good-quality. DNA up to 850 bp [27]; RNA up to 600 bases (data not published yet)	Good-quality. RNA up to 600 bp (data not published yet)
Protein integrity	The protein yield is influenced by the fixation protocol and the age of the sample [47]	Good quality comparable with proteins extracted from fresh frozen tissues [22]	Good quality comparable with proteins extracted from fresh frozen tissues [25]	Good quality comparable with proteins extracted from fresh frozen tissues [26]	Good quality comparable with proteins extracted from fresh frozen tissues [42]	Good quality comparable with proteins extracted from fresh frozen tissues (data not published yet)
Costs	Low	Low	High	High	High	High

In the table, the recommended use, the performance in the preanalytical step, and the effects of each fixative on DNA, RNA, and protein detection are described. Some of the information reported in the table comes from the direct experience of the laboratories involved in the IMPACTS project

detection assays, mRNA quantification by real-time PCR, gene expression profiling…). Among the alcoholic fixatives, an easily accessible fixative that allows performance of the most reliable molecular analyses is the laboratory-prepared methacarn [14, 38, 39]. Due to its toxicity, however, methacarn is not so diffused in pathology laboratories and is generally substituted by other nontoxic alcohol-based solutions. Among the home-made solutions, ethanol is the most adequate, as it allows recovery of high-quality nucleic acids in sufficient quantity for several downstream analyses [31]. Among the commercial fixatives, it has been found that FineFIX, UMFIX, and RCL2 offer better results than formalin in terms of DNA and RNA yield and integrity [25–27, 30, 40, 41].

Similarly, alcohol-based fixatives seem a good alternative to formalin also in proteomics studies. In formalin-fixed specimens, the most critical step is protein extraction and preparation since the fixation process can modify the protein structure. Several proteomics studies have been reported on archival tissues fixed with alternative fixatives and the data obtained so far reveal that alcoholic fixatives can undoubtedly preserve proteins better than conventional formalin treatment. This has been observed both for laboratory prepared solutions, such as ethanol, methacarn, and zinc-based fixatives [14, 22, 24, 40], and for patented fixatives, such as FineFIX, UMFIX, and RCL2 [25–27].

Microwave fixation and/or processing can have variable effects on macromolecule preservation [25, 30]. In any case, temperature energy range must be empirically optimized according to the fixative and the analyzed tissue. In recent proteomics studies we have noticed that the quality of protein lysates, especially from fatty tissues, is remarkably improved using a step of dehydratation-clearing process. It consists of microwaving of biopsies in reagent-grade ethanol for several minutes followed by JFC solution[1] (Milestone, Bergamo, Italy) and wax impregnation, according to Milestone instructions. The use of isopropanol instead of JFC is cheaper but limits the extraction of proteins.

As reported in recent studies, reproducibility of results at the DNA, RNA, and protein levels proves the applicability of LCM in alcohol-fixed tissue specimens, using both laboratory and commercial solutions [27, 39, 42, 32, 43]. Many of the proposed fixatives present characteristics that could be useful at the clinical level in the future, but they need further validation and standardization.

3.4 Technical Considerations that Can Be Adapted to the New Alcohol-Based Fixatives

The reported considerations are based mostly on the experience developed with the use of FineFix.[2]

- DNA and RNA extraction is performed according to the protocols described in Chaps. 7, 8, 9 and 12, with a slight modification in the RNA isolation procedure. As alcohol-based fixatives do not induce protein cross-linking, the Proteinase K digestion step for RNA purification can be carried out in 3 h instead of overnight.
- Quantity assessment of nucleic acids: it can be performed according to the protocol described in the specific chapter (Chap. 16).
- Quality assessment of nucleic acids (see Chap. 17): as for FFPE samples, so for alternative fixatives, the methods based on rRNA integrity assessment are not informative. Fragmentation patterns are similar to those obtained from FFPE samples; for this reason, the use of ribosomal ratio and RIN values is not recommended.
- PCR and quantitative PCR, both endpoint and real time, can be performed with alternative fixatives following the same rules and precautions used when FFPE tissues are considered. The use of FineFIX solution, however, may induce underestimation of the results when gene expression analysis is performed. This phenomenon can be due to the presence of inhibiting components in FineFIX solution that affect reverse transcription step [44].
- Microsatellite Instability Analysis (MSI) analysis: it can be performed according to the protocol described in the specific section of the guidelines (see Chap. 28).
- Methylation analysis: DNA extraction and methylation analysis are performed following the protocols related to FFPE tissues (see Chap. 30 and 31).

---

[1]The JFC is a solution patented by Milestone; it contains a mixture of ethanol, isopropanol, and long-chain hydrocarbon, and is used for consistent and reliable processing of fatty tissues.

[2]If the routine histoprocessor is used to test the performance of an alternative fixative, misleading results could be obtained in case the instrument is contaminated by residual formalin.

• Proteomics: proteins from alcohol-based fixatives can be extracted using the dedicated protocols (see Chap. 39). From the experience of one group involved in the IMPACTS project, it has been noted that the quality of protein lysate is better if, after the microwave fixation, the tissue sample undergoes a step of dehydratation-clearing with JFC solution, instead of isopropanol alone. For all the analyses, the protocols described for FFPE samples can be used.

References

1. Fox CH, Johnson FB, Whiting J, Roller PP (1985) Formaldehyde fixation. J Histochem Cytochem 33(8):845–853
2. Lehmann U, Kreipe H (2001) Real-time PCR analysis of DNA and RNA extracted from formalin-fixed and paraffin-embedded biopsies. Methods 25(4):409–418
3. Macabeo-Ong M, Ginzinger DG, Dekker N, McMillan A, Regezi JA, Wong DT, Jordan RC (2002) Effect of duration of fixation on quantitative reverse transcription polymerase chain reaction analyses. Mod Pathol 15(9):979–987
4. Specht K, Richter T, Muller U, Walch A, Werner M, Hofler H (2001) Quantitative gene expression analysis in microdissected archival formalin-fixed and paraffin-embedded tumor tissue. Am J Pathol 158(2):419–429
5. Stanta G, Bonin S (1998) RNA quantitative analysis from fixed and paraffin-embedded tissues: membrane hybridization and capillary electrophoresis. Biotechniques 24(2): 271–276
6. Becker KF, Metzger V, Hipp S, Hofler H (2006) Clinical proteomics: new trends for protein microarrays. Curr Med Chem 13(15):1831–1837
7. Becker KF, Schott C, Hipp S, Metzger V, Porschewski P, Beck R, Nahrig J, Becker I, Hofler H (2007) Quantitative protein analysis from formalin-fixed tissues: implications for translational clinical research and nanoscale molecular diagnosis. J Pathol 211(3):370–378
8. Antonov J, Goldstein DR, Oberli A, Baltzer A, Pirotta M, Fleischmann A, Altermatt HJ, Jaggi R (2005) Reliable gene expression measurements from degraded RNA by quantitative real-time PCR depend on short amplicons and a proper normalization. Lab Invest 85(8):1040–1050
9. Bonin S, Brunetti D, Benedetti E, Dotti I, Gorji N, Stanta G (2008) Molecular characterisation of breast cancer patients at high and low recurrence risk. Virchows Arch 452(3): 241–250
10. Bonin S, Petrera F, Niccolini B, Stanta G (2003) PCR analysis in archival postmortem tissues. Mol Pathol 56(3):184–186
11. Hwang SI, Thumar J, Lundgren DH, Rezaul K, Mayya V, Wu L, Eng J, Wright ME, Han DK (2007) Direct cancer tissue proteomics: a method to identify candidate cancer biomarkers from formalin-fixed paraffin-embedded archival tissues. Oncogene 26(1):65–76
12. Shen L, Catalano PJ, Benson AB III, O'Dwyer P, Hamilton SR, Issa JP (2007) Association between DNA methylation and shortened survival in patients with advanced colorectal cancer treated with 5-fluorouracil based chemotherapy. Clin Cancer Res 13(20):6093–6098
13. Stanta G, Schneider C (1991) RNA extracted from paraffin-embedded human tissues is amenable to analysis by PCR amplification. Biotechniques 11(3):304, 306, 308
14. Shibutani M, Uneyama C, Miyazaki K, Toyoda K, Hirose M (2000) Methacarn fixation: a novel tool for analysis of gene expressions in paraffin-embedded tissue specimens. Lab Invest 80(2):199–208
15. Srinivasan M, Sedmak D, Jewell S (2002) Effect of fixatives and tissue processing on the content and integrity of nucleic acids. Am J Pathol 161(6):1961–1971
16. Cross SS, Start RD, Smith JH (1990) Does delay in fixation affect the number of mitotic figures in processed tissue? J Clin Pathol 43(7):597–599
17. Douglas MP, Rogers SO (1998) DNA damage caused by common cytological fixatives. Mutat Res 401(1–2):77–88
18. Perlmutter MA, Best CJ, Gillespie JW, Gathright Y, Gonzalez S, Velasco A, Linehan WM, Emmert-Buck MR, Chuaqui RF (2004) Comparison of snap freezing versus ethanol fixation for gene expression profiling of tissue specimens. J Mol Diagn 6(4): 371–377
19. Noguchi M, Furuya S, Takeuchi T, Hirohashi S (1997) Modified formalin and methanol fixation methods for molecular biological and morphological analyses. Pathol Int 47(10):685–691
20. Tyrrell L, Elias J, Longley J (1995) Detection of specific mRNAs in routinely processed dermatopathology specimens. Am J Dermatopathol 17(5):476–483
21. Foss RD, Guha-Thakurta N, Conran RM, Gutman P (1994) Effects of fixative and fixation time on the extraction and polymerase chain reaction amplification of RNA from paraffin-embedded tissue. Comparison of two housekeeping gene mRNA controls. Diagn Mol Pathol 3(3):148–155
22. Lee KY, Shibutani M, Inoue K, Kuroiwa K, U M, Woo GH, Hirose M (2006) Methacarn fixation – effects of tissue processing and storage conditions on detection of mRNAs and proteins in paraffin-embedded tissues. Anal Biochem 351(1):36–43
23. Bonin S, Petrera F, Rosai J, Stanta G (2005) DNA and RNA obtained from bouin's fixed tissues. J Clin Pathol 58(3): 313–316
24. Lykidis D, Van Noorden S, Armstrong A, Spencer-Dene B, Li J, Zhuang Z, Stamp GW (2007) Novel zinc-based fixative for high quality DNA, RNA and protein analysis. Nucleic Acids Res 35(12):e85
25. Stanta G, Mucelli SP, Petrera F, Bonin S, Bussolati G (2006) A novel fixative improves opportunities of nucleic acids and proteomic analysis in human archive's tissues. Diagn Mol Pathol 15(2):115–123
26. Vincek V, Nassiri M, Nadji M, Morales AR (2003) A tissue fixative that protects macromolecules (DNA, RNA, and protein) and histomorphology in clinical samples. Lab Invest 83(10):1427–1435
27. Delfour C, Roger P, Bret C, Berthe ML, Rochaix P, Kalfa N, Raynaud P, Bibeau F, Maudelonde T, Boulle N (2006) Rcl2, a new fixative, preserves morphology and nucleic acid integrity in paraffin-embedded breast carcinoma and microdissected breast tumor cells. J Mol Diagn 8(2):157–169

28. Olert J, Wiedorn KH, Goldmann T, Kuhl H, Mehraein Y, Scherthan H, Niketeghad F, Vollmer E, Muller AM, Muller-Navia J (2001) HOPE fixation: a novel fixing method and paraffin-embedding technique for human soft tissues. Pathol Res Pract 197(12):823–826

29. Benchekroun M, DeGraw J, Gao J, Sun L, von Boguslawsky K, Leminen A, Andersson LC, Heiskala M (2004) Impact of fixative on recovery of mRNA from paraffin-embedded tissue. Diagn Mol Pathol 13(2):116–125

30. Cox ML, Schray CL, Luster CN, Stewart ZS, Korytko PJ, MK KN, Paulauskis JD, Dunstan RW (2006) Assessment of fixatives, fixation, and tissue processing on morphology and RNA integrity. Exp Mol Pathol 80(2):183–191

31. Gillespie JW, Best CJ, Bichsel VE, Cole KA, Greenhut SF, Hewitt SM, Ahram M, Gathright YB, Merino MJ, Strausberg RL, Epstein JI, Hamilton SR, Gannot G, Baibakova GV, Calvert VS, Flaig MJ, Chuaqui RF, Herring JC, Pfeifer J, Petricoin EF, Linehan WM, Duray PH, Bova GS, Emmert-Buck MR (2002) Evaluation of non-formalin tissue fixation for molecular profiling studies. Am J Pathol 160(2):449–457

32. Goldsworthy SM, Stockton PS, Trempus CS, Foley JF, Maronpot RR (1999) Effects of fixation on RNA extraction and amplification from laser capture microdissected tissue. Mol Carcinog 25(2):86–91

33. Paska C, Bogi K, Szilak L, Tokes A, Szabo E, Sziller I, Rigo J Jr, Sobel G, Szabo I, Kaposi-Novak P, Kiss A, Schaff Z (2004) Effect of formalin, acetone, and RNAlater fixatives on tissue preservation and different size amplicons by real-time PCR from paraffin-embedded tissue. Diagn Mol Pathol 13(4):234–240

34. Barrett C, Brett F, Grehan D, McDermott MB (2004) Heat-accelerated fixation and rapid dissection of the pediatric brain at autopsy: a pragmatic approach to the difficulties of organ retention. Pediatr Dev Pathol 7(6):595–600

35. Hsu HC, Peng SY, Shun CT (1991) High quality of DNA retrieved for southern blot hybridization from microwave-fixed, paraffin-embedded liver tissues. J Virol Methods 31(2–3):251–261

36. Su JM, Perlaky L, Li XN, Leung HC, Antalffy B, Armstrong D, Lau CC (2004) Comparison of ethanol versus formalin fixation on preservation of histology and RNA in laser capture micro-dissected brain tissues. Brain Pathol 14(2):175–182

37. Vincek V, Nassiri M, Block N, Welsh CF, Nadji M, Morales AR (2005) Methodology for preservation of high molecular-weight RNA in paraffin-embedded tissue: application for laser-capture microdissection. Diagn Mol Pathol 14(3):127–133

38. Kim JO, Kim HN, Hwang MH, Shin HI, Kim SY, Park RW, Park EY, Kim IS, van Wijnen AJ, Stein JL, Lian JB, Stein GS, Choi JY (2003) Differential gene expression analysis using paraffin-embedded tissues after laser microdissection. J Cell Biochem 90(5):998–1006

39. Uneyama C, Shibutani M, Masutomi N, Takagi H, Hirose M (2002) Methacarn fixation for genomic DNA analysis in micro-dissected, paraffin-embedded tissue specimens. J Histochem Cytochem 50(9):1237–1245

40. Ahram M, Flaig MJ, Gillespie JW, Duray PH, Linehan WM, Ornstein DK, Niu S, Zhao Y, Petricoin EF 3rd, Emmert-Buck MR (2003) Evaluation of ethanol-fixed, paraffin-embedded tissues for proteomic applications. Proteomics 3(4): 413–421

41. Gazziero A, Guzzardo V, Aldighieri E, Fassina A (2009) Morphological quality and nucleic acid preservation in cyto-pathology. J Clin Pathol 62(5):429–434

42. Bellet V, Boissiere F, Bibeau F, Desmetz C, Berthe M, Rochaix P, Maudelonde T, Mange A, Solassol J (2007) Proteomic analysis of RCL2 paraffin-embedded tissues. J Cell Mol Med [Epub]

43. Takagi H, Shibutani M, Kato N, Fujita H, Lee KY, Takigami S, Mitsumori K, Hirose M (2004) Microdissected region-specific gene expression analysis with methacarn-fixed, paraffin-embedded tissues by real-time RT-PCR. J Histochem Cyto-chem 52(7):903–913

44. Dotti I, Bonin S, Basili G, Nardon E, Balani A, Siracusano S, Zanconati F, Palmisano S, De Manzini N, Stanta G (2010) Effects of formalin, methacarn, and FineFix fixatives on RNA preservation. Diagn Mol Pathol 19(2):112–122

45. NCCLS (1999) Quality assurance for immunocytochemis-try. Approved guideline. NCCLS document MM4-AC

46. Godfrey TE, Kim SH, Chavira M, Ruff DW, Warren RS, Gray JW, Jensen RH (2000) Quantitative mRNA expression analysis from formalin-fixed, paraffin-embedded tissues using 5' nuclease quantitative reverse transcription-polymerase chain reaction. J Mol Diagn 2(2):84–91

47. Jiang X, Jiang X, Feng S, Tian R, Ye M, Zou H (2007) Development of efficient protein extraction methods for shotgun proteome analysis of formalin-fixed tissues. J Proteome Res 6(3):1038–1047

Part II

Dissection of Tissue Components

Manual Microdissection

Valentina Faoro and Giorgio Stanta

4

Contents

4.1 Introduction and Purpose

Accuracy and reproducibility of molecular analysis in human tissues are highly dependent on the appropriate selection of starting material. Human tissue specimens are often composed by a heterogeneous mixture of cell types that can variably affect downstream molecular results. In solid tumour samples, for example, neoplastic cells are admixed with stromal tissue cells, inflammatory cells, and blood vessel components that may contaminate the tumoral component. In order to properly separate different populations of cells from a heterogeneous tissue or to reduce the contamination by non-neoplastic cells in a tumour, microdissection of the sample is highly necessary.

Several microdissection methods have been developed to isolate homogeneous cell samples from solid tissues. In this and the following chapter, two types of microdissection from formalin-fixed and paraffin-embedded (FFPE) tissue sections are discussed: manual and laser capture microdissection. In both cases, some precautions should be followed in order to avoid tissue damage and tissue or nucleic acid contamination during tissue processing. Such precautions include cleaning of the instruments with 100% ethanol, clearing of the tissue debris from the work station with xylene, and frequent change of the microtome blades. The FFPE sections are cut onto glass slides and dried overnight at 37°C or for 30–45 min at 60°C. At this point, they are ready to be used or can be stored (protected from the dust).

4.2 Protocol

Manual microdissection is the simplest method for tissue scratching from an unstained slide. Basically, it involves the use of a scalpel blade or needle to scrape

V. Faoro (✉)
Department of Medical, Surgical and Health Sciences,
University of Trieste, Trieste, Italy

G. Stanta
Department of Medical Sciences,
University of Trieste, Cattinara Hospital, Strada di Fiume 447,
Trieste, Italy

G. Stanta (ed.), *Guidelines for Molecular Analysis in Archive Tissues*,
DOI: 10.1007/978-3-642-17890-0_4, © Springer-Verlag Berlin Heidelberg 2011

Fig. 4.1 Donor block (**a**) and the relative punched core (**b**)

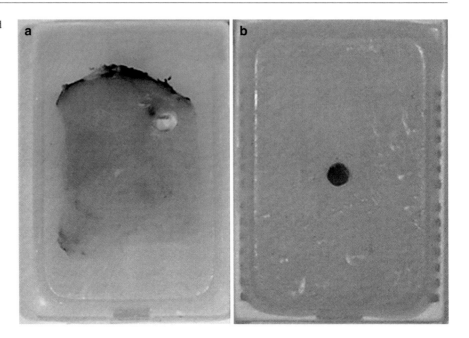

the tissue of interest from the histological section and a standard microscope.

and on the type of the downstream analysis. Usually, 5–20 sections of 5 μm thickness are collected.

4.2.1 Method 1

First, the operator examines the hematoxylin-eosin (H&E)-stained section and marks the area of interest for microdissection (e.g. tumour, with a tumour cell content of more than 80%) with a pen. Then the corresponding unstained consecutive sections are dewaxed[1] with xylene and dehydrated with several graded ethanols. Microdissection is then performed by laying the unstained sections upon the H&E-marked section and scraping the area of interest. Different tools can be used, including any sharp and precise instrument such as a 30-gauge needle or a pointed surgical blade. The number of tissue sections to be microdissected depends on the size of the tissue area, thickness of the sections

4.2.2 Method 2

Manual microdissection is inexpensive in terms of consumables and equipment, but is time-consuming. Recently, the use of the tissue arrayer with needles of 3–5 mm diameter has been introduced as a valid alternative to manual microdissection for molecular analyses in large tissue areas (see Chap. 5). After examination of the (H&E) stained section, the tissue core of interest is punched, embedded in a new paraffin block and used for DNA, RNA and protein extraction (Fig. 4.1). In this case, deparaffinization and tissue digestion are performed directly in the tube, after cutting and collecting 5–20 sections of 5 μm-thickness. This method allows a more precise dissection of the sample, avoiding contaminations with cells of different origin, events that can occur with the previous method. Moreover, in comparison with LCM (see Chap. 6), this method is less time-consuming; in fact, just a few minutes are needed for each punch.

[1]Direct deparaffinization of the tissue slide is used to facilitate the scraping of the sample from the glass. Alternatively, deparaffinization can also be performed after the tissue fragments have been collected in the tube.

Tissue Microarray (TMA)

5

Valentina Faoro and Anna Sapino

Contents

V. Faoro
Department of Medical, Surgical and Health Sciences,
University of Trieste, Trieste, Italy

A. Sapino (✉)
Department of Biomedical Sciences and Human Oncology,
University of Turin, Turin, Italy

5.1 Introduction and Purpose

The Tissue MicroArray procedure was developed by Kononen et al. [1] 10 years ago as a high-throughput tool to investigate a variety of biomarkers on tissue specimens [2]. To construct a TMA, small cores of tissues punched from donor paraffin blocks are transferred into an empty recipient block in arrayed fashion. Using these cores, samples from hundreds of different tissues or patients can be arrayed in a single paraffin block and analyzed by immunohistochemistry, in situ hybridization, or immunofluorescence.

As a research tool, TMAs are used predominantly for the simultaneous investigation of putative prognostic and predictive molecular targets in human cancer tissues [3]. They are also used for the in situ validation of candidate diagnostic markers identified in genomics and proteomics studies [4–7] and for the correlation of staining results with clinical endpoint [8, 9]. Moreover, the so-called progression TMAs are possible, in which cores of a single tissue type are used taking into account different stages of tumor development or different tumor grades. For example, a progression TMA for colon cancer could include normal colon, adenomas, with low and high-grade of dysplasia, as well as carcinomas [10].

TMAs can be also used in experiments aimed to determine whether a protein is expressed or not and to what extent in a wide range of different normal and/or lesional tissues. In addition, this technique can complement other proteomic methods with the advantage of studying the expression pattern of a specific protein with respect to cell compartments.

G. Stanta (ed.), *Guidelines for Molecular Analysis in Archive Tissues*,
DOI: 10.1007/978-3-642-17890-0_5, © Springer-Verlag Berlin Heidelberg 2011

5.1.1 General Considerations Regarding Tissue Microarray Design and Construction

Before sampling the tissues, it is fundamental to define in advance the questions that will be addressed by staining TMAs as it will have an impact on the sampling itself [11]. For example, if the main goal is to characterize the general expression pattern of a specific protein in a tissue, the sampling would be random; if the aim is to compare the expression of a protein between the lesional and the perilesional region of a tumor, the sampling would be done from the right area.

A potential problem in the construction of a TMA is the tissue heterogeneity, resulting in different gene and protein expression in different normal and/or tumor cells. The best way to overcome this intervariability of expression of the target is to take multiple cores of each area of interest from the same sample. Currently no sampling methods are available for standardization, but it seems intuitive that the more samples are taken, the more representative the staining results. Many studies seem to indicate that the results from triplicate TMA cores have up to 98% concordance with the result from full sections [12, 13]. A study by Goethals et al. [14] recommends at least four cores, whereas other authors have achieved more than 95% accuracy with only two cores [15]. There are technical reasons too for including more than one core for each case: some cores could be lost, e.g., during the sectioning. Moreover, in the case of low tumor cell density, doubling the number of cores per case could be necessary.

A particular point of debate refers to the core diameter with respect to tissue sampling. Most tissue arrayer instruments punch with a diameter between 0.6 and 2 mm, which is equivalent to a tissue area of 0.29–3.14 mm². Most of the published studies used the 0.6 mm diameter punches, as it allows preserving more source tissue and including a larger number of cores into the recipient block. Punches with a diameter of more than 0.6 mm are useful for tissues containing a large amount of fatty or connective component or for the study of the whole mucosa thickness, e.g., in the stomach.

Needles with a diameter of 3–5 mm allow a more rapid and precise manual microdissection of large tissue areas or cells of interest (see Chaps. 4 and 6). In this case, the tissue cores can also be used for molecular analyses, such as those performed in some IMPACTS laboratories. Once punched, the single cores are embedded in a new paraffin block, sectioned, and used for DNA, RNA, and protein extraction.

5.2 Protocol

5.2.1 General Considerations Regarding the Layout of Tissue Microarray

Although methods for the layout of a TMA have not been standardized so far, probably because different studies have different requirements, the following components seem to be essential in the TMA design.

To ensure unambiguous orientation and identification of individual cores within the TMA section, it is convenient to add one or more "orientation cores" in a specific position, generally outside the overall geometric margin of the array. Some researchers opted for the introduction of gaps; this empty core positions may help to orient the TMA without any confusion or doubts.

Control tissue cores can be also included in the array, placing them asymmetrically into the grid. These cores may serve both as internal "orientation cores" and as positive or negative internal experimental controls.

The arrangement of the interest cores within the TMA varies according to the type of study and each operator has to identify which design best suits his or her purposes. Ideally cores from the same donor block should not be positioned next to each other, since only random distribution ensures that results from individual cores are not statistically affected by technical blunders. However, random distribution of cores from the same donor block increases the workload and therefore becomes time consuming [11].

Because it is known that immunohistochemistry analyses on full tissue sections may show some staining artifacts at the tissue border, some researchers frame all the TMA with a "protection wall," a row of tissue cores that will not be subsequently analyzed [12].

5.2.2 Equipment

- *TMA Arrayer Instrument*[1]
- *Software interface* for the TMA design
- *Set of two punches of the preferred diameter*: one for the donor block and another one for the recipient block
- *A freshly poured empty paraffin block* (recipient block) of the desired size[2]
- *H&E slides* of the cases of interest on which the desired area for the sampling is marked
- *Donor blocks* of the cases of interest
- *Oven*
- *Glass slides*
- *Microtome*

5.2.3 Method

5.2.3.1 Array Design: Preparation of the Array Pattern

- Using specific software, define the geometrical parameters of the array: the paraffin block size, the needle diameter, the number of spots, and the distance between the spots. If necessary, rows, columns, or groups of empty spots can be introduced to separate the specimens.
- At this point, each donor block is linked to one or more TMA spots in order to create the map that will be followed during the TMA construction. It is sometimes possible to add a note for each donor block, recipient block, or spot that will be present in the final report.

5.2.3.2 TMA Construction: Allocation of Samples in the Array

- In this phase, the software facilitates the completion of the TMA construction. The digital camera is connected, and the homing of the automatic tray[3] and of the needle holder is performed.
- Usually, an operating flowchart guides every step of TMA construction. The steps include:
  1. Definition of the center of the tray.
  2. Insertion of the recipient block.
  3. Checking of the TMA template directly on the recipient block. If the position of the spots is not correct, it can be modified by moving the tray.
  4. Insertion of the first donor block.
  5. Choice of the method for the selection of punch area.[4]
  6. Selection of the punch areas.
  7. Sampling:
     - Prepare the hole in the recipient block
     - Take the sample from the donor block
     - Insert the sample in the prepared hole[5]
- Create the final report of the array in an Excel spread-sheet.

5.2.3.3 Assembling of the TMA

- Once completed, the TMA is placed upside down on a glass slide and into an oven at 40°C for about 30 min to facilitate binding of the donor cores with the paraffin wax of the block itself.
- The glass slide, attached to the TMA block, is used to level the block surface by gently pushing the cores into the block.
- After cooling, the block is ready for sectioning on a microtome[6] and for further analyses (Fig. 5.1).

---

[1]The methodology described here refers to the Galileo TMA CK 3500 Tissue Arrayer and to the IseTMA Software, from the Integrated System Engineering S.r.l. (Galileo TMA CK 3500 Tissue Microarrayer- Operating Manual) for the direct experience of the authors. Other valid arrayers are commercially available, such as the Veridiam Tissue Arrayer, the TMArrayer by Pathology Devices, and others.

[2]Air bubbles may be accidentally created within the paraffin block during the pouring and cooling procedures especially when metal molds are used; the formation of air bubbles can be minimized by using plastic molds.

[3]The automated tray ensures the precise positioning of the paraffin blocks while it is in use.

[4]The overlapping can be manual or digital. In manual overlapping, the reference glass slide is overlapped to the corresponding donor block and manually aligned with it looking at the monitor. In digital overlapping, the digital image of the reference slides is overlapped and aligned to the live donor block image.

[5]Repeat "Selection of the punch areas" and "Sampling" for each donor block, according to the design plan.

[6]The TMA block should be cut only by expert technicians as the TMA section needs to be cut and picked up from the hot water bath with great care to avoid distortion prior to aligning it in parallel with the edge of the glass slide to facilitate analyses.

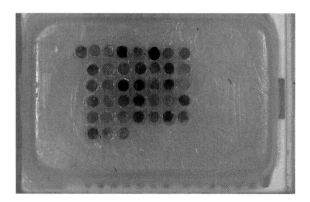

Fig. 5.1 Representative picture of a TMA block in which 1.5 mm cores have been inserted from selected donor blocks

5.2.3.4 Evaluation of the Results

Generally, TMA sections are read serially using a conventional light microscope. However, keeping track of the position of each core can sometimes become very difficult when many spots are present in one slide. Open source types of software have been developed and are available, for example from Stanford University (http://genome-www.stanford.edu/TMA/) and Johns Hopkins University, (http://tmaj.pathology.jhmi.edu/) to make digital scanning and image for the evaluation of the results.

References

1. Kononen J, Bubendorf L, Kallioniemi A, Barlund M, Schraml P, Leighton S, Torhorst J, Mihatsch MJ, Sauter G, Kallioniemi OP (1998) Tissue microarrays for high-throughput molecular profiling of tumor specimens. Nat Med 4(7): 844–847

2. Yan P, Seelentag W, Bachmann A, Bosman FT (2007) An agarose matrix facilitates sectioning of tissue microarray blocks. J Histochem Cytochem 55(1):21–24

3. Egervari K, Szollosi Z, Nemes Z (2007) Tissue microarray technology in breast cancer HER2 diagnostics. Pathol Res Pract 203(3):169–177

4. Bubendorf L, Kolmer M, Kononen J, Koivisto P, Mousses S, Chen Y, Mahlamaki E, Schraml P, Moch H, Willi N, Elkahloun AG, Pretlow TG, Gasser TC, Mihatsch MJ, Sauter G, Kallioniemi OP (1999) Hormone therapy failure in human prostate cancer: analysis by complementary DNA and tissue microarrays. J Natl Cancer Inst 91(20):1758–1764

5. Dube V, Grigull J, DeSouza LV, Ghanny S, Colgan TJ, Romaschin AD, Siu KW (2007) Verification of endometrial tissue biomarkers previously discovered using mass spectrometry-based proteomics by means of immunohistochemistry in a tissue microarray format. J Proteome Res 6(7):2648–2655

6. Hewitt SM (2006) The application of tissue microarrays in the validation of microarray results. Meth Enzymol 410:400–415

7. Skacel M, Siva A, Xu B, Tubbs RR (2007) From array to array: confirmation of genomic gains and losses discovered by array-based comparative genomic hybridization utilizing fluorescence in situ hybridization on tissue microarrays. J Mol Histol 38(2):135–140

8. Kallioniemi OP, Wagner U, Kononen J, Sauter G (2001) Tissue microarray technology for high-throughput molecular profiling of cancer. Hum Mol Genet 10(7):657–662

9. Venkataraman G, Ananthanaranayanan V (2005) Tissue microarrays: potential in the Indian subcontinent. Indian J Cancer 42(1):9–14

10. Chen WC, Lin MS, Zhang BF, Fang J, Zhou Q, Hu Y, Gao HJ (2007) Survey of molecular profiling during human colon cancer development and progression by immunohistochemical staining on tissue microarray. World J Gastroenterol 13(5):699–708

11. Parsons M, Grabsch H (2009) How to make a tissue microarray. Diagn Histopathol 15(3):142–150

12. Hoos A, Cordon-Cardo C (2001) Tissue microarray profiling of cancer specimens and cell lines: opportunities and limitations. Lab Invest 81(10):1331–1338

13. Sapino A, Marchiò C, Senetta R, Macrì L, Cassoni P, Ghisolfi G, Cerrato M, D'Ambrosio E, Bussolati G (2006) Routine assessment of prognostic factors in breast cancer using a multicore tissue microarray procedure. Virchow Arch 449(3):288–296

14. Goethals L, Perneel C, Debucquoy A, De Schutter H, Borghys D, Ectors N, Geboes K, McBride WH, Haustermans KM (2006) A new approach to the validation of tissue microarrays. J Pathol 208(5):607–614

15. Camp RL, Charette LA, Rimm DL (2000) Validation of tissue microarray technology in breast carcinoma. Lab Invest 80(12):1943–1949

Laser Capture Microdissection (LCM)

6

Elvira Stacher, Hannelore Kothmaier, Iris Halbwedl, and Helmut H. Popper

Contents

E. Stacher (✉), H. Kothmaier, I. Halbwedl, and H.H. Popper
Research Unit for Molecular Lung and Pleura Pathology,
Institute of Pathology, Medical University of Graz, Graz,
Austria

6.1 Introduction and Purpose

LCM represents the automation of manual microdissection and requires the use of laser technology-based instruments [1–3].

The following protocol specifically provides a method to perform laser capture microdissection with the Veritas System by Arcturus on formalin-fixed and paraffin-embedded (FFPE) tissues for further DNA extraction, but also systems from other companies can be used with efficient results.

6.2 Protocol

6.2.1 Reagents

Note: Reagents from specific companies are reported here, but reagents of equal quality purchased from other companies may be used:

- *Xylene* (Merck, 1086812500)
- *100% Ethanol* (Merck, 1000.983.2511)
- *90%, 70%, 50% Ethanol*
- *Sterile filtered* ddH$_2$O
- *Proteinase K* (Sigma, P-6556; 20 mg/ml in sterile filtered ddH$_2$O, stored at −20°C)
- *TE buffer pH 9* (DNA Minikit, buffer AE; Qiagen, Germany)
- *Papanicolaou 1a Harris hematoxylin* (Merck, 1092530500)

G. Stanta (ed.), *Guidelines for Molecular Analysis in Archive Tissues*,
DOI: 10.1007/978-3-642-17890-0_6, © Springer-Verlag Berlin Heidelberg 2011

6.2.2 Equipment

Note: Equipment from specific companies is reported here, but others with similar characteristics can be used:

- *PEN membrane glass slides* (2 µm; Microdissect, Germany, MDG3P40W)
- *Stratagene Autocrosslinker*
- *Laser microdissection system*, Arcturus, USA
- *Arcturus CapSure Macro LCM Caps* (Plastic Carriers; LCM0211)
- *PCR machine* (e.g., GeneAmp PCR System 9600, Applied Biosystems)
- *0.2 ml PCR-tubes* (MicroAmp, N801–0840, Applied Biosystems)
- *Heat-sterilized glass staining jars*
- *Sterile pipette tips*
- *Glass microscopy slides, ethanol-cleaned* (e.g., Menzel)
- *Light microscope*

6.2.3 Method

This protocol is structured into four parts:

1. Slide preparation
2. Deparaffinization and staining
3. Laser capture microdissection (LCM)
4. Digestion

6.2.3.1 Slide Preparation

- Cross-link PEN membrane slides with the Stratagene Autocrosslinker.
- Freshly prepare five to ten consecutive sections from the FFPE tissue (7 µm thickness) on special membrane slides. The first and the last section should be prepared on glass slides, deparaffinized, and stained with hematoxylin-eosin (H.E.). H.E.-stained slides are used to find the area of interest.
- Dry the sections for LCM at room temperature overnight or for at least 3 h at 37°C.

6.2.3.2 Deparaffinization and Staining

- Put sections in heat-sterilized staining jars with
  - Xylene for 10 min × 3
  - 100% ethanol for 5 min × 2
  - 90% ethanol for 5 min
  - 70% ethanol for 5 min
  - 50% ethanol for 5 min
  - ddH_2O for 5 min
- Stain the deparaffinized sections on the special membrane slides with sterile filtered, undiluted Papanicolaou 1a Harris hematoxylin solution for a few seconds[1] (using the syringe with the sterile filter) and rinse with ddH_2O several times (using a filtered pipette tip).[2]
- Dry the sections completely at 37°C for approximately 6–7 h.

6.2.3.3 Laser Capture Microdissection

- Use the H.E. slides and simple light microscopy to find the desired areas.
- General settings at the Veritas system: use low energy UV laser of the Veritas instrument for cutting the selected area and IR laser for capturing the desired cells on Arcturus Macrocaps. For detailed setting information it is strongly recommended to study the manufacturer's guidelines.

6.2.3.4 Digestion

- Freshly prepare digestion buffer:
  10 µl TE buffer pH 9
  0.5 µl Proteinase K
  2 µl ddH_2O[3]

---

[1] For the laser sections, a short-time staining with hematoxylin gives the best results for visualization in the instrument (max 30 s.).

[2] Do not stain with eosin; this will affect the DNA quality! Slight staining of nuclei by hematoxylin will help to sort out tumor or dysplasia from normal cells.

[3] ddH_2O is used only to prevent evaporation during the long period of digestion.

- 12.5 μl
- Peel off the membrane from the cap with a filter tip and put it into a PCR tube[4] with the 12.5 μl digestion buffer.
- Digest for 72 h at 55°C in the PCR machine.
- Each day add 2 μl ddH$_2$O and 0.5 μl fresh Proteinase K.
- After 72 h, inactivate Proteinase K by heating at 99°C for 10 min in a PCR machine.
- Spin down the condensate and immediately store the digested material in the supernatant at −20°C until amplification (all 10 μl of the digestion are used).

6.2.4 Troubleshooting

6.2.4.1 Inefficient Microdissection

- Sections have to be placed centrally on membrane slides; otherwise, the instrument may not reach all areas.
- Increase laser power to ensure adequate melting or cutting.
- The UV laser has to be focused exactly to guarantee a small cutting trace with low energy.
- Sections should be completely dried before laser capture microdissection in order to make it possible for the IR laser to capture the cells.

6.2.4.2 Tissue Remnants on the Slide After Microdissection

The tissue section is too thick or uneven: in such case, cut sections at 5–8 μm for IR capture LCM or 2–15 μm for UV laser cutting.

6.2.4.3 Decreased DNA Yield or Inadequate Amplification

- Inadequate removal of paraffin from FFPE tissue sections: deparaffinization should be prolonged. Deparaffinize in three changes of xylene up to 15 min each.
- Proteinase K is not inactivated completely. Proteinase K inactivation should be performed at least at 99°C for 10 min.
- Double the number of microdissected cells and optimize downstream analysis.

6.2.4.4 Contamination of Samples

- Always take care to avoid DNA contamination. A digestion blank is strongly recommended.

References

1. Espina V, Milia J, Wu G, Cowherd S, Liotta LA (2006) Laser capture microdissection. Methods Mol Biol 319:213–229
2. Espina V, Wulfkuhle JD, Calvert VS, VanMeter A, Zhou W, Coukos G, Geho DH, Petricoin EF 3rd, Liotta LA (2006) Laser-capture microdissection. Nat Protoc 1(2):586–603
3. Frost AR, Eltoum IE, Siegal GP (2001) Laser capture microdissection. Curr Protoc Mol Biol, Chap 25:Unit 25A 21

---

[4]Tubes should be UV cross-linked before use. UV irradiation (cross-linking) is a decontaminating step that removes exogenous DNA from tubes, surfaces, pipettes…

Extraction, Purification, Quantification and Quality Assessment of Nucleic Acids

DNA Extraction from Formalin-Fixed Paraffin-Embedded (FFPE) Tissues

7

Serena Bonin, Patricia J.T.A. Groenen, Iris Halbwedl, and Helmut H. Popper

Contents

S. Bonin (✉)
Department of Medical, Surgical and Health Sciences,
University of Trieste, Cattinara Hospital, Strada di Fiume 447,
Trieste, Italy

P.J.T.A. Groenen
Department of Pathology, Radboud University
Nijmegen Medical Centre, Nijmegen, The Netherlands

I. Halbwedl and H.H. Popper
Research Unit for Molecular Lung and Pleura Pathology,
Institute of Pathology, Medical University of Graz,
Graz, Austria

7.1 Introduction and Purpose

This protocol provides a method of obtaining DNA suitable for PCR analyses from formalin-fixed and paraffin-embedded (FFPE) tissue specimens [1–4], even of autopsy origin [5, 6]. This procedure is based mainly on deparaffinization of tissues (as optional step) and a proteolytic digestion with Proteinase K. The proteolysis step is fundamental to degrade proteins and generate pure DNA. The time required for the whole procedure is 4 days.

Commercial kits are also available for DNA extraction from FFPE; some of these are specifically dedicated to archive tissues (i.e., QIAamp DNA FFPE Tissue Kit). However, some minor modifications could be useful to achieve better results depending on the type of tissues; requirements for the specific molecular test should be taken into consideration [7].

7.2 Protocol

7.2.1 Reagents

Note: Reagents from specific companies are reported here, but similar reagents from other providers could be used:

- *Xylene* (Fluka or Sigma-Aldrich)
- *Absolute, 90% and 70% Ethanol* (Sigma)
- *20 mg/ml Proteinase K* (stock solution): (Sigma P2303) Dissolve 100 mg of Proteinase K in 5 ml of autoclaved 50% glycerol diluted in sterile H_2O.[1]

---

[1]The solubilization of Proteinase K in 50% sterile glycerol maintains the solution fluid at −20°C with a better preservation of the enzymatic activity.

G. Stanta (ed.), *Guidelines for Molecular Analysis in Archive Tissues*,
DOI: 10.1007/978-3-642-17890-0_7, © Springer-Verlag Berlin Heidelberg 2011

Store at −20°C. The stock solution of 20 mg/ml should be diluted to a final concentration of 1 mg/ml Proteinase K in digestion buffer

- *10× Digestion Buffer*[2] (stock solution): 500 mM Tris HCl pH 7.5, 10 mM EDTA, 1 M NaCl, 5% Tween 20
- *1× Digestion Buffer*: 50 mM Tris HCl pH 7.5, 1 mM EDTA, 100 mM NaCl, 0.5% Tween 20. Complete the solution with Proteinase K (1 mg/ml = final concentration) just before use
- *Phenol-Tris buffered pH 8*[3]*/CHCl₃ 50:50*: Mix 1 part of buffered phenol with 1 part of chloroform. Top the organic phase with 1× TE buffer (about 1 cm height) and allow the phase to separate. Store at 4°C in a light-tight bottle
- *Phenol-Tris buffered pH 8/CHCl₃–isoamyl alcohol 50:49:1*: Mix 48 ml of Chloroform with 2 ml of isoamyl alcohol. Mix 1 part of buffered phenol with 1 part of chloroform–isoamyl alcohol. Top the organic phase with 1× TE buffer (about 1 cm high) and allow the phase to separate. Store at 4°C in a light-tight bottle
- *Iso-propanol or EtOH/Sodium acetate*
- *1 mg/ml Glycogen in water*
- *10× TE buffer*: 100 mM Tris pH 8, 10 mM EDTA pH 8

7.2.2 Equipment

- *Disinfected*[4] *adjustable pipettes*, range: 2–20 µl, 20–200 µl, 100–1,000 µl
- *Nuclease-free aerosol-resistant pipette tips*
- *1.5 ml tubes* (autoclaved)
- *Single-packed toothpicks*
- *Sterile or disposable tweezers*
- *Microtome, with new blade*

- *Centrifuge* suitable for centrifugation of 1.5 ml tubes at 13,200 or 14,000 rpm
- *Thermoblock*
- *Thermomixer* (e.g., Eppendorf)
- *SpectroPhotometer*

7.2.3 Method

7.2.3.1 Sample Preparation

- If possible, cool the paraffin blocks at −20°C or on dry ice in aluminium foil in order to cut the sections.
- Using a clean, sharp microtome blade[5], cut two to ten sections of 5–10 µm thickness depending on the size of the sample. Discard the first section and displace the other ones in 1.5 ml tubes, using a sterile toothpick or tweezers (depending on the section size). Use some sections from a paraffin block without included tissue, treated together with other samples, for negative control analysis.

7.2.3.2 Deparaffinization[6] (Optional)

- Add 1 ml of xylene,[7] vortex for 10," and then maintain the tube at room temperature for approximately 5 min.[8]
- Spin the tube for 5 min at maximum speed (14,000 rpm) in a microcentrifuge and then carefully remove and discard the supernatant using a micropipette or a glass Pasteur pipette.[9]
- Repeat wash with a fresh aliquot of xylene.

---

[2]It is possible to digest the proteins using the following buffer: PCR buffer 1× final (10 mM Tris–HCl pH 8.3, 50 mM KCl) and Proteinase K, 1 mg/ml final. The use of this buffer, without EDTA and detergent, is suggested to avoid the possible inhibition of PCR reaction by the omitted reagents.

[3]We strongly recommend purchase of saturated phenol pH 8 from a commercial manufacturer.

[4]Clean the pipettes with alcohol or another disinfectant and leave them under the UV lamp for 10 min. Alternatively, it is possible to autoclave the pipette depending on the provider instructions.

[5]Clean the microtome with xylene.

[6]Deparaffinization step could be completely skipped; alternatively, it could be performed by adding 300 µl of mineral oil to the tube containing the section and incubating at 90°C for 20 min to dissolve the wax [8].

[7]When working with xylene, avoid breathing fumes. It is better to perform the deparaffinization step under a fume hood.

[8]Wear gloves when isolating and handling DNA to minimize the contamination with exogenous nucleases. Use autoclaved pipette tips and 1.5 ml microcentrifuge tubes.

[9]Xylene is harmful; the wasted xylene must be collected in a chemical waste container and discharged according to the local hazardous chemical disposal procedures.

- Wash the pellet by adding 1 ml of absolute ethanol. Flick the tubes to dislodge the pellet and then vortex the tubes for 10 s.
- Leave at room temperature for approximately 5 min.
- Spin the tube for 5 min at maximum speed (14,000 rpm) in a microcentrifuge and then carefully remove and discard the supernatant.
- Repeat washes using 90% and 70% ethanol.
- After removing 70% ethanol, allow the tissue pellet to air dry in a thermoblock at 37°C for about 30 min.

7.2.3.3 Proteolytic Digestion and DNA Extraction

- Add to the tissue pellet 150–300 µl of digestion buffer 1x supplemented with Proteinase K at final concentration of 1 mg/ml. The amount of digestion buffer depends on the tissue amount. The digestion buffer must cover the tissue pellet completely.[10]
- Incubate in the thermomixer for 48–72 h[11] at 55°C, shaking moderately. For longer digestion, Proteinase K can be added again every 24 h.
- Add 1 volume of phenol[12]-buffered pH 8.0 Tris/CHCl$_3$ /isoamyl alcohol (50:49:1 v/v/v).[13] Mix well by inverting the tube, and leave on ice for 10–20 min.
- Centrifuge at 14,000 rpm at 4°C for 20 min. The mixture will separate into a lower organic phase, an interphase, and an upper aqueous phase.
- Transfer the upper phase into a new tube, add 1 volume of CHCl$_3$, mix well for 5 min, and centrifuge at 14,000 rpm at 4°C for 20 min.

- Transfer the supernatant in a new tube containing 5 µl of glycogen solution (1 mg/ml stock) as precipitation carrier. Carefully avoid transferring the interphase containing proteins.
- Precipitate overnight at –20°C with 1 volume of iso-propanol or 2.5 volumes of EtOH supplemented with 0.1 volumes of Sodium acetate 3 M pH 7.
- Centrifuge at 14,000 rpm at 4°C for 20 min and discard the supernatant.
- Wash the pellet with 200 µl of 70% ethanol without resuspending the pellet to wash away the remaining salts.
- Air dry the pellet and resuspend the DNA pellet in the appropriate amount of TE buffer 1×. Store the DNA solution at –20°C.
- For DNA measurement, pipette 199 µl of sterile water into a fresh tube and add 1 µl of DNA extract (dilution factor = 200). Determine the DNA concentration photometrically at 260 and 280 nm (see Chap. 16, Sect. 16.2.1 for more details).[14]

7.2.4 Troubleshooting

- If the DNA yield is low, you may have lost the DNA pellet; in such case, repeat the entire process of extraction.
- If the pellet is not visible after centrifugation, the precipitation could have been incomplete because of the absence of a precipitation carrier. Add 5 µl of glycogen 1 mg/ml, and leave at –20°C overnight to complete precipitation.
- If DNA is absent, a nuclease contamination could have occurred. In such case, repeat the extraction using freshly made reagents.

---

[10]If the pellet is firmly lodged at the bottom of the tube, it is possible to dislodge it in the digestion buffer using a sterile toothpick.

[11]Longer digestion time (at least 48 h) increases the yield of the DNA.

[12]Phenol is very toxic and should be handled in a fume hood; the wasted phenol must be collected with hazardous chemical waste.

[13]The extraction can also be performed with 1volume of phenol (Tris saturated)-chloroform-(50:50, v/v). Phenol is an inhibitor of PCR reaction, because of Taq Polymerase inactivation. A single chloroform-isoamyl alcohol (24:1, v/v) extraction could be performed after the phenol (Tris saturated)-chloroform-isoamyl alcohol extraction in order to completely remove phenol traces.

---

[14]The concentration of dsDNA expressed in µg/µl is obtained as follows: $[DNA] = A_{260} \times$ dilution factor $\times 50 \times 10^{-3}$ (see Chap. 16). A clean DNA preparation should have a A_{260}/A_{280} ratio of 1.5–2. This ratio is decreased by the presence of proteins, oligo-, and polysaccharides. Concentration estimation can also be affected by phenol contamination, as phenol absorbs strongly at 260 nm and therefore can mimic higher DNA yield and purity.

References

1. Gilbert MT, Haselkorn T, Bunce M, Sanchez JJ, Lucas SB, Jewell LD, Van Marck E, Worobey M (2007) The isolation of nucleic acids from fixed, paraffin-embedded tissues-which methods are useful when? PLoS ONE 2(6):e537
2. Lassmann S, Gerlach UV, Technau-Ihling K, Werner M, Fisch P (2005) Application of BIOMED-2 primers in fixed and decalcified bone marrow biopsies: analysis of immuno-globulin H receptor rearrangements in B-cell non-Hodgkin's lymphomas. J Mol Diagn 7(5):582–591
3. Lehmann U, Kreipe H (2001) Real-time PCR analysis of DNA and RNA extracted from formalin-fixed and paraffin-embedded biopsies. Methods 25(4):409–418
4. Pauluzzi P, Bonin S, Gonzalez Inchaurraga MA, Stanta G, Trevisan G (2004) Detection of spirochaetal DNA simultaneously in skin biopsies, peripheral blood and urine from patients with erythema migrans. Acta Derm Venereol 84(2):106–110
5. Bonin S, Petrera F, Niccolini B, Stanta G (2003) PCR analysis in archival postmortem tissues. Mol Pathol 56(3):184–186
6. Bonin S, Petrera F, Stanta G (2005) PCR and RT-PCR analysis in archivial postmortem tissues. In: Encyclopedia of diagnostic genomics and proteomics. Marcel Dekker, New York
7. Bonin S, Hlubek F, Benhattar J, Denkert C, Dietel M, Fernandez PL, Hofler G, Kothmaier H, Kruslin B, Mazzanti CM, Perren A, Popper H, Scarpa A, Soares P, Stanta G, Groenen PJ (2010) Multicentre validation study of nucleic acids extraction from FFPE tissues. Virchows Arch 457(3):309–317
8. Lin J, Kennedy SH, Svarovsky T, Rogers J, Kemnitz JW, Xu A, Zondervan KT (2009) High-quality genomic DNA extraction from formalin-fixed and paraffin-embedded samples deparaffinized using mineral oil. Anal Biochem 395(2):265–267

DNA Extraction from Formalin-Fixed Paraffin-Embedded Tissues (FFPE) (from Small Fragments of Tissues or Microdissected Cells)

8

Serena Bonin, Patricia J.T.A. Groenen, Iris Halbwedl, and Helmut H. Popper

Contents

S. Bonin (✉)
Department of Medical, Surgical and Health Sciences,
University of Trieste, Cattinara Hospital, Strada di Fiume 447,
Trieste, Italy

P.J.T.A. Groenen
Department of Pathology, Radboud University Nijmegen
Medical Centre, Nijmegen, The Netherlands

I. Halbwedl and H.H. Popper
Research Unit for Molecular Lung and Pleura
Pathology, Institute of Pathology, Medical
University of Graz, Graz, Austria

8.1 Introduction and Purpose

This protocol was proposed further to the comparison with the original methodologies for DNA extraction from formalin-fixed and paraffin-embedded (FFPE) tissues used by the participants in the IMPACTS project [1–4]. This procedure provides a home-made method of obtaining DNA suitable for PCR analyses especially from small specimens. The described method is dedicated mainly to microdissected tissues. In order to reduce the loss of DNA, the extraction step with phenol-chloroform and alcohol precipitation is skipped. The protocol for Proteinase K deactivation is based on temperature denaturation. The time required for the whole procedure is 3 days.

Commercial kits are also available for DNA extraction from FFPE, and some of these are specifically dedicated to archive tissues (i.e., QIAamp DNA FFPE Tissue Kit). However, some minor modifications could be useful to achieve better results depending on the type of tissue and requirements for the specific molecular test should be taken into consideration [5].

8.2 Protocol

8.2.1 Reagents

Note: Reagents from specific companies are reported here, but similar reagents of other companies can be used:

- *Xylene* (Fluka or Sigma-Aldrich)
- *Absolute ethanol* (Sigma), 90% and 70% ethanol

G. Stanta (ed.), *Guidelines for Molecular Analysis in Archive Tissues*,
DOI: 10.1007/978-3-642-17890-0_8, © Springer-Verlag Berlin Heidelberg 2011

- *20 mg/ml Proteinase K* (stock solution): (Sigma P2303) Dissolve 100 mg of Proteinase K in 5 ml of autoclaved 50% glycerol diluted in sterile H_2O.[1] Store at −20°C. The stock solution is 20 mg/ml, and the final concentration of Proteinase should be 1 mg/ml in digestion buffer
- *10X PCR Buffer without MgCl₂*: 500 mM KCl, 100 mM Tris pH 8.3
- *Digestion Buffer*: 1X PCR buffer without $MgCl_2$ supplemented with Proteinase K 1 mg/ml final concentration just before use

8.2.2 Equipment

- *Disinfected[2] adjustable pipettes*, range: 2–20 µl, 20–200 µl, 100–1,000 µl
- *Nuclease-free aerosol-resistant pipette tips*
- *1.5 ml tubes* (autoclaved)
- *Single-packed toothpicks*
- *Sterile or disposable tweezers*
- *Microtome, with new blade*
- *Centrifuge* suitable for centrifugation of 1.5 ml tubes at 13,200 or 14,000 rpm
- *Thermoblock*
- *Thermomixer* (e.g., Eppendorf)
- *SpectroPhotometer*

8.2.3 Method

8.2.3.1 Sample Preparation

- Cool the paraffin blocks at −20°C or on dry ice in aluminum foil in order to cut the sections.
- Using a clean, sharp microtome blade, cut two to ten sections of about 5 µm thickness. Discard the first section and displace the others in a clean 1.5 ml tube using a sterile toothpick or tweezers. Alternatively,

the tissues can previously be laser capture microdissected directly from the sections (see Chaps. 4 and 6). Use some sections from the paraffin block without the included tissue, but processed together with other samples, for negative control analysis.

8.2.3.2 Deparaffinization (Optional[3])

- Add 1 ml of xylene,[4] vortex for 10" and then maintain the tube at room temperature for approximately 5 min.[5]
- Spin the tube for 5 min at maximum speed (14,000 rpm) in a microcentrifuge and then carefully remove and discard the supernatant using a micropipette or a glass Pasteur pipette.[6] Repeat wash with a fresh aliquot of xylene.
- Wash the pellet adding 1 ml of absolute ethanol. Flick the tubes to dislodge the pellet and then vortex the tubes for 10 s. Leave at RT for approximately 5 min.
- Spin the tube for 5 min at maximum speed (14,000 rpm) in a microcentrifuge and then carefully remove and discard the supernatant.
- Repeat washes using once 90% and 70% ethanol.
- After removing 70% ethanol allow the tissue pellet to air dry in a thermoblock at 37°C for about 30 min.

8.2.3.3 Proteolytic Digestion and DNA Extraction

- Add to the tissue pellet 20–100 µl of 1× digestion buffer supplemented with Proteinase K at the final concentration of 1 mg/ml. The amount of

---

[1]The solubilization of Proteinase K in 50% sterile glycerol maintains the solution fluid at −20°C with a better preservation of the enzymatic activity.

[2]Clean the pipettes with alcohol or another disinfectant and leave them under the UV lamp for at least 10 min. Alternatively, it is possible to autoclave the pipette depending on the provider instructions.

[3]It is possible to bypass the deparaffinization step by adding the digestion buffer directly to the cut sections. In this case, it is better to add a short incubation time of 5 min at 65°C to melt the paraffin rapidly [6].

[4]When working with xylene, avoid breathing fumes; it is better to perform the deparaffinization step under a fume hood.

[5]Wear gloves when isolating and handling DNA to minimize the contamination with exogenous nucleases. Use autoclaved pipette tips and 1.5 ml microcentrifuge tubes.

[6]Xylene is harmful; the wasted xylene must be collected in a chemical waste container and discharged according to the local hazardous chemical disposal procedures.

digestion buffer depends on the section size. The digestion buffer must cover the tissue pellet[7] completely.

- Incubate for 48–72 h[8] at 55°C, shaking moderately. For longer digestion, Proteinase K can be added every 24 h.
- Inactivate Proteinase K by heating at 95°C for 10 min.
- Centrifuge at room temperature for 10 min at maximum speed (14,000 rpm); transfer the supernatant to a new tube and store at −20°C.
- For DNA measurement, pipette 199 μl of sterile water into a fresh tube and add 1 μl of DNA extract (dilution factor = 200). Determine the DNA concentration photometrically at 260 and 280 nm (see Chap. 16 for more details).[9]

8.2.4 Troubleshooting

- If DNA yield is low or DNA is not detected by UV measurement, decrease the dilution factor and measure the concentration again.
- If no DNA has been extracted, a nuclease contamination could have occurred. In such case, repeat the extraction using freshly made reagents.

References

1. Gilbert MT, Haselkorn T, Bunce M, Sanchez JJ, Lucas SB, Jewell LD, Van Marck E, Worobey M (2007) The isolation of nucleic acids from fixed, paraffin-embedded tissues-which methods are useful when? PLoS ONE 2(6):e537
2. Lassmann S, Gerlach UV, Technau-Ihling K, Werner M, Fisch P (2005) Application of BIOMED-2 primers in fixed and decalcified bone marrow biopsies: analysis of immunoglobulin H receptor rearrangements in B-cell non-Hodgkin's lymphomas. J Mol Diagn 7(5):582–591
3. Lehmann U, Kreipe H (2001) Real-time PCR analysis of DNA and RNA extracted from formalin-fixed and paraffin-embedded biopsies. Methods 25(4):409–418
4. Pauluzzi P, Bonin S, Gonzalez Inchaurraga MA, Stanta G, Trevisan G (2004) Detection of spirochaetal DNA simultaneously in skin biopsies, peripheral blood and urine from patients with erythema migrans. Acta Derm Venereol 84(2): 106–110
5. Bonin S, Hlubek F, Benhattar J, Denkert C, Dietel M, Fernandez PL, Hofler G, Kothmaier H, Kruslin B, Mazzanti CM, Perren A, Popper H, Scarpa A, Soares P, Stanta G, Groenen PJ (2010) Multicentre validation study of nucleic acids extraction from FFPE tissues. Virchows Arch 457(3): 309–317
6. Coombs NJ, Gough AC, Primrose JN (1999) Optimisation of DNA and RNA extraction from archival formalin-fixed tissue. Nucleic Acids Res 27(16):e12

---

[7] If the pellet is firmly lodged at the bottom of the tube, it is possible to dislodge it in the digestion buffer using a sterile toothpick.

[8] Long digestion (at least 48 h) increases the yield of the DNA.

[9] The concentration of dsDNA expressed in μg/μl is obtained as follows: $[DNA] = A_{260} \times$ dilution factor $\times 50 \times 10^{-3}$. A clean DNA preparation should have a A_{260}/A_{280} ratio of 1.5–2. This ratio is decreased by the presence of proteins, oligo- and polysaccharides.

Fast Protocol for DNA Extraction from Formalin-Fixed Paraffin-Embedded Tissues

9

Falk Hlubek and Andreas Jung

Contents

9.1 Introduction and Purpose

This protocol provides a fast method to obtain a crude extract of genomic DNA from microdissected or entire sections of FFPE tissue for PCR analysis. It is based on the method described by Higuchi [1] with modifications in the deparaffinization steps and in the procedure for recovering the tissue. For the described home-made method, approximately 1 cm^2 tissue (3–5 μm thick sections) is required. Sectioning is followed by deparaffinization, tissue microdissection when needed, and proteolytic hydrolysis with Proteinase K. There are also available commercial kits for extraction of DNA from FFPE (i.e., Qiagen, Roche, Ambion…), to this purpose visit the companies' websites for instruction.

The time required for the complete procedure is about 1.5 h and an additional overnight incubation.

9.2 Protocol

9.2.1 Reagents

Note: Reagents from specific companies are reported here, but might be substituted by reagents of comparable quality from other vendors.

- *Xylene* (ultrapure) (AppliChem, Fluka or Sigma-Aldrich)
- *Absolute ethanol* (AppliChem, Sigma-Aldrich)
- *25 mg/ml Proteinase K, stock solution* (Sigma P2308): Dissolve 25 mg of Proteinase K in 1 ml of dd H$_2$O

F. Hlubek (✉) and A. Jung
Department of Pathology, LMU, Munich, Germany

[1]Alternatively, the ready-to-use Proteinase K solution (Qiagen 19133) can be used; it has a concentration of about 20 mg/ml (store in aliquots at +4°C).

G. Stanta (ed.), *Guidelines for Molecular Analysis in Archive Tissues*,
DOI: 10.1007/978-3-642-17890-0_9, © Springer-Verlag Berlin Heidelberg 2011

(final concentration of Proteinase K is 3.26 mg/ml in digestion buffer). Store the solution in aliquots at −20°C. Do not refreeze after thawing.[1]

- *1× Hydrolysis buffer*: 50 mM Tris HCl pH 8.5, 1 mM EDTA pH 8.0, 0.5% (v/v) Tween 20. Store the buffer in aliquots at −20°C. Do not refreeze after thawing.

9.2.2 Equipment

- *Clean*[2*] *adjustable pipettes,* range: 1–10 µl, 10–100 µl, 100–1,000 µl
- *Nuclease-free aerosol-resistant pipette tips*
- *2 ml Safelock and 1.5 ml microreaction tubes (nuclease free)*
- *Microtome, with new blades*
- *Cooling plate for cooling down the paraffin-embedded tissue*
- *Centrifuge* suitable for centrifugation of 2 ml microreaction tubes at ~17,900 × g
- *Thermoblock*
- *Thermomixer* (e.g., Eppendorf)
- *UV-light Photometer (wavelength [λ]: 260 and 280 nm)*

9.2.3 Method

- Cool the paraffin blocks on a cooling plate or at −20°C in order to cut the sections.
- In order to avoid DNase cross-contaminations use a new, clean microtome blade for every paraffin block and clean the microtome with xylene. Cut 3–5 µm-thick sections together representing at least 1 cm^2 area of tissue and mount them on microscope slides. As a negative control an open microreaction tube containing Hydrolysis buffer may be put in the area of the cutting. It is closed after the cutting procedure and run in parallel with the other microreaction tubes containing tissue material.

- Deparaffinize sections twice in xylene for 10 min each and twice in absolute ethanol for 10 min each before air drying.
- Moisten the tissue with the Hydrolysis buffer and scratch off the tissue using another sterile microscope slide in a way one would prepare blood smears. Transfer the moist tissue clump into a 2 ml Safelock microreaction tube.[3] If microdissection is necessary, the tissue might be microdissected by scraping off the area of interest from the section using a sterile scalpel blade and transferring it into a new 2 ml Safelock microreaction tube. It might be helpful to moisten the tip of the blade so that the scratched-off tissue will adhere to it.
- Add 200 µl Hydrolysis buffer. The Hydrolysis buffer should cover the tissue completely. If the tissue adheres to the wall of the microreaction tubes, briefly spin it down.
- Add 30 µl Proteinase K (25 mg/ml) to a final concentration of 3.26 mg/ml.
- Incubate overnight (at least 16 h); but for microdissected tissue, incubate for a maximum of 5 h at 56°C while shaking (450 rpm).
- Inactivate Proteinase K by incubation at 95°C for 10 min.
- Centrifuge for 10 min at 14,000 rpm at room temperature.
- Transfer the supernatant into a new microreaction tube (0.5 or 1.5 ml tube) and discard the pellet.
- Determine the DNA concentration photometrically at 260 nm[4] and use the digestion buffer as blank. The A260/A280 ratio will be low because the lysate is a crude DNA extract containing cellular proteins. When working with microdissected tissue fragments it might be difficult or even impossible to obtain reliable absorbtion values at A260.
- Use the DNA extract for PCR analysis, or store it at −20°C.

---

[2]Clean pipettes with DNase Away™ to avoid DNase and DNA contamination. Alternatively, it is possible to clean pipettes by first using mild detergent containing aqueous solutions, followed by application of antiseptic alcohol solution (e.g., 70% (v/v) ethanol) or another disinfectant and then leaving them under UV light for at least 10 min.

[3]Depending on the Thermomixer, smaller microreaction tubes may be used.

[4]The concentration of dsDNA expressed in µg/µl is obtained as follows: $[DNA] = A_{260} \times$ dilution factor $\times 50 \times 10^{-3}$. A clean DNA preparation should have a A_{260}/A_{280} ratio of 1.5–2. This ratio is decreased by the presence of proteins, oligo- and polysaccharides. Concentration estimation can be also affected by phenol contamination, as phenol absorbs strongly at 260 nm and therefore can mimic higher DNA yield and purity.

9.2.4 Troubleshooting

- If the yield of DNA is low or DNA is absent by UV measurement, decrease the dilution factor and measure the concentration again.
- If DNA is absent, a nuclease contamination could have occurred. In such case, clean the workspace and pipettes with DNase Away™ and repeat the extraction using freshly made reagents. These hints may not apply when working with tissue fragments isolated with the help of microdissection. In these cases, the DNA contents may not be detectable in the crude lysates or by photometry.

Reference

1. Higuchi R (1989) Simple and rapid preparation of samples for PCR. PCR technology; principles and applications for DNA amplification. Stockton, New York

DNA Extraction from Blood and Forensic Samples

10

Solange Sorçaburu Cigliero, Elisabetta Edalucci, and Paolo Fattorini

Contents

S.S. Cigliero and P. Fattorini (✉)
Department of Medicine, Surgery and Health,
University of Trieste, Trieste, Italy

E. Edalucci
Department of Nuclear Medicine, Azienda U-O di Trieste,
Trieste, Italy

10.1 Introduction and Purpose

DNA is usually extracted from fresh blood. However, successful individual-specific DNA profiles are routinely obtained from any biological source (saliva, hair, semen, etc.) containing nucleated cells, even from the cellular debris left on a touched object [1]. All this makes DNA analysis an irreplaceable tool for personal identification in Forensic Medicine and allows, in the Clinical Laboratory, the resolution of real or suspected specimen mislabelling. In the same manner, genetic data can be gathered from aged samples, such as skeletal remains or museum specimens [2]. Nevertheless, since a higher number of PCR cycles is usually required to produce amplicons from such samples, the risk of contamination has to be always considered [1–3]. Thus, rigorous precautions have to be adopted alongside the extraction procedures both to prevent and identify the exogenous contamination which can be inadvertently introduced. Reagents, disposables, pipettes, gloves, etc. can be important sources of contamination and even the operator can contaminate the sample by his/her DNA (by breathing, for example). For all these reasons, particular precautions have to be adopted when handling forensic/aged samples. Moreover, the size of the forensic specimen is usually very small so that the risk of mistyping due to exogenous contamination increases exponentially [1, 2].

10.2 Precautions

The usual precautions adopted in each laboratory are enough in case of DNA extraction from fresh blood, while particular precautions need to be introduced when handling both forensic and aged specimens.

The particular precautions suggested for DNA extraction from minute/aged samples are the following:

- DNA extraction should be performed in a separate room dedicated solely for this purpose.
- Reagents, pipettes and disposables employed in that room should be dedicated solely to DNA extraction.
- Pipettes have to be cleaned often with a solution of bleach in water (10% v/v).
- Reagents are intended to be autoclaved or UV treated.
- Disposables should be UV treated.
- Nuclease-free aerosol-resistant pipette tips are compulsory.
- Avoid handling large volumes of working solutions: it is preferable to handle small-volume solutions.
- Wear disposable latex gloves (to be changed often), mask and white coat. The white coat should be left in the extraction room and dedicated solely for this purpose. In particular cases, wear a surgical cap and a mono-use white coat.
- Discard the tips into a mono-use plastic container to which 10% of bleach in water has been added.
- Do not extract the forensic sample (e.g. hair) and the reference samples (e.g. fresh blood) simultaneously.
- Before opening the Eppendorf tubes containing DNA, centrifuge them briefly to spin down any trace of the sample from the tube cover: your thumb could be an important source of cross-contamination.
- After use, clean the working desk and the inside part of the centrifuge with 10% of bleach.
- Metallic instruments (forceps, scissors etc.) can be sterilized on the Bunsen flame.
- Alongside the extraction procedure, always introduce at least one blank extraction control (BEC). Most simply, you have to perform a step-by-step extraction from a lysis buffer in which no biological components have been added (remember that the BEC validates your PCR-based results; see also Sect. 10.5.
- Record in a database the genetic profiles of all the samples analysed in your laboratory as well as the genotypes of the operators.

10.3 DNA Isolation from Blood Samples

This protocol is suitable for EDTA/blood. This protocol allows recovery of up to 30–40 μg of DNA, so it is suitable for long-term replicate uses of the samples (population genetics, for example). If the sample is a clot, put a small part of it into an Eppendorf tube and start from step 7 of Method 10.4.1.3.

10.3.1 Reagents

- *Absolute ethanol*: Fluka (02,860) and ethanol (70%) in water (v/v).
- *10 mg/ml Proteinase K* (stock solution): Sigma (P2308). Dissolve 100 mg of Proteinase K in 10 ml of autoclaved 50% glycerol. Store at −20ºC.
- *3 M Sodium acetate solution*: Fluka (71,196). Dilute to 0.2 M with water (Milli Q) to provide the working solution.
- *Sodium dodecyl sulphate* (SDS): Sigma (L-71,725). Prepare a 10% (w/v) solution in water. Weigh the SDS in a fume cupboard, wearing the mask.
- *8-Hidroxy-quinoline*: Sigma (H 6,752).
- *Phenol*: Sigma (P9346).
- *Chloroform* (CHCl$_3$): Fluka (25,670).
- *Isoamyl alcohol* (3-methylbutanol): Sigma (I9392).
- *Phenol/CHCl$_3$/isoamyl alcohol* (25:24:1 v/v/v) solution. Mix 25 ml of Phenol, 24 ml of Chloform, 1 ml of Isoamyl alcohol, and 50 mg of 8-Hydroxy-quinoline. Remember that these substances are toxic: follow the recommendations reported on the bottles. The mixture is then equilibrated by forming an emulsion with an equal volume of 10 mM Tris pH 7.5. Allow the phases to separate, remove the upper phase and repeat the process. Store at 4°C in a light-tight bottle.

10.3.2 Equipment

- *Disinfected adjustable pipettes* with the following ranges: 0.5–10 μl, 20–200 μl, and 100–1,000 μl
- *Nuclease-free aerosol-resistant pipette* tips
- *Eppendorf tubes*
- *Centrifuge* suitable for centrifugation of Eppendorf tubes at full speed (12,000–14,000 rpm)

10.3.3 Method

Red Cells Lysis

1. Flip the blood tubes several times to ensure homogeneity. Transfer a 0.7 ml aliquot into an Eppendorf tube.

2. Adjust to 1.5 ml with sterile water; then centrifuge the tubes at full speed (12,000–14,000 rpm) for 1 min in a micro-centrifuge.

3. Discard 1 ml from the supernatant of each tube without disturbing the leukocyte pellet.

4. Add 1 ml of sterile water to each tube and gently flip them. Centrifuge as described above. Remove and discard the supernatant.

5. Add to each leukocyte pellet 375 µl of 0.2 M Na-Acetate pH 7.0, 25 µl of 10% SDS, and 15 µl of Proteinase K solution.

6. Resuspend the pellet by vortexing. Incubate at 37°C for at least 1 h (overnight incubation is suggested for a higher recovery of DNA).

Phenol/Chloroform/isoamyl alcohol purification

7. Spin down the condensate from the tube walls.

8. Add 200 µl of the Phenol/CHCl$_3$/Isoamyl alcohol (25:24:1 v/v/v) solution and shake for about 10–20 s to obtain an emulsion.

9. Centrifuge the tubes for 2 min at full speed. The upper aqueous phase[1] is transferred into a new Eppendorf tube.

10. Precipitate the DNA by adding 1 ml of 100% ethanol. Close the tube and gently invert it until the DNA is visible as a clot.

11. Spin down the precipitate. Remove and discard the supernatant without disturbing the pellet.

12. Resuspend the pellet in 200 µl of 0.2 M Na-Acetate. Incubate in a water bath at 55°C for about 10 min or until suspended.

13. Precipitate the DNA with 500 µl of 100% ethanol as described in step 10 and repeat step 11.

14. Wash the sample by adding 1 ml of 70% ethanol and repeat point 11. Dry the pellet at 37 °C in an oven.

15. Add 200–400 µl of sterile water. Redissolve the sample at 37 °C for about 2–6 h. Store at −20°C until use.

10.4 DNA Isolation from Forensic Samples

Forensic analysis requires procedures which enable the isolation of genomic DNA from a big variety of biological samples usually adhering to solid (wood, glass, metal, plastic) or soft substrata (cloth, paper, cardboard, carpet, etc.). In addition, in some circumstances the biological specimens can be so tiny that they are not visually appreciable.

One of most relevant concern in handling all these samples is represented by the chemical composition of the substrata themselves which can contain unknown substances inhibiting *Taq I* polymerase, so leading to PCR failure. Empirically, the smaller the amount of the treated substratum, the higher the chance of a successful outcome of the PCR.

The real case-work offers a virtually infinite set of substrata. However, from a practical point of view, they can be divided into:

- Absorbing substrata (AS): cloth, paper, wood, etc.
- Non-absorbing substrata (NAS): metal, plastic, nylon, etc.

As a basic strategy, while for AS it is preferable to put a small part of the whole specimen into the extraction solution, for NAS it is more convenient to remove the biological component from the substratum before DNA extraction. DNA extraction from bones, soft tissues, saliva swabs and post coital swabs will be considered separately.

10.4.1 DNA Extraction from Absorbing Substrata

This protocol is suitable for blood stains, semen stains,[2] saliva stains, cigarette butts,[3] etc. It is always recommended that at least half of the specimen is stored for confirming tests.

10.4.1.1 Reagents

Note: In addition to Reagents of Sect. 10.3.1

- *Physiological solution*: (145 mM NaCl)
- *1× Lysis buffer (LB)*: (2% SDS, 100 mM NaCl, 40 mM DTT, 10 mM Tris pH 7.5, 10 mM Na$_2$EDTA pH 8.0). Aliquot small volumes of this solution

---

[1]Always take care not to remove any of the interface material or the lower organic phase.

[2]For a differential lysis see Sect. 10.4.7

[3]For cigarette butts, remove the outer part of the filter using a blade.

(5–10 ml) and store them at −20°C. Before use, melt at 37°C

- *CHCl₃/Isoamyl alcohol* (24:1 v/v) solution. Mix 24 ml of Chloroform and 1 ml of Isoamyl alcohol. The mixture is equilibrated by forming an emulsion with 10 mM Tris pH 7.5 (equal volume). Allow the phases to separate, remove the upper phase and repeat the process. Store at 4°C in a light-tight bottle
- *Glycogen*: 10 mg/ml in water (Bioline, BIO-37,077)

10.4.1.2 Equipment

Note: In addition to Equipment of Sect. 10.3.2

- *Sterile surgical blades and tweezers*

10.4.1.3 Method

1. Cut a piece of the stain (about 10–15 mm²) on a Petri dish and put it into an Eppendorf tube.[4]
2. Add 1 ml of physiological solution.
3. Incubate the sample at room temperature for about 30 min, with occasional shaking.
4. Using a tip remove the substratum, squeezing it well.
5. Centrifuge the sample at full speed for 5 min (12,000–14,000 rpm).
6. Remove the supernatant without disturbing the pellet.
7. Resuspend the pellet in 0.5–0.75 ml of LB and 20–30 µl of Proteinase K.
8. Incubate at 37 °C for at least 3–6 h. Centrifuge briefly.
9. Add 300 µl of Phenol/CHCl₃/Isoamyl alcohol (25:24:1 v/v/v).
10. Vortex for about 10 s.
11. Centrifuge for 2 min at full speed (12,000–14,000 rpm).
12. Transfer the supernatant in a new tube.
13. Repeat steps 9–12.
14. Add 300 µl of CHCl₃/Isoamyl alcohol(24:1)
15. Vortex and centrifuge as described previously.

16. Transfer the supernatant in a new tube.
17. Precipitate the samples by adding 1/10 volume of 3 M Na-acetate pH 7.5, 2.5 volumes of absolute ethanol and 1 µl of (10 mg/ml) glycogen. Leave at −20°C for at least 3 h.
18. Centrifuge for 15 min at full speed[5].
19. Discard the supernatant by a tip.
20. Wash by adding 1 ml of 70% ethanol; invert the tube a few times and centrifuge at full speed for 30 s.
21. Discard the supernatant as above.
22. Dry the DNA pellet in an oven at 37°C.
23. Add appropriate volume (20–50 µl) of sterile water, resuspend the sample and store it at −20°C until use.

10.4.2 DNA Extraction from Non-absorbing Substrata

This protocol is suitable for DNA extraction from any kind of biological material left on non-absorbing substrata such as plastic, nylon, glass, etc.

10.4.2.1 Reagents

Note: In addition to Reagents of Sect. 10.4.1.1

- *2× Lysis buffer* (4% SDS, 200 mM NaCl, 80 mM DTT, 20 mM Tris pH 7.5, 20 mM Na₂EDTA pH 8.0)

10.4.2.2 Equipment

Note: In addition to Equipment of Sect. 10.3.2
None

10.4.2.3 Method

1. Using a tipped pipette, dissolve the stain on the surface using 150 µl of water (see footnote 4).

---

[4]If the stain is actually small (or not visible), put the substratum directly into 500 µl of LB to which 30 µl of Proteinase K has been added and start from point 8 of this method. Remember to remove the substratum before phenol purification.

[5]When inserting the Eppendorf tube into the rotor, mark (or remember) the position of the cover hinge. This will help you in identifying the location of the pellet.

2. Transfer this solution into an Eppendorf tube.
3. Repeat step 1 on the surface of the substratum.
4. To about 300 µl of the recovered solution, add an equal volume of 2× Lysis buffer and 30 µl of Proteinase K.
5. Proceed as described from step 8 onwards in Method Sect. 10.4.1.

10.4.3 DNA Extraction from Saliva Swabs

In the forensic practice, it is a validated procedure to collect a saliva swab as reference sample. This is a safe and robust protocol which allows obtaining enough DNA to perform tens of genetic tests.

10.4.3.1 Reagents

Note: In addition to Reagents of Sect. 10.4.1.1
   None

10.4.3.2 Equipment

Note: In addition to Equipment of Sect. 10.3.2
   None

10.4.3.3 Method

1. Isolate the swab from the stick used to collect the saliva from the mouth and put it into an Eppendorf tube.[6]
2. Add 1 ml of physiological solution and incubate at room temperature for 5–10 min, stirring occasionally to remove the cells from the swab.
3. With a sterile tip, remove the cotton swab from the stick, squeezing it well.
4. Centrifuge for 2 min at full speed and discard the supernatant.

5. Add to the pellet 500 µl of 0.2 M Na-Acetate, 25 µl of 10% SDS and 30 µl of Proteinase K (10 mg/ml). Resuspend the pellet. Incubate at 37°C for at least 2–3 h.
6. Purify the sample by adding 250 µl of phenol/chloroform/isoamyl alcohol (25:24:1) as usual.
7. Precipitate by adding 1 ml of 100% ethanol and 1 µl of (10 mg/ml) glycogen. Leave at −20°C for at least 3 h.
8. Centrifuge at full speed for 10 min. Discard the supernatant and wash the pellet with 1 ml of 70% ethanol without disturbing the pellet.
9. Centrifuge for 10 s.
10. Discard the ethanol without disturbing the pellet.
11. Air dry and resuspend in an appropriate volume (20–40 µl) of water. Store at −20°C until use.

10.4.4 DNA Extraction from Soft Tissues

Each biological tissue contains DNA suitable for genetic typing. However, from decomposed bodies it is preferable to perform DNA extraction from connective tissues, as ligaments, cartilages, nails [4], etc.

10.4.4.1 Reagents

Note: In addition to Reagents of Sect. 10.4.1.1
   None

10.4.4.2 Equipment

Note: In addition to Equipment of Sect. 10.3.2

• Ten millilitre tubes

10.4.4.3 Method

1. Put the sample (about 20–40 mg) in a 10 ml tube containing 5 ml of physiological solution.[7]
2. Shake briefly and discard the solution.

---

[6]Use no more than half a swab and store the remaining part at −20°C.

[7]If the sample is mummified, before this step it is preferable to incubate the sample in water at room temperature for at least 12-18 h.

3. Repeat steps 1 and 2 twice.
4. Place the sample on a Petri dish.
5. Cut the sample into small pieces.
6. Put the fragments into an Eppendorf tube.
7. Add 0.5–0.75 ml of 1 X Lysis buffer and 40 µl of Proteinase K.
8. Incubate at 37°C for 2–3 h.[8]
9. Centrifuge briefly.
10. Add 20 µl of Proteinase K and proceed as described from step 8 onwards in Method "DNA extraction from absorbing substrata"

10.4.5 DNA Extraction from Hairs

Hair roots are a suitable source of nuclear DNA while only mitochondrial DNA (mDNA) typing is expected from hair shafts. However, an extensive SNP analysis from a large ≈4,000-year-old-permafrost-preserved hair tuft was recently reported [6].

10.4.5.1 Reagents

Note: In addition to Reagents of Sect. 10.4.1.1
   None

10.4.5.2 Equipment

Note: In addition to Equipments of Sect. 10.3.2
   None

10.4.5.3 Method

1. Identify the hair root by a magnifying lens.
2. Cut about 8–10 mm of that end with a blade.
3. Place the sample into an Eppendorf tube.
4. Add 0.5–0.75 ml of 1 X Lysis buffer and 40 µl of Proteinase K.
5. Proceed as described from step 8 onwards in Method "DNA extraction from absorbing substrata" Sect. 10.4.1.

---

[8]As even a temperature of 37°C causes DNA damage [3], working at room temperature is suggested for ancient samples [5].

10.4.6 DNA Extraction from Bone

Bones usually provide the last source of genetic material of any individual. Before DNA extraction, their surface needs to be cleaned carefully to remove both soil and other several sources of contamination. In addition, a decalcification step by the chelating Na_2EDTA permits a more efficient recovery of the genetic material still present in the sample [7]. Long bones (femur, for example) are usually preferable.[9]

10.4.6.1 Reagents

Note: In addition to Reagents of Sect. 10.4.1.1

• *EDTA solution*: Na_2EDTA 0.5 M pH 8.0

10.4.6.2 Equipment

Note: In addition to Equipments of Sect. 10.3.2

• *A large base (φ about 4–5 cm) tube*
• *A saw* (sterilized by UV radiation)
• *Glass paper* (sterilized by UV radiation)
• *Ultra-filtration devices*: Amicon® Ultra 100 kDa filter units (Millipore; cod. UFC 810,024 for volumes <4 ml and cod. UFC510, 024 for volumes <0.5 ml)
• *Centrifuge*: for UFC 810, 024, a suitable centrifuge allowing 4,000 rpm is required

10.4.6.3 Method

1. Remove the external surface of a section of the bone by the glass paper.
2. Slice a section of about 3–5 mm thickness with the saw.
3. Put the section in the large base tube and add about 20 ml of EDTA solution.
4. Incubate at room temperature for 18–48 h, with gentle shaking.
5. Remove the bone section and put it into a Petri dish.
6. With a blazer, remove small fragments of the inner part (medulla) of the section.

---

[9]The same protocol can be used for teeth. In this case, after cleaning the surface using a sterile drill, the tooth has to be divided by a longitudinal sawing and one of the two parts processed as described for bones.

7. Put them into an Eppendorf tube.
8. Add 0.5–0.75 ml of 1 X Lysis buffer and 40 µl of Proteinase K.
9. Incubate at 37°C for 2–3 h.[10]
10. Centrifuge briefly. Add 20 µl of Proteinase K and incubate at 37°C for another 8–12 h.
11. Proceed from step 9 to step 16 of Method of Sect. 10.4.1.
12. Purify the sample by ultra-filtration following the manufactures protocols (at least three washes are recommended).

10.4.7 Differential Lysis Between Sperm and Epithelial Cells

Vaginal or anal swabs can contain a mixture of sperm and epithelial cells. Although the employment of Y-specific STR markers allows the identification of the male compound, it is however preferable to separate the two genetic fractions. This is usually performable by a preferential lysis of the epithelial cells whose cellular membrane is more sensitive to the SDS/Proteinase K treatment. Indeed, by this procedure the more robust sperm cells can be preferentially recovered by centrifugation.

10.4.7.1 Reagents

Note: In addition to Reagents of Sect. 10.4.1.1
   None

10.4.7.2 Equipment

Note: In addition to Equipment of Sect. 10.3.2
   None

10.4.7.3 Method

Differential Lysis

1. Isolate the swab from the stick (see footnote 6) and put it into an Eppendorf tube.

2. Add 1 ml of physiological solution and incubate at room temperature for 15 min, mixing occasionally to remove the cells from the substratum.
3. With a sterile tip remove the cotton swab squeezing it well.
4. Centrifuge for 3 min at full speed and discard the supernatant without disturbing 50 µl of the pellet.[11]
5. Resuspend the pellet in 500 µl of 1× LB supplemented with 15 µl Proteinase K (10 mg/ml). Mix gently. Incubate for at least one hour at 56°C in order to lyse the epithelial cells, but for less than 2 h to minimize the lysis of the sperm cells.
6. Centrifuge for 5 min.
7. Transfer all the supernatant, except 50 µl of the pellet, into a clean Eppendorf tube (SN tube)[12]
8. Resuspend the pellet gently in 200 µl of physiological solution.
9. Centrifuge for 5 min at full speed.
10. Transfer the supernatant into the SN tube leaving 50 µl of the pellet in the original tube (OT).
11. Repeat step 8 twice by adding 200 µl of physiological solution.

DNA Extraction from the Pellet (Sperm Fraction)

Proceed from step 7 in Method of Sect. 10.4.1.3 using the pellet of the original tube (OT).

DNA Extraction from the Supernant (Epithelial Fraction of the SN Tube)

Add 30 µl of Proteinase K and 20 µl of 10% SDS to half volume of the supernatant and proceed from step 8 in Method of Sect. 10.4.1.3. Store the remaining supernatant at −20°C.

10.4.8 DNA Extraction from Urine

Also urine is a source of DNA suitable for identification purposes. Note, however, that a high amount of bacterial contamination is expected.

---

[10]As even a temperature of 37°C causes DNA damage [3], working at room temperature is suggested for ancient samples [5].

[11]The pellet can be composed both by sperm and apithelial cells. Thus, it is strongly recommended to analyze 1-2 µl microscopically.

[12]The supernatant (SN) is employed for DNA extraction from the epithelial fraction.

10.4.8.1 Reagents

Note: In addition to Reagents of Sect. 10.4.1.1
   None

10.4.8.2 Equipment

Note: In addition to Equipment of Sect. 10.3.2
   None

10.4.8.3 Method

1. Transfer 1 ml of urine into an Eppendorf tube.
2. Centrifuge at full speed for 2 min.
3. Remove the supernatant.
4. Add to the pellet 1 ml of physiological solution.
5. Resuspend the pellet gently and centrifuge at full speed for 2 min.
6. Remove the supernatant.
7. Proceed from step 7 in Method of Sect. 10.4.1

10.5 Remarks and Troubleshooting

By the employment of the protocols described above, your samples should be ready for PCR amplification. If the extraction protocol worked successfully on a suitable biological sample, the amplification of a locus with low molecular weight (no more than 200–300 bp) through 30–32 cycles of PCR should give the following results at the checking mini-gel (see Chap. 17, Sect. 17.1):

- Positive PCR control; +
- Negative PCR control: −
- Blank extraction control (BEC): −
- Unknown sample (e.g. your sample): +

If the PCR controls (both + and −) are consistent, note that a positive signal in "blank extraction control" (BEC) indicates that contamination occurred alongside the extraction procedure. In this case, it is mandatory to perform a new extraction by freshly prepared solutions and new consumables. However, genotype the amplicons obtained from the BEC in any case: this could help in the identification of the source of contamination. In addition, these undesirable results are often represented by non-specific PCR products, so a

correct interpretation of the genetic profile of the unknown sample could be however performed.

Regrettably, it is quite common that aged/forensic samples do not provide amplicons at all (− at checking mini-gel). This unsuccessful outcome of amplification is usually due to one of these two main reasons: (1) no suitable template; (2) inhibition of the PCR.

No Suitable Template

Successful amplification outcomes require minimal amounts of suitable template [3]. Nevertheless, the genetic material obtained from aged/forensic samples can be both degraded (e.g. of low molecular weight) and chemically damaged [2, 8]. This means that no (or low copy number of) entire templates are available for PCR amplification. In addition, a high degree of exogenous (fungal/bacterial) contamination can be present. All these features make the sample not or scarcely sensitive to PCR amplification. If this is the case (see Sect. 10.5.1) both a higher number of PCR cycles and a higher amount of template could be enough. As last chance, a new extraction can be performed employing a much higher amount of the original specimen.

Inhibition of the PCR

A forensic or aged biological sample can be contaminated by several compounds which are co-purified with the nucleic acids in spite of the accuracy of the protocol employed. Out of these contaminating substances, several (for example, humic acid, tannins, urea, etc.) are known to act as inhibitors of the DNA polymerases. Sometimes, the presence of such substances is revealed simply by a visual check, being the sample heavily coloured or cloudy. In other cases, a particular fluorescence is appreciable at UV transillumination (Fig. 10.1).

More often, no particular feature characterises these contaminated samples, so a logical approach has to be adopted to distinguish an actual inhibition of the polymerisation [9, 10] from an unsuitable template (no or low copy amount of entire templates).

If inhibition is identified (see the following section), several strategies can be carried out. The first attempt could be a further treatment of the sample by Proteinase K treatment, phenol extraction and ethanol

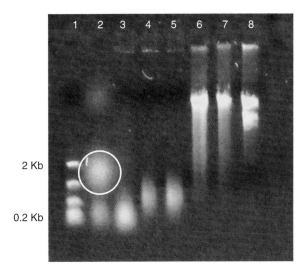

2 Kb

0.2 Kb

Fig. 10.1 UV transillumination of a 1% agarose gel (containing EtBr) after electrophoresis. The circle in lane 2 shows a grey-blue fluorescence contaminating the DNA preparation from a 2,100 year-old mummified tissue. This unknown contaminating substance was removed by ultra-filtration (lane 3). Also note the low molecular weight of this sample. Lanes 4 and 5: low molecular weight control DNA; lanes 6 and 7: high molecular weight control DNA; lane 1: easy ladder (Bioline); lane 8: λ/Bam HI

precipitation. Nevertheless, this is sample-consuming and often unsuccessful. Therefore, the employment of centrifugal filter devices is more convenient because these allow purification of the samples (see Equipment of Sect. 10.4.6). Recently, the employment of *Taq I* polymerases, which are less sensitive to inhibition, has also been described [11].

10.5.1 Distinguishing Between No Suitable Template and Inhibition of the PCR

The detection of PCR inhibitors can be performed by several methods, among which Real-Time PCR is preferred [9, 10]. Although this approach is both sensitive and accurate, it requires expensive equipments and often the availability of commercial kits. Here we describe a simple and inexpensive PCR-based method routinely employed in our Laboratory as screening assay. It is based upon the presence of an internal positive control (IPC) which is able to point out the presence of PCR inhibitors within the unknown sample. In addition, since repeated *Alu* sequences are the target of this PCR assay, a high sensitivity is also assured.

The described method allows the synthesis of a 262 bp long fragment [12] from a sub-cellular amount (<1 pg) of human control DNA.

10.5.1.1 Primers, Additional Reagents and Standards

- *Primers*:P1:5'GCCTGTAATCCCAGCACTTT3';P: 5'GAGACAGGGTCTCGCTCTG3'.
- *Mineral Oil*[13]: (Sigma, M5904).
- *High molecular weight human DNA* (CTRL sample) at the concentration of 10 pg/μl (as determined at DO_{260}/DO_{280}).

10.5.1.2 Method

PCRs are performed in a final volume of 15 μl through 28 cycles of denaturation at 94°C for 30 s, annealing at 60°C for 30 s and extension at 72°C for 30 s.

Set up the samples as follows:

- For *n* samples to test, prepare 13 μl × (2 *n* + 3) of a PCR mix containing 115 nM of each primer, 230 μM of dNTP, 1.7 mM of $MgCl_2$, 1.15 × buffer and 1.15 U of *TaqI* polymerase. In a standard assay (supposing that a single unknown sample has to be analysed) the protocol of amplification should be the following, as described in Table 10.1.

Assuming the consistency of the results from tubes 1 and 2, negative results both in tube 3 and 4 indicate PCR inhibition while PCR failure only in tube 3 shows

Table 10.1 Description of the amplification protocol

Tube	PCR mix (μl)	H₂O (μl)	CTRL sample (μl)	Unknown (μl)
1 (+PCR)	13	1	1	–
2 (–PCR)	13	2	–	–
3 Unknown	13	1	–	1
4 Competition assay	13	–	1	1

---

[13]In some thermal-cyclers mineral oil must be added to the samples to prevent their evaporation. However, when handling forensic/ancient samples it is always preferable to employ it to prevent dispersions during PCR cycles.

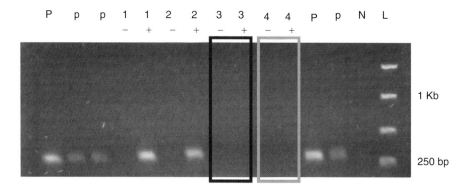

Fig. 10.2 Agarose gel electrophoresis of the *Alu* assay performed to detect PCR inhibition (30 cycles). P: positive PCR control (10 pg of high molecular weight DNA); p: positive control (1 pg of high molecular weight DNA); N: negative PCR control; 1, 2, 3 and 4 indicate four aged forensic samples amplified as alone (–) and in the presence (+) of the internal positive control (P). In samples 1 and 2 amplifications occurred only when added with the internal positive control. This indicates no or low copy numbers of entire templates. For samples 3 and 4 (indicated by the rectangles) amplification never occurred. In these cases, the presence of inhibitors has to be considered. L: molecular weight ladder

unsuitable template (no or low copy number of entire template) (Fig. 10.2).

- Mix well and dispense 13 μl of this solution into the PCR tubes (discarding the remaining volume of about 13 μl).
- Cover this solution by a drop of mineral oil.
- Add the water into the tubes that require it.
- Finally add the DNA.
- Run the cycles as described above.
- Analyse 4 μl of each reaction by electrophoresis through a 2% agarose gel containing EtBr as usual.

References

1. Alaeddini R, Walsh SJ, Abbas A (2010) Forensic implications of genetic analyses from degraded DNA – a review. Forensic Sci Int Genet 4:148–157
2. Paabo S, Poinar H, Serre D, Jaenicke-Despres V, Hebler J, Rohland N, Kuch M, Krause J, Vigilant L, Hofreiter M (2004) Genetic analyses from ancient DNA. Annu Rev Genet 38:645–679
3. Eckert KA, Kunkel TA (1991) in McPherson MJ, Quirke P, Taylor GR (Eds) PCR: a practical approach IRL Press, Oxford, 225–244
4. Allouche M, Hamdoum M, Mangin P, Castella V (2008) Genetic identification of decomposed cadavers using nails as DNA source. Forensic Sci Int Genet 3(1):46–49
5. Rohland N, Hofreiter M (2007) Ancient DNA extraction from bones and teeth. Nat Protoc 2(7):1756–1762
6. Rasmussen M, Li Y, Lindgreen S, Pedersen JS, Albrechtsen A, Moltke I, Metspalu M, Metspalu E, Kivisild T, Gupta R, Bertalan M, Nielsen K, Gilbert MT, Wang Y, Raghavan M, Campos PF, Kamp HM, Wilson AS, Gledhill A, Tridico S, Bunce M, Lorenzen ED, Binladen J, Guo X, Zhao J, Zhang X, Zhang H, Li Z, Chen M, Orlando L, Kristiansen K, Bak M, Tommerup N, Bendixen C, Pierre TL, Gronnow B, Meldgaard M, Andreasen C, Fedorova SA, Osipova LP, Higham TF, Ramsey CB, Hansen TV, Nielsen FC, Crawford MH, Brunak S, Sicheritz-Ponten T, Villems R, Nielsen R, Krogh A, Wang J, Willerslev E (2010) Ancient human genome sequence of an extinct palaeo-eskimo. Nature 463(7282):757–762
7. Loreille OM, Diegoli TM, Irwin JA, Coble MD, Parsons TJ (2007) High efficiency DNA extraction from bone by total demineralization. Forensic Sci Int Genet 1(2):191–195
8. Fattorini P, Marrubini G, Ricci U, Gerin F, Grignani P, Cigliero SS, Xamin A, Edalucci E, La Marca G, Previderè C (2009) Estimating the integrity of aged DNA samples by CE. Electrophoresis 30(22):3986–3995
9. King C, Debruyne R, Kuch M, Schwarz C, Poinar H (2009) A quantitative approach to detect and overcome PCR inhibition in ancient DNA extracts. Biotechniques 47(5):941–949
10. Kontanis EJ, Reed FA (2006) Evaluation of real-time PCR amplification efficiencies to detect PCR inhibitors. J Forensic Sci 51(4):795–804
11. Hedman J, Nordgaard A, Rasmusson B, Ansell R, Radstrom P (2009) Improved forensic DNA analysis through the use of alternative DNA polymerases and statistical modeling of DNA profiles. Biotechniques 47(5):951–958
12. Previderè C, Micheletti P, Perossa R, Grignani P, Fattorini P (2002) Molecular characterisation of the nucleic acids recovered from aged forensic samples. Int J Legal Med 116(6):334–339

Restoration and Reconstruction of DNA Length

11

Serena Bonin and Federica Tavano

Contents

11.1 Introduction and Purpose

Among the archive tissues, autopsy tissues may represent an important resource for the study of rare diseases, neuropathology, cardiopathology, or molecular epidemiology, because of the possibility to analyze both pathological and normal tissues not available from other sources. The tissue fixation step is usually performed in a buffered formaldehyde solution, in the dark for about 24 h before paraffin wax embedding. This procedure is not usually followed for autopsy tissues, which are generally fixed for longer periods of time. Several factors, such as postmortem interval, the type of fixative, and the fixation time could affect the quality and utilization of nucleic acids from autoptic archive tissues. For example, the extensive DNA degradation that is often found in autoptic archival tissues (on average, fragments less than 100 bases long are produced) restricts PCR amplification to very short sequences. This protocol provides a method to partially restore DNA using a pre-PCR treatment, which fills single-strand breaks. This method allows the amplification of longer sequences ranging 300 bases without any modification to the usual DNA extraction procedure [1, 2]. This method could be applied even to DNA extracts from Bouin's fixed tissues [2].

11.2 Protocol

11.2.1 Reagents

- *Pre-PCR DNA restoration treatment solution*: 10 mM Tris–HCl (pH 8.3), 1.5 mM $MgCl_2$, 2% Triton X-100, and 200 µM of each dNTP
- *Taq DNA Polymerase* (e.g., Amersham)

S. Bonin (✉) and F. Tavano
Department of Medical, Surgical and Health Sciences,
University of Trieste, Cattinara Hospital, Strada di Fiume 447,
Trieste, Italy

G. Stanta (ed.), *Guidelines for Molecular Analysis in Archive Tissues*,
DOI: 10.1007/978-3-642-17890-0_11, © Springer-Verlag Berlin Heidelberg 2011

11.2.2 Equipment

- *Disinfected[1] adjustable pipettes,* range: 2–20 µl, 20–200 µl, 100–1,000 µl
- *Autoclaved PCR tubes* (0.2 ml)
- *Thermoblock*
- *Thermomixer* (e.g., Eppendorf)

11.2.3 Method

- Incubate DNA samples for 1 h at 55°C in 100 µl of Pre-PCR DNA restoration treatment solution.
- After this incubation, add 1 unit of Taq DNA Polymerase and perform DNA polymerization at 72°C for 20 min.

---

[1]Clean the pipettes with alcohol or another disinfectant and leave them under the UV lamp for almost 10 min. Alternatively, it is possible to autoclave the pipettes according to the provider instructions.

- Store the treated samples at –20°C until they are processed.
- Just before PCR amplification, proceed with the denaturation step: incubate 10 µl of restored DNA solution at 95°C for 5 min and then immediately chill on ice.
- Add the PCR solution to the sample and proceed with the amplification.

References

1. Bonin S, Petrera F, Niccolini B, Stanta G (2003) PCR analysis in archival postmortem tissues. Mol Pathol 56(3): 184–186
2. Bonin S, Petrera F, Rosai J, Stanta G (2005) DNA and RNA obtained from Bouin's fixed tissues. J Clin Pathol 58(3): 313–316

RNA Extraction from Formalin-Fixed Paraffin-Embedded Tissues

12

Serena Bonin and Giorgio Stanta

Contents

S. Bonin (✉)
Department of Medical, Surgical and Health Sciences,
University of Trieste, Cattinara Hospital,
Strada di Fiume 447, Trieste, Italy

G. Stanta
Department of Medical Sciences,
University of Trieste, Cattinara Hospital,
Strada di Fiume 447, Trieste, Italy

12.1 Introduction and Purpose

This protocol provides a method of obtaining RNA suitable for reverse transcription and PCR analyses from formalin-fixed and paraffin-embedded (FFPE) specimens [1–4], even of autopsy origin [5]. This protocol is based mainly on deparaffinization of tissues and a proteolytic digestion with Proteinase K. The basic principle of RNA extraction is very similar to that for DNA but the methods used differ in many specific aspects. RNA is more liable than DNA and more susceptible to nuclease degradation. Furthermore, RNases are more resistant to protein denaturants. Care should be taken to avoid the accidental introduction of RNases from external sources. The time required for the whole procedure is 3 days.

Commercial kits and semi or fully automated systems have been described for RNA extraction from FFPE [6]. Furthermore, some kits, commercially available, are specifically dedicated to archive tissues (i.e., Kit-Qiagen RNeasy FFPE, Roche High Pure RNA). However, using the same kit, different laboratories have shown remarkable differences in RNA yield. Small adjustments in the manufacturer's instructions (e.g., Proteinase K digestion time) can cause interlaboratory variability in both RNA quantity and quality. These observations are consistent with earlier reports demonstrating that different results were obtained by different groups using the same commercial kits [7].

12.2 RNA Handling

Special precautions must be taken when working with RNA because this macromolecule is extremely instable. Its fragility is partly due to the ribonuclease

G. Stanta (ed.), *Guidelines for Molecular Analysis in Archive Tissues*,
DOI: 10.1007/978-3-642-17890-0_12, © Springer-Verlag Berlin Heidelberg 2011

activity (RNases are ubiquitous in living organisms and exceptionally stable even after autoclaving) that can easily contaminate RNA preparations. Moreover, RNA is thermodynamically less stable than DNA because the 2′-OH group on the ribose ring promotes the hydrolytic reaction. Such RNA degradation is increased at high temperatures and pH. The following good laboratory practices (GLP) are suggested:

- Wipe bench surfaces, tube holders etc. with 100% ethanol to remove contaminating microbes and their RNases.
- Wear gloves all the time and change them frequently; try to avoid touching surfaces that could be contaminated by RNases.
- Use sterile and RNase-free materials. It is important to underline that autoclaving will not remove RNases. Sterile, disposable plasticware should preferably be used if guaranteed RNase free.
- Employ glassware, plasticware, pipettors, and buffers dedicated only for RNA experiments. This equipment should be kept in dedicated RNase-free areas.
- Prepare all solutions with DEPC-treated water and in RNase-free glassware.
- During RNA isolation and reverse transcription, use inhibitors of RNases.
- Store open solutions or buffers in small working aliquots; discard each aliquot after use.
- Store RNA at −80°C in DEPC-treated water. Under these conditions, no degradation of RNA is detectable after 1 year. Prepare disposable RNA aliquots in order to reduce multiple freezing-thawing.
- When handling RNA, keep it on ice and put it back at −80°C as soon as possible.
- It is advisable to keep RNase-free reagents separately, and their containers must never be touched without gloves.

General laboratory glassware or plasticware can be cleaned by soaking them for at least 2 h in a 10% solution of hydrogen peroxide or 1% SDS followed by thorough washing with DEPC-treated water. Alternatively, glassware could be put in oven at 250°C for at least 2 h.

After DEPC treatment, water is essentially free from RNases and can be used to prepare solutions or to rinse any items to be used for RNA isolation (e.g., apparatus for electrophoresis). Every solution for RNA extraction should be treated with 0.1% DEPC or made with DEPC-treated H$_2$O. Note, however, that DEPC reacts quickly with amines and so it cannot be used to treat solutions containing Tris. Tris buffers should be made using Tris prepared in DEPC-treated H$_2$O.

It is better to weigh all the reagents by tapping directly from the bottle on the technical balance covered by a new aluminium foil, without touching the powder with any instrument. Alternatively, it is possible to use a stainless steel spatula or spoons that have been pre-treated in an oven at 250° for at least 2 h.

12.3 Protocol

12.3.1 Reagents

All reagents should be RNase-free or DEPC-treated and dedicated exclusively to RNA analysis:

- *DEPC-treated water (DEPC H$_2$O)*: Add 1 ml of diethylpyrocarbonate (DEPC) (0.1% final concentration) to 1 l of sterile water and incubate overnight at 37°C, then autoclave[1]
- *Xylene* (e.g., Fluka or Sigma-Aldrich)
- *Absolute ethanol* (e.g., Sigma), 90% and 70% ethanol
- *20 mg/ml Proteinase K (stock solution)*: (e.g., Sigma P2303) Dissolve 100 mg of Proteinase K in 5 ml of autoclaved 50% glycerol diluted in DEPC H$_2$O.[2] Store at −20°C
- *1 × Digestion buffer to be completed*: 1.6 M Guanidine Thiocyanate[3], 30 mM Tris HCl pH 7.5, 0.72% N-Lauril Sarcosine. Store at room temperature in an aluminium-wrapped bottle. Complete the solution just before use with 0.2% 2-Mercaptoethanol (final concentration)[4] and 6 mg/ml Proteinase K (final concentration)

---

[1]DEPC is a carcinogen and should be handled with care under a fume hood.

[2]The solubilization of Proteinase K in 50% sterile glycerol keeps the solution fluid at −20°C, with a better preservation of the enzymatic activity.

[3]Guanidinium thiocyanate is a very strong protein denaturant, which is used to denature all the cellular proteins including RNases. It is harmful, so it must be handled under a fume hood.

[4]2-Mercaptoethanol is toxic by inhalation, ingestion, and through skin contact, and is a severe eye irritant. Use only with adequate ventilation, and wear gloves and safety glasses. Wasted 2-Mercaptoethanol must be placed in a chemical waste container.

- *Phenol[5]-H$_2$O[6]/CHCl$_3$ 70:30 or Phenol-citrate buffered pH 4.3/CHCl$_3$ 70:30*: Mix 7 parts of buffered or H$_2$O phenol with 3 parts of chloroform. Top the organic phase with DEPC H$_2$O or 0.1 M citrate buffer pH 4.3 (about 1 cm high) and allow the phase to separate. Store at 4°C in a light-tight bottle
- *Iso-propanol or EtOH/0.8 M LiCl*
- *1 mg/ml glycogen* in DEPC H$_2$O as a precipitation carrier

12.3.2 Equipment

- *Disinfected[7] adjustable pipettes*, range: 2–20 μl, 20–200 μl, 100–1,000 μl
- *Nuclease-free aerosol-resistant pipette tips*
- *1.5 or 2 ml tubes* (autoclaved)
- *Single-packed toothpicks*
- *Sterile or disposable tweezers*
- *Microtome with new blades*
- *Centrifuge* suitable for centrifugation of 1.5 ml tubes at 13,200 or 14,000 rpm
- *Thermoblock*
- *Thermomixer* (e.g., Eppendorf)
- *Spectrophotometer*

12.3.3 Method

12.3.3.1 Sample Preparation

- Cool the paraffin blocks at −20°C or on dry ice in aluminium foil in order to cut the sections.[8]

---

[5]We strongly recommend purchase of saturated phenol from a commercial manufacturer.

[6]The pH of Phenol/H$_2$O is around 5, because phenol makes the water acidic during the process.

[7]Clean the pipettes with alcohol or another disinfectant and leave them under the UV lamp for at least 10 min. Alternatively, it is possible to autoclave the pipette depending on the provider instructions.

[8]Wear gloves and change them frequently over the entire procedure. Do not talk while handling the sample to avoid introduction of saliva drops into the tubes. If possible, perform the extraction under a laminar flow hood. After the deparaffinization steps, the tubes must be kept on ice.

- Using a clean, sharp microtome blade,[9] cut two to ten sections of 5–10 μm thickness, depending on the size of the sample. Discard the first section and displace the others in a tube, using a sterile toothpick or tweezers (depending on the section size). Use some sections from a paraffin block without tissue, treated together with the other specimens as negative control analysis.[10]

12.3.3.2 Deparaffinization (Optional [11])

- Add 1 ml of xylene,[12] vortex for 10″ and then maintain the tube at room temperature (RT) for approximately 5 min.[13]
- Spin the tube for 5 min at maximum speed (14,000 rpm) in a microcentrifuge and then carefully remove and discard the supernatant using a micropipette or a glass Pasteur pipette.[14]
- Repeat wash with a fresh aliquot of xylene.
- Wash the pellet adding 1 ml of absolute ethanol. Flick the tubes to dislodge the pellet and then vortex the tubes for 10 s. Leave at RT for approximately 5 min.
- Spin the tube for 5 min at maximum speed (14,000 rpm) in a microcentrifuge and then carefully remove and discard the supernatant.
- Repeat washes using 90% and 70% ethanol.
- After removing 70% ethanol, allow the tissue pellet to air dry in a thermoblock at 37°C for about 30 min.

---

[9]Clean the microtome with xylene.

[10]RNA extraction from sections previously adhered on slides is also possible. In this case, displace the slides in a jar for histochemistry and deparaffinize the tissues by incubating twice with xylene for 10 min. After the EtOH washings bring the samples to water and scrape the sections from the slide by the use of a needle and transfer it into a tube with the digestion buffer as described in point 10. All the equipments used must be RNase-free.

[11]It is possible to bypass the deparaffinization step by adding the digestion buffer directly to the cut sections. In such case, it is better to add a short incubation time of 5 min at 65°C to rapidly melt the paraffin. Paraffin remains on the top of the solution [8].

[12]When working with xylene, avoid breathing fumes; it is better to perform the deparaffinization step under a fume hood.

[13]Wear gloves when isolating and handling RNA to minimize contamination with exogenous nucleases. Use autoclaved pipette tips and 1.5 ml microcentrifuge tubes.

[14]Xylene is harmful; the wasted xylene must be collected in a chemical waste container and discharged according to the local hazardous chemical disposal procedures.

12.3.3.3 Proteolytic Digestion and RNA Extraction

- Add to the tissue pellet 150–300 µl of digestion buffer supplemented with 2-mercaptoethanol (0.2% final concentration) and Proteinase K at final concentration of 6 mg/ml. The amount of digestion buffer depends on the section size. The digestion buffer must cover the tissue pellet[15] completely.
- Incubate in a thermomixer overnight[16] at 55°C, shaking moderately. For longer digestion time (>24 h), Proteinase K can be added every 24 h.
- Add 1 volume of phenol[17]-citrate buffered pH 4.3/ CHCl$_3$ (7:3 ratio)[18] and mix well by gently inverting the tube or vortexing.[19]
- Incubate the suspension in ice for 20 min.
- Centrifuge at 14,000 rpm at 4°C for 20 min. The mixture separates into a lower organic phase,[20] an interphase, and an upper aqueous phase.
- Transfer the upper phase to a new tube containing 5 µl of glycogen solution (1 mg/ml stock solution) as precipitation carrier. Carefully avoid transferring the protein-containing interphase.
- Precipitate with 1 volume of iso-propanol or 3 volumes of EtOH-LiCl (0.1 volume of 8 M LiCl and 2–3 volumes of absolute ethanol)[21] overnight at –20°C.
- Centrifuge at 14,000 rpm at 4°C for 20 min and discard the supernatant.
- Wash the pellet with 200 µl of 70% ethanol without resuspending the pellet to wash the remaining salts.
- Air dry the pellet and resuspend the RNA pellet in the appropriate amount of DEPC-treated water.
- Store the RNA solution at –70°C in aliquots.[22]
- For RNA measurement, pipette 199 µl of sterile water into a fresh tube and add 1 µl of RNA extract. Determine the RNA concentration photometrically at 260 and 280 nm[23] (see Chap. 16, Sect. 16.2.1).

12.3.4 Troubleshooting

- If the yield of RNA is low, you may have lost the RNA; in such case, repeat the entire process of extraction.
- If the pellet is not visible after centrifugation, the precipitation could have been incomplete because of the absence of a precipitation carrier. Add 5 µl of glycogen 1 mg/ml, and leave at –20°C overnight to complete precipitation.
- If the yield of RNA is low by UV measurement decrease the dilution factor to check the concentration and repeat the measuring.

---

[15]If the pellet is firmly lodged at the bottom of the tube, it is possible to dislodge it in the digestion buffer using a sterile toothpick.

[16]Longer duration of digestion (up to 48 h) increases the yield of the RNA, because it promotes a greater cross-link reversal. Optional incubation in the digestion buffer at 70°C for 20 min after the overnight digestion facilitates the disruption of cross-links, resulting in improved quantity and quality of RNA.

[17]Phenol is very toxic and should be handled in a fume hood. The wasted phenol must be collected with hazardous chemical waste.

[18]The extraction could be performed with 1 volume of phenol (H$_2$O saturated)-chloroform in the same ratio (70/30). Alternatively it is also possible to use chloroform-isoamyl alcohol (24:1, v/v) instead of pure chloroform in the extraction. Usually, isoamyl alcohol is added to the chloroform to prevent foaming. Phenol is an inhibitor of PCR reaction, because of Taq Polymerase inactivation. A single chloroform-isoamyl alchol (24:1, v/v) extraction could be performed after the phenol (H$_2$O saturated)-chloroform extraction in order to completely remove phenol traces.

[19]Alternatively, it is possible to perform the extraction using a monophasic commercial solution (i.e., Trizol, RNazol…). If this procedure is chosen, do not proceed with the successive steps of this protocol, but follow the manufacturer's instruction.

[20]After the aqueous phase has been transferred into a new tube, it is possible to extract DNA from the organic phase and interphase by adding an equal volume of Tris 50 mM pH 8–8.5. At this point, follow the protocol described in the chapter "DNA extraction from FFPE tissues." See also other specific chapters dedicated to DNA extraction.

[21]As LiCl is highly soluble in ethanol-containing solutions, the salt is not coprecipitated with the nucleic acid even at –70°C.

[22]It is better to store RNA extracts in small aliquots in order to prevent multiple thawing/freezing that may degrade the nucleic acid.

[23]The concentration of RNA expressed in µg/µl is obtained as follows: $[RNA] = A260 \times dilution\ factor \times 40 \times 10^{-3}$; for example, when diluting 1 µl RNA in 199 µl sterile water, the dilution factor is 200. A clean RNA preparation should have a A260/A280 ratio of 1.5–2. This ratio is decreased by the presence of proteins, oligo- and polysaccharides. Concentration estimation can be also affected by phenol contamination. Also phenol strongly absorbs at 260 nm and therefore can mimic higher DNA yield and purity.

• If RNA is absent, an RNase contamination could have occurred. In such case, it is better to check all the solutions and materials used along the extraction procedure, and repeat the entire process.

References

1. Gilbert MT, Haselkorn T, Bunce M, Sanchez JJ, Lucas SB, Jewell LD, Van Marck E, Worobey M (2007) The isolation of nucleic acids from fixed, paraffin-embedded tissues-which methods are useful when? PLoS ONE 2(6):e537

2. Godfrey TE, Kim SH, Chavira M, Ruff DW, Warren RS, Gray JW, Jensen RH (2000) Quantitative mRNA expression analysis from formalin-fixed, paraffin-embedded tissues using 5' nuclease quantitative reverse transcription-polymerase chain reaction. J Mol Diagn 2(2):84–91

3. Lehmann U, Kreipe H (2001) Real-time PCR analysis of DNA and RNA extracted from formalin-fixed and paraffin-embedded biopsies. Methods 25(4):409–418

4. Stanta G, Bonin S, Perin R (1998) RNA extraction from formalin-fixed and paraffin-embedded tissues. Methods Mol Biol 86:23–26

5. Bonin S, Petrera F, Stanta G (2005) PCR and RT-PCR analysis in archivial postmortem tissues. In: Encyclopedia of diagnostic genomics and proteomics. Marcel Dekker, New York, pp 985–988

6. Bohmann K, Hennig G, Rogel U, Poremba C, Mueller BM, Fritz P, Stoerkel S, Schaefer KL (2009) RNA extraction from archival formalin-fixed paraffin-embedded tissue: a comparison of manual, semiautomated, and fully automated purification methods. Clin Chem 55(9):1719–1727

7. Bonin S, Hlubeck F, Benhattar J, Denkert C, Dietel M, Fernandez PL, Höfler G, Kothmaier H, Kruslin B, Mazzanti CM, Perren A, Popper H, Scarpa A, Soares P, Stanta G, Groenen PJTA (2010) Multicentre validation study of nucleic acids extraction from FFPE tissues. Virchows Arch 457(3):309–317

8. Coombs NJ, Gough AC, Primrose JN (1999) Optimisation of DNA and RNA extraction from archival formalin-fixed tissue. Nucleic Acids Res 27(16):e12

RNA Extraction from Decalcified and Non-decalcified Formalin-Fixed Paraffin-Embedded Tissues

13

Marco Alberghini, Stefania Benini, Gabriella Gamberi, Stefania Cocchi, and Licciana Zanella

Contents

13.1 Introduction and Purpose

This protocol provides a method for obtaining RNA suitable for RT-PCR analysis from decalcified and non-decalcified formalin-fixed and paraffin-embedded (FFPE) tissue specimens. Decalcification causes RNA fragmentation and it is generally believed that this damage precludes successful RT-PCR. For the best performance in calcified tissue, it is necessary to choose an appropriate protocol to decalcify the tissue in order to perform histological, immunological, and molecular studies. The protocol below describes the decalcification procedure using formic acid and nitric acid [1]. For decalcified and non-decalcified FFPE tissue samples, we have standardized an RNA extraction protocol based on a commercial kit (i.e., Pinpoint slide RNA isolation systemII, Zymo Research Corp) [2–4]. This kit for nucleic acid isolation is specifically developed for archive tissues. Some modifications have been made in order to obtain good-quality RNA, depending on the type, volume, and treatment of the tissue [5].

The RNA extraction procedure is based mainly on deparaffinization, tissue isolation, and proteolytic digestion with Proteinase K. The time required for the entire procedure is approximately 6 h.

13.2 Protocol

13.2.1 Reagents

Listed below are the products we use, but similar reagents from other manufacturers may be selected:

- *Nitric acid 65% solution* (Fluka or Sigma-Aldrich)
- *Formic acid 98% solution* (Fluka or Sigma-Aldrich)

M. Alberghini (✉), S. Benini, G. Gamberi, S. Cocchi, and L. Zanella
Pathology Department, Istituto Ortopedico Rizzoli, Bologna, Italy

G. Stanta (ed.), *Guidelines for Molecular Analysis in Archive Tissues*,
DOI: 10.1007/978-3-642-17890-0_13, © Springer-Verlag Berlin Heidelberg 2011

- *Xylene* (Fluka or Sigma-Aldrich)
- *Absolute, 96% and 70% ethanol* (Fluka or Sigma-Aldrich)
- *20 mg/ml Proteinase K (stock solution)*: (Zymo Research Corp). Dissolve 5 mg of Proteinase K in 250 µl of storage glycerol Buffer. Store at −20°C.[1]
- *RNAse Away*[2] (Invitrogen) ready-to-use solution for eliminating RNase and DNA contamination from labware
- *Sterile H₂O*

13.2.2 Equipment

- *Disinfected adjustable pipettes,*[3] range: 0.1–2 µl, 2–10 µl, 20–200 µl, 100–1,000 µl
- *Nuclease-free aerosol-resistant pipette tips*
- *1.5 ml tubes* (autoclaved)
- *Sterile or disposable tweezers*
- *Microtome, with new blade*
- *Orbital shaker*
- *Refrigerated MicroCentrifuge* suitable for centrifugation of 1.5 ml tubes
- *Waterbath*
- *Spectrophotometer*
- *Clean Micro slides with ground edges* (76×26 mm)
- *Glass bowl*
- *Rack for slides*

13.2.3 Method

13.2.3.1 Bone Fixation and Bone Decalcification Protocol

- For a large bone specimen, cut the sample in a large coronal or axial section using the band saw.[4]

- Fix the calcified sections in 4% neutral buffered formalin[5] at room temperature for 24–48 h (for biopsy specimens the time of fixation is approximately 6 h)
- Submerge the samples in the decalcifying solution[6] at room temperature for 7–10 days depending on size and calcification level (for biopsy specimens, the decalcification time is about 1–3 h)
- Wash the samples in tap water to remove the decalcifying solution.
- Dissect the specimen into smaller samples, then proceed with routine paraffin-embedding tissue protocol.

13.2.3.2 Sample Preparation

- If possible, cool the paraffin blocks at −20°C or on ice in order to facilitate cutting of the sections.[7]
- Using a clean, sharp microtome blade,[8] cut two to five sections of 6–8 µm thickness depending on the size of the sample. Discard the first section and displace the other sections on a floating-out bath of clean deionized water at 40°C.
- Mount the sections on clean glass slides using sterile tweezers.
- Dry the sections at 60°C for 10 min.
- The first sections are used for RNA extraction; while the last section is hematoxylin-eosin-stained and used to check the morphology in order to select the area with adequate cellularity and to compare it with the unstained sections.
- Mark the selected area of interest for RNA isolation with a diamond tip.

[1]The solubilization of Proteinase K in glycerol Buffer keeps the solution fluid at −20°C, with a better preservation of the enzymatic activity.

[2]Apply RNAse Away over the surface of glassware or plasticware to be treated. Unwanted RNase and DNA contamination are eliminated.

[3]Clean the pipettes with RNAse Away or leave them under the UV lamp for about 10 min to prevent contaminations. Alternatively, the pipette can be autoclaved, if possible, according to the manufacturer's specifications.

[4]This step should be performed by two operators under a fume hood.

[5]Avoid breathing fumes when working with formalin. The wasted formalin must be collected in a chemical waste container and discharged according to the local hazardous chemical disposal procedures.

[6]For 1 l of decalcifying solution, mix 51 ml of 98% formic acid (5% final concentration), 30.8 ml of 65% nitric acid (2% final concentration), and 918.2 ml of distilled water. Formic acid and nitric acid are corrosive solutions. Wear gloves and handle the solutions under the chemical hood.

[7]Wear gloves when isolating and handling RNA or reagents for RNA isolation to minimize the contamination with exogenous RNAses. Use autoclaved pipette tips and 1.5 ml microcentrifuge tubes.

[8]Clean the microtome and blade with RNAse Away. It is recommended that the blade be changed after the cutting of each paraffin block to avoid any potential contamination.

13.2.3.3 Deparaffinization

- Submerge the slides in clean xylene at room temperature for 1 h changing the xylene once after approximately 30 min[9].
- Hydrate the slides by washing progressively for 2 min in clean ethanol 100%, 96%, 75%, and then in pure water[10].
- When the tissue has been decalcified, incubate the hydrated slides in 0.001N EDTA pH 8.0 for 40 min in orbital shaker at room temperature.
- Wash the slides in distilled water for 5 min. Repeat this step.
- Air dry the sections, leaving the slide at room temperature for approximately 10 min or under a hood for 2–3 min.

13.2.3.4 Tissue Isolation and Digestion

- Apply the Pinpoint solution to the selected area on the slide to remove the tissue region. The amount of solution depends on the tissue area[11]
- Allow the Pinpoint solution to dry completely at room temperature.[12] When it is dry, it appears as a blue film embedding the tissue and cells underneath.
- Remove the embedded tissue from the slide[13] and transfer to a sterile tube.
- Centrifuge briefly to locate the tissue sample at the bottom of the tube
- Add to the recovered tissue, 25 µl of digestion buffer 1× supplemented with Proteinase K at a final concentration of 1 mg/ml and mix gently. The amount of

digestion buffer depends on the amount of tissue. The digestion buffer must cover the tissue pellet completely.
- Incubate the tube in a waterbath at 55°C for 4 h[14].
- Centrifuge the tube briefly at the end of incubation.

13.2.3.5 RNA Extraction

- Add 50 µl of RNA extraction Buffer and mix well with a micropipette.
- Add 75 µl of 100% ethanol to the tube. Vortex lightly.
- Transfer the mixture to a Column in a 2 ml collection tube[15]
- Spin the column at 10,000 rpm in a microcentrifuge for 1 min.
- Add 200 µl of RNA wash buffer to the column and centrifuge at 10,000 rpm for 1 min. Repeat this step.
- Transfer the column into a new RNase-free 1.5 ml tube.
- Add 10 µl of distilled water directly to the membrane. Wait for 2 min.
- Spin the column at 10,000 rpm in a microcentrifuge for 1 min to elute the RNA[16]
- For RNA measurement, pipette 121 µl of sterile water into a fresh tube and add 4 µl of RNA extract (dilution factor = 25). Determine the RNA concentration photometrically at 260 and 280 nm (see Chap. 16, Sect. 16.2.1).[17]

13.2.4 Troubleshooting

- *RNA degradation.* RNA is highly susceptible to RNase digestion, we encourage the use of freshly prepared sections. If a sample cannot be processed immediately, store it at ≤70°C or submerge it in 96% ethanol at −20°C. Processing of tissue sections stored for 1 month or more at room temperature is not recommended.

---

[9]Avoid breathing fumes when working with xylene. It is better to perform the deparaffinization step under a chemical hood. Xylene is harmful; the wasted xylene must be collected in a chemical waste container and discharged according to the local hazardous chemical disposal procedures.

[10]Make sure that ethanol dilutions used to RNA isolation procedure, are performed in RNase-free water. This step is performed in an orbital shaker (shaking moderately).

[11]Use a sterile pipette tip or a glass pasteur to gently spread a small amount of Pinpoint solution over the selected tissue region. Generally, about 0.5 µl of Pinpoint solution is used per mm^2 of tissue area. Usually, one drop of Pinpoint is adequate for 25 mm^2 of tissue area. Four drops on tissue with appropriate cellularity (using one to four slides) allow good results.

[12]Leave the slides for about 30–45 min; if left under the chemical hood, 10–15 min are sufficient.

[13]Use a sterile blade or scalpel to cut and remove the embedded section from the slide. Transfer the sample to a 1.5 ml tube.

[14]Vortex the tube every 30 min to improve the digestion.

[15]Use one Column for each extraction tube.

[16]The isolated RNA can be used directly for RT-PCR amplification, or it can be stored at −70°C for future use.

[17]The concentration of RNA expressed in µg/µl is obtained as follows: [RNA] = A260×dilution factor×40×10^{-3} (see Chap. 16). An appropriate RNA preparation from FFPE tissue should have a A260/A280 ratio of 1.40–1.80. The ratio variability is linked to the presence of proteins and oligo-, polysaccharides.

- *Insufficient RNA*. Make sure an appropriate sampling area is selected for processing. Select an area of the tissue that will contain ≥50 cells. Increase the sampling area if the tissue type contains few cells (e.g., fatty tissue, fibrous or cartilaginous tissue). The sampling size can vary from 1 mm² to over 100 mm².
- *DNA contamination*. Traces of fragmented DNA may be present in the eluted RNA fraction. DNA-free RNA can be obtained with subsequent DNase I treatment.

References

1. Shibtat Y, Fujita S, Takahashi H, Yamaguchi A, Koji T (2000) Assessment of decalcifying protocols for detection of specific RNA by non-radioactive in situ hybridization in calcified tissues. Histochem Cell Biol 113:153–159

2. Weizacher FV, Labeit S, Koch HK, Oehlert W, Blum HE (1991) A simple and rapid method for the detection of RNA in formaliln-fixed, paraffin-embedded tissue by PCR amplification. Biochem Biophys Res Commun 174:176–180

3. Greer CE, Lund JK, Manos M (1991) PCR amplification from paraffin-embedded tissue: recommendations on fixatives for long-term storage and prospective studies. PCR Methods Appl 1:46–50

4. Bonin S, Hlubek F, Benhattar J, Denkert C, Dietel M, Fernandez PL, Hofler G, Kothmaier H, Kruslin B, Mazzanti CM, Perren A, Popper H, Scarpa A, Soares P, Stanta G, Groenen PJ (2010) Multicentre validation study of nucleic acids extraction from FFPE tissues. Virchows Arch 457(3): 309–317

5. Mangham DC, Williams A, KcMullan DJ, McClure J, Sumathi VP, Grimer RJ, Davies AM (2006) Ewing's sarcoma of bone: the detection of specific transcripts in a large, consecutive series of formalin-fixed, decalcified, paraffin-embedded tissue samples using the reverse transcriptase-polymerase chain reaction. Histopathology 48: 363–376

RNA Temperature Demodification

14

Serena Bonin and Giorgio Stanta

Contents

S. Bonin (✉)
Department of Medical, Surgical and Health Sciences,
University of Trieste, Cattinara Hospital,
Strada di Fiume 447, Trieste, Italy

G. Stanta
Department of Medical Sciences,
University of Trieste, Cattinara Hospital,
Strada di Fiume 447, Trieste, Italy

14.1 Introduction and Purpose

The extraction of useful RNA from formalin-fixed and paraffin-embedded (FFPE) tissues is often compromised for the extraction efficacy. Moreover, RNA in FFPE is not completely available for reverse transcriptase-polymaerase chain reaction (RT-PCR) reactions because it is resistant to extraction due to cross-linking with proteins [1, 2]. RNA, as DNA, is modified in FFPE tissues by the presence of methylol addition. Prolonged fixations could favour further reactions with the above-mentioned groups, resulting in irreversible artifacts. Masuda et al. [1] demonstrated that all four bases showed addition of mono-methylol (CH_2OH) in formalin, with different rates, ranging from 40% for adenine to 4% for the less reactive uracil. The presence of the methylol group doesn't allow the reverse transcription reaction, but the presence of the majority of these groups could be removed, by simply heating the RNA extracts in formalin-free buffers. Several methods have been reported to demodify RNA obtained from FFPE [1, 3, 4]; here we report two simple temperature treatments derived from Masuda [1] and Li [3]. It is also possible to use more sophisticated methods for RNA demodification, such as the one described by Oberli [4]. The latter is based on a chemical treatment with NH_4Cl, followed by a heating step at 94°C . The starting point of the protocol described hereafter will be RNA extracts from FFPE tissues (See Chap. 12 for more details).

From our experience, by the use of these simple methods it is possible to decrease the Ct values in real-time RT-PCR by about two cycles when compared to untreated RNA extracts. Moreover, the best results are achieved by the use of "Method I" (described below): resuspension of the RNA in TE buffer 1× at pH 7.5. Nevertheless the demodification treatment seems to improve the RNA recruitment only in old samples.

G. Stanta (ed.), *Guidelines for Molecular Analysis in Archive Tissues*,
DOI: 10.1007/978-3-642-17890-0_14, © Springer-Verlag Berlin Heidelberg 2011

The methods described hereafter could be added as routine procedure to the RNA extraction protocols especially for very old samples.

14.2 Precautions

Glassware, plasticware, micropipette, and reagents must be kept in specifically dedicated RNase-free areas. Sterile, disposable plasticware should preferably be used. General laboratory glassware or plasticware should be presoaked in 0.1% DEPC-treated H_2O for 2 h at 37°C. The DEPC-treated items should be rinsed thoroughly with DEPC H_2O and then autoclaved. Alternatively, glassware could be put in oven at 250°C for at least 2 h.

After DEPC treatment the water is essentially free from RNases and can be used to prepare solutions or to rinse any items to use for RNA isolation. Every solution for RNA extraction should be treated with 0.1% DEPC or made with DEPC H_2O. Note, however, that DEPC reacts quickly with amines and so it cannot be used to treat solutions containing Tris. Tris buffers should be made using Tris prepared in already DEPC-treated H_2O.

It is better to weigh all the reagents by tapping directly from the bottle on the technical balance covered by a new aluminium foil, without touching the powder with any instrument. Alternatively, it is possible to use stainless steel spatula or spoons that have been previously treated in oven at 250° for at least 2 h.

14.3 Protocol

14.3.1 Reagents

Note: All reagents should be RNase-free or DEPC-treated and used exclusively for RNA analysis.

- DEPC-treated water (H_2O DEPC): Add 1 ml of diethylpirocarbonate (DEPC) (0.1% final concentration) to 1 l of sterile water and incubate overnight at 37°C; then autoclave[1]
- TE buffer 1× pH 7.5: 10 mM Tris-HCl (pH 7.5) and 1 mM EDTA (pH 8.0). Stock solution 10×. For this purpose it is possible to use both pH 7.5 or pH 8 TE

buffers; however, in our experience we obtained better results by using TE buffer pH 7.5
- TE buffer 10×: 100 ml 1 M Tris-HCl pH 7.5, 20 ml 500 mM EDTA pH 8.0, 880 ml DEPC water. Autoclave

14.3.2 Equipment

- Disinfected[2] adjustable pipettes, range: 2–20 µl, 20–200 µl, 100–1,000 µl
- Nuclease-free aerosol-resistant pipette tips
- 1.5 or 2 ml tubes (autoclaved)
- Thermoblock

14.3.3 Method I

Note: Derived from Masuda [1] with minor modifications; it can be used in already extracted RNA.

1. RNA pellets from standard extraction procedure[3] (See Chap. 12 for details) should be resuspended in about 20 µl of TE buffer 1× pH 7.5.
2. Incubate in a thermoblock at 70°C for 20 min.[4]
3. Chill on ice.
4. Spin down briefly to collect the drops.
5. Store the RNA solution at –70°C in aliquots[5].
6. For RNA measurement, pipette 199 µl of sterile water into a fresh tube and add 1 µl of RNA extract. Determine the RNA concentration photometrically at 260 and 280 nm[6] (See Chap. 16).

---

[1]DEPC is a carcinogen and should be handled with care under a fume hood.

[2]Clean the pipettes with alcohol or another disinfectant and leave them under the UV lamp for at least 10 min. Alternatively, it is possible to autoclave the pipette depending on the provider instructions.

[3]For this demodification method, it is not mandatory to use the RNA extraction protocol described in this book; any extraction procedure, even commercial kits, can be used.

[4]It is possible to protract this step up to 1 h; however, in our experience, the best results have been obtained by heating at 70°C for 20 min.

[5]It is better to store RNA extracts in small aliquots to prevent multiple thawing/freezing, which may degrade the nucleic acid.

[6]The concentration of RNA expressed in µg/µl is obtained as follows: [RNA]= A_{260} × dilution factor × 40×10^{-3}; for example, when diluting 1 µl RNA in 199 µl sterile water, the dilution factor is 200. A clean RNA preparation should have a A_{260}/A_{280} ratio of 1.5–2.0. This ratio is decreased by the presence of proteins, phenol and oligo-, polysaccharides.

14.3.4 *Method II* [3]

Note: It can be used during RNA extraction procedures using the crude RNA extract after the proteolysis step.

1. Crude RNA extracts (after the proteolysis step) in Proteinase K buffer[7] (See Chap. 12 for details) should be incubated in a thermoblock at 70°C for 20 min.
2. Chill on ice.
3. Spin down briefly to collect the drops on the tube walls and proceed with protocol for RNA extraction to purify RNA extracts.

14.3.5 *Troubleshooting*

- If the yield of RNA is low by UV measurement decrease the dilution factor to check the concentration and repeat the measuring.

- If RNA is absent, an RNase contamination could have occurred. In such case, it is better to check all the solutions and materials used along the extraction procedure, and repeat the entire process.

References

1. Masuda N, Ohnishi T, Kawamoto S, Monden M, Okubo K (1999) Analysis of chemical modification of RNA from formalin-fixed samples and optimization of molecular biology applications for such samples. Nucleic Acids Res 27(22): 4436–4443
2. Srinivasan M, Sedmak D, Jewell S (2002) Effect of fixatives and tissue processing on the content and integrity of nucleic acids. Am J Pathol 161(6):1961–1971
3. Li J, Smyth P, Cahill S, Denning K, Flavin R, Aherne S, Pirotta M, Guenther SM, O'Leary JJ, Sheils O (2008) Improved RNA quality and Taqman pre-amplification method (preamp) to enhance expression analysis from formalin fixed paraffin embedded (FFPE) materials. BMC Biotechnol 8:10
4. Oberli A, Popovici V, Delorenzi M, Baltzer A, Antonov J, Matthey S, Aebi S, Altermatt HJ, Jaggi R (2008) Expression profiling with RNA from formalin-fixed, paraffin-embedded material. BMC Med Genet 1:9

---

[7]For this demodification method, it is not mandatory to use the proteinase K digestion buffer described in this book; any proteinase K buffer could be used, even the ones provided in commercial kits or commercial solutions.

MicroRNA Extraction from Formalin-Fixed Paraffin-Embedded Tissues

15

Roberto Cirombella and Andrea Vecchione

Contents

R. Cirombella and A. Vecchione (✉)
University of Rome "La Sapienza," Department of Clinical and
molecular medicine, Division of Pathology, Ospedale
Sant'Andrea, Rome, Italy

15.1 Introduction and Purpose

Formalin-fixed paraffin-embedded (FFPE) tissues are valuable samples for the study of human cancer, since they are generally retrieved with extensively documented clinico-pathological histories. Isolating nucleic acids from archived samples has the potential to unlock a wealth of additional information and could facilitate the study of human cancer at the molecular level.

While a standard method of preservation using formalin is ideal for maintaining the tissue structure and preventing putrefaction, they pose challenges for the molecular analyses of these samples. Nucleic acids become trapped and modified through protein–nucleic acid and nucleic acid–nucleic acid cross-links. RNA isolated from FFPE samples is often fragmented to a random range of sizes and chemically modified to a degree that is incompatible with many molecular analysis techniques.

MicroRNAs (miRNAs) are a class of 20–25 nucleotide-long noncoding RNAs that modulate gene expression and play a crucial role in many cellular processes, tumorigenesis included [1]. Recently identified, miRNAs have a diagnostic and prognostic potential for various diseases, most notably for cancer. In particular, miRNA expression profiling has been associated with the diagnosis, progression, and prognosis of many human tumors [2]. To realize such a potential, expression analysis of miRNA expression using archived FFPE samples is one of the key components [3]. Microarray-based hybridization has proven to be a powerful technique for miRNA profiling in these samples. Currently, the locked nucleic acid (LNA)-based miRNA array represents one of the most sensitive platforms for miRNA array expression analysis in both FFPE samples and fresh tissues [3].

G. Stanta (ed.), *Guidelines for Molecular Analysis in Archive Tissues*,
DOI: 10.1007/978-3-642-17890-0_15, © Springer-Verlag Berlin Heidelberg 2011

Several technical issues must be considered when using FFPE for miRNA expression analysis. The primary challenge is the extraction of total RNA that appropriately retains the small RNAs. Most methods for extracting RNA from FFPE samples have been optimized for recovery of significantly longer RNAs. Extensive cross-linking of miRNA with proteins during fixation makes miRNA more resistant to extraction. Enzyme degradation, which occurs before and during the fixation process, as well as chemical degradation, results in decreased miRNA yield and integrity. Finally, formalin is responsible for forming mono-methylol adducts with bases of nucleic acids, in particular with adenine. This covalent modification reduces the efficiency of reverse transcription in qRT-PCR and negatively affects the performance of miRNA samples in other downstream applications. However, recent studies in FFPE have shown that miRNAs are more accessible to expression analysis than mRNA, making them good candidates for biomarker discovery. For example, Siebolts et al. [2,3] showed in their studies that miRNA accessibility is not affected by prolonged formalin fixation and seems to be comparable to snap-frozen material. Moreover, several studies support the notion that miRNAs may be more stable or more easily recovered than mRNAs, due to their small size. The longer an RNA molecule is, the greater the probability that a cross-link still exists after the Proteinase K digestion [4]. So the extraction of a small molecule seems more accessible than the larger one. This result determines the clear advantage of using miRNA for biomarker discovery compared to mRNA [4,5].

Several kits for the total RNA isolation from FFPE are available:

- "RecoverAll Total Nucleic Acid Isolation Kit" (Ambion, Austin, TX) for extraction of total RNA from archived fixed tissues. This protocol allows the robust and reproducible recovery of an approximately twofold higher yield of total RNA, miRNA included [5]
- "Rneasy FFPE kit" (Quiagen, Valencia, CA); a supplementary protocol can be used for copurification of total RNA and miRNA from FFPE [5]
- "PureLink FFPE RNA Isolation kit" (Invitrogen, Carlsbad, CA)

- "High Pure miRNA isolation kit" (Roche, Indianapolis, IN)

The recovery of miRNA can be detected by means of Real-Time Quantitative RT-PCR, after a reverse transcription reaction. For example, the "TaqMan MicroRNA reverse transcriptase kit" (Applied Byosistems, California, CA) and the "TaqMan MicroRNA assay kit" (Applied Byosistems, California, CA) can be used in order, following the manufacturer's instructions [2].

As an alternative, it is possible to investigate the miRNA expression profiles of FFPE samples with Locked nucleic acid (LNA)-based miRNA array analysis using:

- "miRCURY LNA microRNA Array labeling kit" (Exiqon Inc.) for the labelling
- "miRCURY LNA microRNA Array" (Exiqon Inc.) for detection.

Among the currently available methods for miRNA extraction from FFPE, our laboratory selected the commercial High Pure miRNA isolation Kit (Roche, Indianapolis, IN), introducing some minor modifications.

15.2 Protocol

15.2.1 Reagents

Note: Reagents from specific companies are reported here, but similar reagents from other providers could be used:

- *High Pure miRNA isolation Kit* (Roche, cat. No. 05 080 576 001). It contains:

  - *Paraffin Tissue Lysis Buffer*
  - *Proteinase K*, 20 mg/ml (Roche, stock solution): Dissolve 100 mg of Proteinase K in 4.5 ml of nuclease-free, sterile, double-distilled water. Store at −20°C
  - *Binding buffer*: For tissue and cell disruption. For one sample, mix 80 μl of Binding Buffer and 320 μl of nuclease-free, sterile, double-distilled water (or Elution Buffer) in a sterile RNase-free

tube to prepare 20% Binding Buffer. Contains guanidine-thiocyanate[1]
- *Binding enhancer*
- *Elution buffer*: Nuclease-free, sterile, double-distilled water
- *Wash buffer*: Add 40 ml absolute ethanol to wash buffer. It removes large polynucleotides, salts, and proteins
- *Xylene* (Fluka or Sigma-Aldrich)
- *Absolute ethanol* (Sigma)
- *10% SDS solution* (Sigma): Added to the lysis buffer

15.2.2 Equipment

- *High pure filter tubes* (two bags with 50 columns for processing up to 700 µl sample volume) and *Collection Tubes* (two bags with 50 polypropylene tubes [2 ml]) contained in the High Pure miRNA isolation Kit (Roche)
- *Microtome, with new blade*
- *Disinfected adjustable pipettes*, range: 2–20 µl, 20–200 µl, 100–1,000 µl
- *Centrifuge* suitable for centrifugation of 1.5 ml tubes at 13,200 or 14,000 rpm
- *Thermoblock*
- *Thermomixer* (i.e. Eppendorf)
- *Spectrophotometer*

---

[1]Guanidine-thiocyanate in Binding Buffer is an irritant. Always wear gloves and follow standard safety precautions to minimize contact when handling. Do not let these buffers touch your skin, eyes, or mucous membranes. If contact does occur, wash the affected area immediately with large amounts of water; otherwise, the reagent may cause burns. If you spill the reagent, dilute the spill with water before wiping it up. Never store or use the Binding Buffer near human or animal food. Always wear gloves and follow standard safety precautions when handling these buffers. Guanidine-thiocyanate in Binding Buffer can form toxic gases when combined with bleach or acid. If a spilled sample containing this solution is potentially infectious, do not directly add bleach for decontamination.

15.2.3 Method

General precautions should be taken when working with RNA:

- Wear gloves at all times
- Use nuclease-free aerosol-resistant pipette tips
- Use autoclaved 1.5 ml tubes
- Use sterile or disposable tweezers

15.2.3.1 Sample Preparation

- Put the tissue block on a cold plate.
- Using a clean, sharp microtome blade, cut two to ten[2] sections of 5–10 µm thickness depending on the size of the sample. Displace the section in 1.5 ml tubes, using a sterile toothpick or tweezers (depending on the section size). Use some sections from a paraffin block without included tissue, treated together with other samples, for negative control analysis.

15.2.3.2 Deparaffinization

- Add 800 µl Xylene, incubate for 5 min, and mix by shaking.
- Add 400 µl ethanol abs. and mix. Centrifuge for 2 min at 14,000 rpm and discard supernatant.
- Add 1 ml ethanol abs. and mix by shaking and centrifuge for 2 min at 14,000 rpm and discard supernatant.
- Invert tube and blot briefly on a paper towel to get rid of residual ethanol. Dry the tissue pellet for 10 min at 55°C.
- Proceed with miRNA isolation from FFPE tissue.

---

[2]Kit requires only one tissue section of 5–10 µm thickness.

15.2.3.3 Proteolytic Digestion and miRNA Extraction[3]

- Add 100 µl Paraffin Tissue Lysis Buffer, 16 µl 10% SDS, and 40 µl Proteinase K working solution to each deparaffinized sample as described above and vortex three times for 5 s.
- To increase yield, incubate at 55°C for at least 3 h (up to overnight) shaking moderately. For longer digestion, proteinase K can be added again every 24 h.
- Add 325 µl Binding Buffer and vortex briefly.
- Add 120 µl Binding Enhancer, and vortex three times for 5 s.
- Combine the High Pure filter tube with a 2 ml collection tube and pipette the lysate into the upper reservoir.[4]
- Centrifuge for 30 s at 14,000 rpm in a microcentrifuge and collect the flowthrough.
- Add 205 µl Binding Enhancer and vortex three times for 5 s.
- Combine the High Pure filter tube with a collection tube and pipette the whole mixture from the previous step into the upper reservoir.
- Centrifuge for 30 s at 14,000 rpm, and discard the flowthrough. These two steps can be repeated, in order to load the column with additional sample material (do not overload the column).
- Add 500 µl Wash Buffer working solution and Centrifuge for 30 s at 14,000 rpm, discard the flowthrough.
- Add 300 µl Wash Buffer working solution and as described in the previous step. Centrifuge for 30 s at 14,000 rpm, discard the flowthrough.
- Centrifuge at 14,000 rpm for 1 min, in order to dry the filter fleece completely.
- Place the High Pure filter tube into a fresh 1.5 ml microcentrifuge tube, add 100 µl Elution Buffer[5], and incubate for 1 min at 15–25°C.
- Centrifuge for 1 min at 14,000 rpm.

---

[3]It is possible to modify this protocol in order to use only one column; however, this may lead to higher DNA contamination.

[4]To minimize the pipetting steps, a 2 ml microcentrifuge tube can be used instead of the collection tube.

[5]To increase RNA concentration of your sample, elution with 50 µl Elution Buffer is possible.

15.2.4 Troubleshooting

If the amount of miRNA is too low, the causes may be different:

- Kit stored under nonoptimal conditions. Store kit at +15°C to +25°C at all times upon arrival.
- High levels of RNase activity. Be careful to create an RNase-free working environment. Process sample immediately or store at 80°C until it can be processed. Use eluted RNA directly in downstream procedures or store immediately at 80°C.
- Insufficient disruption or homogenization. Add more 20% Binding Buffer and repeat the homogenization step to reduce viscosity.
- Incomplete elution. Elute RNA with 2 volumes of Elution Buffer (50 µl each). Make sure to centrifuge after each addition of Elution Buffer.

References

1. Inui M, Martello G, Piccolo S (2010) MicroRNA control of signal transduction. Nat Rev Mol Cell Biol 11(4): 252–263
2. Siebolts U, Varnholt H, Drebber U, Dienes HP, Wickenhauser C, Odenthal M (2009) Tissues from routine pathology archives are suitable for microRNA analyses by quantitative PCR. J Clin Pathol 62(1):84–88
3. Xi Y, Nakajima G, Gavin E, Morris CG, Kudo K, Hayashi K, Ju J (2007) Systematic analysis of microRNA expression of RNA extracted from fresh frozen and formalin-fixed paraffin-embedded samples. RNA 13(10):1668–1674
4. Li J, Smyth P, Flavin R, Cahill S, Denning K, Aherne S, Guenther SM, O'Leary JJ, Sheils O (2007) Comparison of miRNA expression patterns using total RNA extracted from matched samples of formalin-fixed paraffin-embedded (FFPE) cells and snap frozen cells. BMC Biotechnol 7:36
5. Doleshal M, Magotra AA, Choudhury B, Cannon BD, Labourier E, Szafranska AE (2008) Evaluation and validation of total RNA extraction methods for microRNA expression analyses in formalin-fixed, paraffin-embedded tissues. J Mol Diagn 10(3):203–211

Quantification of Nucleic Acids

16

Isabella Dotti and Serena Bonin

Contents

16.1 Introduction and Purpose

Several procedures are available for quantification of nucleic acids. Since the data obtained with different methods are not directly comparable [1], it's essential to choose the same approach when a comparison of data from different reports has to be done. Some of the following methods give information of both quantity and quality of nucleic acids. All the reported methods are more or less suitable for nucleic acids extracted from formalin-fixed and paraffin-embedded (FFPE) tissues. It should also be taken into account that when RNA concentration is measured, it involves mostly ribosomal RNA that represents 85% of the total RNA in a cell (mRNA is present at 1–5%).

Methods for nucleic acid quantification have been divided according to whether quantification relies on spectrophotometric measurement, reading in presence of a fluorescent dye, or on real-time amplification.

16.2 Methods

16.2.1 Spectrophotometric Quantification

The spectrophotometer is the most common device used to determine the quantity of the extracted nucleic acids. Concentration is estimated through absorbance reading. Measurements of the absorbance are recorded at different wavelengths to calculate DNA/RNA concentration and the presence of contaminants. The following absorbance lengths are usually analyzed:

I. Dotti (✉) and S. Bonin
Department of Medical, Surgical and Health Sciences,
University of Trieste, Cattinara Hospital, Strada di Fiume 447,
Trieste, Italy

G. Stanta (ed.), *Guidelines for Molecular Analysis in Archive Tissues*,
DOI: 10.1007/978-3-642-17890-0_16, © Springer-Verlag Berlin Heidelberg 2011

- *A260 nm*: Both DNA and RNA concentrations are read at this wavelength, therefore this measurement cannot distinguish between the two nucleic acids. This value is used in the following formula to obtain the nucleic acid concentration (Lambert–Beer law):

$$C_{\mu g/\mu l} = A \times \text{dil. factor} \times \varepsilon \times 10^{-3} \quad (16.1)$$

where A is the absorbance, dilution factor (dil. factor) is the ratio between the total volume used for the measurement and the volume of sample, ε (molar extinction coefficient) is a physical constant that is unique for each substance and describes the amount of absorbance at 260 nm (A_{260}) of 1 mole/l of nucleic acid solution measured in a 1 cm path-length cuvette. The molar extinction coefficient is 50 for double-stranded DNA (dsDNA), 40 for RNA, and 33 for oligonucleotides. Estimation of nucleic acid concentration can be affected by phenol contamination. Since phenol strongly absorbs at 260 nm, it can falsely increase DNA yield and purity.

- *A260/230 nm ratio*: It gives the level of contamination from copurified organic compounds (sugars, heparin, guanidine isothiocyanate...) that could inhibit downstream experiments. A260/230 ratio >1.8 for both DNA and RNA is indicative of a "pure" nucleic acid preparation. This type of contamination is common in biological samples.
- *A260/280 nm ratio*: It gives the level of contamination from proteins, salts, and other copurified reagents. A260/280 nm ratio >1.8 for both DNA and RNA is indicative of a "pure" nucleic acid preparation.

16.2.1.1 Conventional Spectrophotometer

It requires that the DNA/RNA sample is diluted before quantification. Depending on the expected concentration, 2–3 μl of sample diluted in 100–500 μl of H_2O is normally used.

Major disadvantages of this method are the poor sensitivity (the lower limit is generally 0.5–1 μg nucleic acid) and the interferences in signal reading by contaminating components such as nucleotides, proteins, and salts present in the solution.

16.2.1.2 NanoDrop (e.g., ND-3300, NanoDrop Technologies, USA)[1]

The NanoDrop is a modern spectrophotometer for highly sensitive quantification of DNA and RNA (proteins included). A major advantage of this system is that it requires very low sample consumption, which is very useful when using FFPE, and it is very reliable because no sample dilutions are necessary.[2] As a conventional spectrophotometer, it gives information about nucleic acid concentration (but it can also distinguish between DNA and RNA) and about the presence of contaminants. The results are acquired on a PC software programme (Fig. 16.1).

16.2.2 Quantification with a Fluorescent Dye

This approach relies on the measurement of the fluorescence emitted by a dye when it is intercalated in the DNA or RNA filaments. Emitted fluorescence is proportional to DNA/RNA sample concentration. The choice of the method depends on the nature of the material that should be measured.

16.2.2.1 Ethidium Bromide Gel-Based Assays

This approach can be used for quantification of both PCR product bands and genomic DNA (less commonly for RNA), but it is not the best choice when a precise quantification is required. For PCR product quantification, the assay can be easily performed by electrophoresis of the sample in an agarose gel stained with ethidium bromide (see Sect. 17.2 and Appendix A). The agarose percentage depends on the length of the fragment that has to be resolved.[3] Quantification can be carried out by

---

[1]See http://www.nanodrop.com/ for details.

[2]The measurement is performed by pipetting 1 μl of nucleic acid solution directly onto the pedestal of the instrument.

[3]A 0.7% gel shows a good resolution of large fragments (5–10 kb), while a 2% gel will show a good resolution for small fragments (0.2–1 kb). For smaller fragments a vertical polyacrylamide gel is more appropriate (see Appendix A and Appendix B for more details).

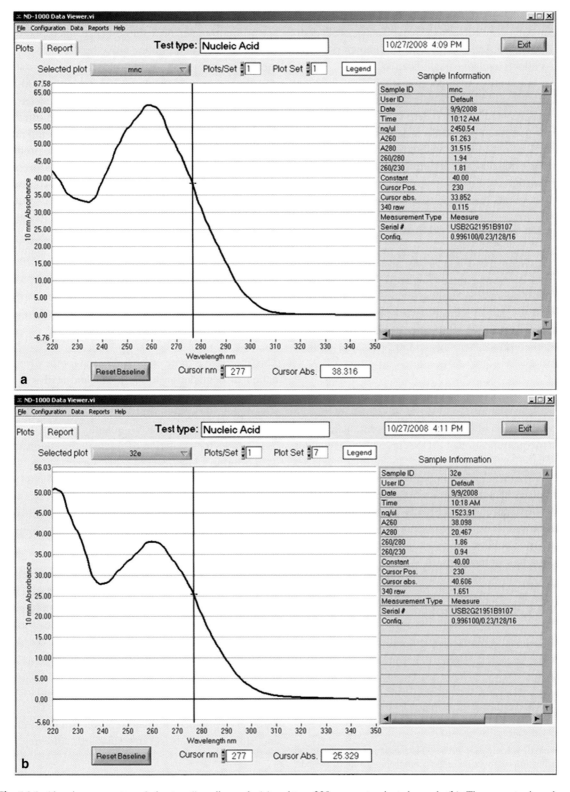

Fig. 16.1 Absorbance spectra relative to a "pure" sample (**a**) and to a 230 nm-contaminated sample (**b**). These spectra have been acquired using the software of the PC connected to the NanoDrop instrument

acquiring the gel image with a standard gel imaging instrument and calculating the band intensity with dedicated software. The comparison with known reference DNAs allows a relative quantification.

An alternative method, especially for genomic DNA, is to spot the DNA sample directly on an agarose gel prepared in a small Petri capsule together with four or five reference DNAs at known concentration. This gel must be treated with ethidium bromide solution and evaluated at a UV transilluminator by direct comparison.

16.2.2.2 SYBR Green I Method[4]

SYBR Green I dye is a highly sensitive fluorescent stain for detecting single-stranded DNA (ssDNA) and dsDNA in agarose and polyacrylamide gels. It can also detect RNA, but with much lower sensitivity. This dye is approximately 25–100 times more sensitive than ethidium bromide staining and much less mutagenic.

16.2.2.3 PicoGreen Method[5]

PicoGreen is an ultrasensitive fluorescent nucleic acid stain for quantification of dsDNA in solution. It enables the quantification of less than 25 ng/ml of dsDNA. using a standard spectrofluorometer or a fluorescence microplate. This method is ~500-fold more sensitive than absorbance measurements at 260 nm and the concentration measurement is not affected by contaminants (nucleotides, single-strand nucleic acids, proteins, organic compounds...).

16.2.2.4 RiboGreen Method[6]

RiboGreen is an ultrasensitive fluorescent nucleic acid stain for measuring RNA concentration in solution.

The use of this reagent requires the availability of a spectrofluorometer or a fluorescence microplate and the preparation of a titration curve before each experiment. The method is ~1,000-fold more sensitive than absorbance measurements at 260 nm and ~200-fold more sensitive than ethidium bromide-based assays, allowing detection of 1 ng RNA/ml. Signal intensity of RiboGreen can be affected by several compounds that commonly contaminate nucleic acid preparations, so it should be advisable to make a preliminary quantification by spectrophotometer to ensure the purity of the sample. It does not discriminate between DNA and RNA; for this reason, a treatment with DNase should be performed before RNA quantification. For practical details, follow the manufacturer's instructions.

16.2.2.5 Agilent 2100 Bioanalyzer[7]

This approach is more commonly used for the determination of degradation levels in nucleic acid samples (see Chap. 17). However, this microfluidic capillary electrophoresis system is also a very sensitive tool for calculation of DNA and RNA quantity. The instrument uses a laser for excitation of intercalating fluorescent dyes offering a high level of sensitivity.

16.2.3 Quantification by Real-time PCR

Real-time PCR can be used as an alternative method for the concentration estimation of both DNA and RNA when the starting material is extremely poor. This quantification procedure is based on the absolute quantification of specific target sequences (DNA or cDNA sequence depending on the origin of the starting material) [2]. The absolute quantitation relies on a curve generated from serially diluted standards of known concentrations or number of copies. The standard curve produces a linear relationship between Ct and initial amounts of the standard, allowing the

---

[4]See Roche datasheet at the website: https://www.roche-applied-science.com/pack-insert/1988131a.pdf. Equal reagents from other companies may be used.

[5]Provided by Molecular Probes. Datasheet is available at the website: http://probes.invitrogen.com/media/pis/mp07581.pdf. Equal reagents from other companies may be used.

[6]See Molecular Probes, web site: http://probes.invitrogen.com/media/pis/mp11490.pdf.

[7]See http://www.chem.agilent.com/Scripts/PDS.asp?lPage=51 for more details.

determination of the concentration/number of copies of the unknown target sequences based on their Ct values. This method assumes that all standards and samples have approximately equal amplification efficiencies. The standard used for the curve can be a fragment of dsDNA, ssDNA, or complementary RNA (cRNA) containing the target sequence. The inner standard can be obtained by in vitro transcription (cRNA), by cloning into a plasmid (DNA), or by purification of a PCR product. The primer pair for the target sequence should be designed in order to produce an amplicon no longer than 70 bases (for general information on real-time PCR set up see Chap. 25).

References

1. Bustin SA (2005) Real-time, fluorescence-based quantitative PCR: a snapshot of current procedures and preferences. Expert Rev Mol Diagn 5(4):493–498
2. Wong ML, Medrano JF (2005) Real-time PCR for mRNA quantitation. Biotechniques 39(1):75–85

Integrity Assessment of Nucleic Acids

17

Isabella Dotti and Serena Bonin

Contents

I. Dotti (✉) and S. Bonin
Department of Medical, Surgical and Health Sciences,
University of Trieste, Cattinara Hospital,
Strada di Fiume 447, Trieste, Italy

17.1 Introduction and Purpose

The level of degradation of nucleic acids, and especially of RNA, is a very important parameter to assess assay amenability when using fixed and paraffin-embedded (FPE) samples. The reason for this is that it's not a very well predictable phenomenon and its variable effects may false the results of qualitative and quantitative analysis on DNA and RNA. This variability depends on both biological and technical factors (type of tissue, fixation parameters, sample processing...). The degradation levels can be measured by different approaches and they are chosen according to the nature of the nucleic acid.

17.2 Agarose Gel Electrophoresis for DNA

Indicative integrity of genomic DNA can be checked on a 0.4% agarose gel with ethidium bromide and the level of degradation can be graded according to the electrophoretic migration of sample DNA in comparison to a known molecular weight marker (usually λDNA/HindIII fragments). It can be run with or without previous enzymatic digestion depending on the expected level of degradation and the consequent length of DNA fragments present in the sample. In the first case, the degradation rate can be defined by comparing the band pattern between an intact DNA and the unknown sample; in the second case, the index of degradation is determined by the presence of a smearing pattern. For more details on the gel preparation, see also Appendix A.

G. Stanta (ed.), *Guidelines for Molecular Analysis in Archive Tissues,*
DOI: 10.1007/978-3-642-17890-0_17, © Springer-Verlag Berlin Heidelberg 2011

17.3 Denaturing Agarose Gel Electrophoresis for RNA

The quality of mRNA from fresh samples (cultured cell lines, fresh frozen tissue samples…) can be easily assessed by running total RNA in a denaturing gel stained with ethidium bromide. This method relies on the assumption that rRNA quality and quantity reflect those of the underlying mRNA population. RNA integrity evaluation is based on the intensity ratio between the 28S and 18S bands (2:1 ratio is considered an index for intact RNA) and on the observation of the rRNA degradation pattern. Anyway, it is not clear how rRNA degradation actually reflects the quality of the underlying mRNA population, since in certain conditions rRNA can turn over more rapidly than mRNA [1]. Besides, visual assessment of the 28S:18S rRNA ratio on agarose gels is not an objective parameter because the appearance of rRNA bands is affected by the electrophoretic conditions, the amount of loaded RNA, and the saturation of ethidium bromide fluorescence.

This approach is not suitable when RNA integrity assessment is performed on archival tissues because large amounts of input RNA are needed and the ribosomal bands are rarely detectable due to the high levels of RNA degradation.

17.3.1 Reagents

All reagents should be RNase free[1] and DEPC treated

- *DEPC H_2O*
- *10% SDS or alternatively 1 M NaOH or H_2O_2*
- *Agarose powder electrophoresis grade*
- *37% Formaldehyde*
- *10X MOPS buffer*: 0.2 M MOPS, 0.05 M sodium acetate pH 6, 0.01 M EDTA
- *RNA loading buffer.*[2] For 1.5 ml: 720 µl formamide, 160 µl 10X MOPS buffer, 260 µl 37%

formaldehyde, 80 µl 1% blue bromophenol, 100 µl 80% glycerol, 180 µl DEPC H_2O
- *Ethidium Bromide, 10 mg/ml (stock solution).*[3] Stock solution of EtBr should be stored at 4°C in a dark bottle
- *Running buffer*: 1/10 of 37% formaldehyde in volume, 1X MOPS

17.3.2 Equipment

- *Disinfected adjustable pipettes,*[4] range: 2–20 µl, 20–200 µl, 100–1,000 µl
- *Nuclease-free aerosol-resistant pipette tips*
- *0.5 or 1.5 ml tubes* (autoclaved)
- *Centrifuge* suitable for centrifugation/spinning at 13,200 or 14,000 rpm
- *Thermoblock*
- *Electrophoretic apparatus* with spacers
- *Power supply*
- *UV transilluminator*

17.3.3 Method

- Wash the electrophoretic apparatus and spacers with 1% SDS (alternatively it is possible to use 1 M NaOH or 7% H_2O_2), cover and leave for 2 h under a fume hood. Wash with distilled H_2O, followed by sterile H_2O and finally DEPC H_2O.
- Prepare the 1% agarose gel by melting 0.5 g agarose in 42.3 ml DEPC H_2O. When cooled, add 5 ml 10X MOPS and 2.7 ml 37% formaldehyde. Pour the solution in the apparatus.[5]
- To prepare the RNA samples, use 5–50 µg total RNA. Mix 2–10 µl of RNA solution with 5 µl of RNA loading buffer. Adjust the volume with DEPC H_2O to a final volume of 20 µl.

---

[1]Sterile, disposable plasticware should be preferably used because it is RNase free. If general laboratory glassware or plasticware is used, it should be presoaked in 0.1% DEPC-treated H_2O for 2 h at 37°C. The DEPC treated-items should be rinsed thoroughly with DEPC H_2O and then autoclaved. Alternatively, glassware can be put in an oven at 250°C for at least 2 h.

[2]It can be stored at 4°C for a maximum of 2 weeks or at −20°C for a maximum of 2 months.

[3]EtBr is a potentially carcinogenic compound. Always wear gloves. Used EtBr solutions must be collected in containers for chemical waste and discharged according to the local hazardous chemical disposal procedures.

[4]Clean the pipettes with a disinfectant (e.g. Meliseptol®rapid) and leave them under the UV lamp for at least 10 min. Alternatively, it is possible to autoclave the pipette depending on the provider instructions.

[5]Formaldehyde is carcinogenic; for this reason, gels should be prepared and run under a fume hood.

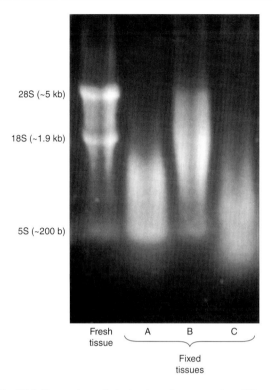

Fig. 17.1 Denaturing gel electrophoresis representing different rRNA degradation patterns. RNA from fresh frozen tissue is almost intact (ribosomal bands are clearly visible) while RNA from fixed tissues (formalin – *A*, methacarn – *B* and commercial FineFIX® – *C*) is highly degraded (only a smear is present)

- Denature RNA for 5′ at 65°C and chill on ice.
- Add 1 µl ethidium bromide previously diluted to 200 µg/ml and load the samples into the gel.
- Separate the products at 70 V in 1X running buffer until the dye reaches three-fourths of the gel length.
- Check the gel under the UV light.[6] Two main bands corresponding to 28S and 18S rRNA should be detected. Acquire the intensity of the bands by a computer-based system. A 2:1 ratio between the two bands is considered an index of intact RNA. The presence of a smear and/or the absence of the ribosomal bands is an index of RNA degradation (Fig. 17.1).

---

[6]If the gel must be submitted to further analysis (e.g., Northern blotting), deposit it on a clean plastic wrap before checking on the transilluminator.

17.4 PCR Amplification of DNA Fragments of Increasing Length

DNA quality from archival samples can be assessed by PCR amplification selecting fragments of increasing length. One commercial method that can be used for this purpose is the "Specimen Control Size Ladder" that is provided by InVivo Scribe Technologies. The kit consists of a master mix composed of a buffered magnesium solution, deoxynucleotides, and multiple primer pairs that allow multiplex amplification of five DNA fragments ranging from 100 to 600 base pairs. With the "Specimen Control Size Ladder" – Unlabeled both a conventional agarose and acrylamide gel can be run to visualize the results (Fig. 17.2).

17.5 RT-PCR Amplification of mRNA Fragments of Increasing Length

The setting up of this method is very easy. Messenger RNA integrity is normally evaluated by RT-PCR amplification of a housekeeping gene selecting a ladder of mRNA fragments of increasing length. Reverse

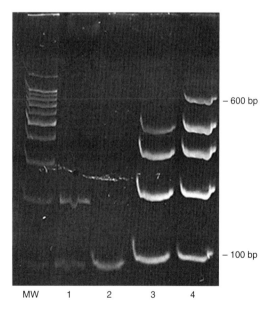

Fig. 17.2 Electrophoretic run of DNA samples at different degradation levels amplified using the "Specimen Control Size Ladder" kit. In this example, sample 2 is the most degraded, sample 4 is the best preserved

Fig. 17.3 Polyacrylamide gel electrophoresis representing different mRNA degradation levels. (**a**) Intact RNA (isolated from a fresh frozen tissue) (**b**) Almost degraded RNA (extracted from a FFPE tissue). RT-PCR amplification has been performed using a ladder of mRNA fragments ranging from 77 to 651 bases

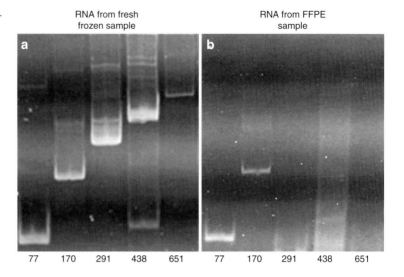

transcription is normally performed using MMLV enzyme combined with the random-hexamer priming strategy. The specific primer pairs used in the amplification step are designed for fragments of progressive length, preferably keeping either primer constant (Fig. 17.3). The length of the amplicons can be chosen arbitrarily, but is usually between 200 and 3,000 bp for DNA, and between 60 and 1,000 bp for RNA [2]. The major advantage of mRNA integrity determination is that the results are independent from rRNA integrity. This method can be useful in FFPE samples as it requires small amounts of starting material. (For the setting of this experiment follow the rules reported in Chaps. 19 and 25).

17.6 3′/5′ mRNA Integrity Assay

This method is specific to assess mRNA degradation levels and represents a variation of the previous approach ("RT-PCR amplification of mRNA fragments of increasing length"). It measures the integrity of the selected target gene (usually a highly expressed housekeeping gene such as GAPDH and beta-actin), which is expected to reflect the integrity of the entire mRNA subpopulation in the sample [3, 4]. In this assay reverse transcription is performed with oligo-dT and Superscript II/III. A real-time PCR reaction is carried out for three 60–80 bp-long amplicons along the transcript using three differently labeled TaqMan probes. Usually, amplicons are located at 100, 400–500, and 800–1,000 bases from the 3′ end of the gene, corresponding to the so-called 3′, intermediate and 5′ fragment, respectively.

The quantitative ratios between the Ct values of the 3′ and 5′ amplicons reflect the degradation level and the efficiency of the RT enzyme to proceed along the transcript. The higher the RNA fragmentation, the lower the RT efficiency, and consequently, the PCR yield. A 3′:5′ ratio of about 1 indicates high RNA integrity, whereas progressive ratios indicate increasing degradation (Fig. 17.4). For details regarding the setting of the TaqMan-based experiments and interpretation of the results, see Chap. 25, Sect. 25.2.

17.7 Agilent Bioanalyzer 2100[7]

This microfluidic capillary electrophoresis system by Agilent Technologies is based on fluorescent dyes binding to nucleic acid and allows the checking of not only DNA and RNA integrity, but also nucleic acid quantity (as already reported in Chap. 16, Sect. 16.2.2.5). When RNA integrity is assessed, three types of data analysis can be performed by this instrument. The first is the visual analysis of the electropherogram ("visual method"). It assigns a categorization of the RNA degradation levels (e.g., from 1 to 5) [5], taking into account the ribosomal peak height, the baseline flatness, and the presence of additional peaks between the ribosomal peaks (Fig. 17.5). Alternatively, two automatically calculated parameters can be used. The first one is the 28S/18S ratio ("Ratio method"). A ribosomal

---

[7]See http://www.chem.agilent.com/Scripts/PDS.asp?lPage=51 for more details

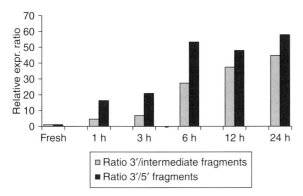

Fig. 17.4 Quantitative ratios calculated from the Ct values associated with the 3′ fragment and the two fragments located at increasing length from the 3′ end of the gene. Ratios of RNA isolated from fresh frozen sample usually correspond to 1 and progressively increase in RNA extracted from formalin-fixed samples (in this example cultured cell samples were progressively formalin fixed from 1 to 24 h)

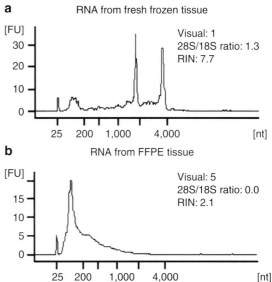

Fig. 17.5 Electropherograms representing different levels of RNA degradation. (**a**) Nearly intact RNA (isolated from fresh frozen tissue) (**b**) Highly degraded RNA (isolated from FFPE tissue). These traces can be used to evaluate RNA integrity by visual assessment (operator-dependent), 28S/18S ratio and RIN number (automatically calculated by the software)

band ratio of (28S:18S) between 0.7 and 2.5 is considered to be typical of good-quality RNA. The second one is the RIN (RNA Integrity Number). It takes into account the whole electropherogram, including in the analysis the total RNA ratio (the fraction of the area in the region of 18S and 28S compared to the total area under the curve), the fast region (the area between 18S peak and the marker), and the marker height.[8] The RIN range is between 1 and 10 (RIN 1 corresponds to completely degraded RNA, whereas a RIN 10 is achieved for an intact RNA sample). Values equal or higher than 6 are associated with good-quality RNA [6]. The Agilent Bioanalyzer system offers several advantages. It requires a very small amount of starting material, allowing assaying DNA and RNA quality even in small samples. Furthermore, interpretation of integrity is standardized and for this reason it is comparable among different laboratories. However, this approach appears more suitable for fresh frozen samples than for archival specimens because of the high degradation levels of nucleic acids isolated from fixed and embedded material. When RNA integrity is assessed in these specimens, 28S/18S ratios are usually around 0 and RIN are usually between 1 and 3.5 [1]. These low values, however, do not preclude the use of nucleic acids isolated

from archival samples in PCR-based assays. For details about sample preparation and loading, follow the manufacturer's instructions.

References

1. Dotti I, Bonin S, Basili G, Nardon E, Balani A, Siracusano S, Zanconati F, Palmisano S, De Manzini N, Stanta G (2010) Effects of formalin, methacarn and finefix fixatives on RNA preservation. Diagn Mol Pathol 19(2):112–122
2. Stanta G, Mucelli SP, Petrera F, Bonin S, Bussolati G. A novel fixative improves opportunities of nucleic acids and proteomic analysis in human archive's tissues. Diagn Mol Pathol. 2006 Jun;15(2):115–23
3. Nolan T, Hands RE, Bustin SA. Quantification of mRNA using real-time RT-PCR. Nat Protoc. 2006;1(3):1559–82
4. Penland SK, Keku TO, Torrice C, He X, Krishnamurthy J, Hoadley KA, Woosley JT, Thomas NE, Perou CM, Sandler RS, Sharpless NE. RNA expression analysis of formalin-fixed paraffin-embedded tumors. Lab Invest. 2007 Apr;87(4):383–91
5. Imbeaud S, Graudens E, Boulanger V, Barlet X, Zaborski P, Eveno E, Mueller O, Schroeder A, Auffray C (2005) Towards standardization of quality assessment using user-independent classifiers of microcapillary electrophoresis traces. Nucleic Acids Res 33(6):e56
6. Schroeder A, Mueller O, Stocker S, Salowsky R, Leiber M, Gassmann M, Lightfoot S, Menzel W, Granzow M, Ragg T (2006) The RIN: an integrity number for assigning integrity values to measurements. BMC Mol Biol 7:3

---

[8]The marker is an internal standard that is used to align the ladder analysis with the individual sample analysis. The RNA ladder is a mixture of six RNA transcripts of well-defined size and total concentration that is run first from the ladder well. Both marker and ladder are provided in the kit.

DNase Treatment of RNA

18

Isabella Dotti and Serena Bonin

Contents

18.1 Introduction and Purpose

A frequent cause of false positives in RT-PCR-based assays is the amplification of genomic DNA (gDNA) contaminating RNA preparations. Because of PCR sensitiveness, a single copy of a genomic sequence can be theoretically detected. This phenomenon is exacerbated when tested genes present pseudogenes at the DNA level, which have sequences similar to mRNA (information about pseudogenes can be obtained in the web[1]). For this reason DNA removal by DNase digestion is often a necessary step. Moreover, what has been observed is a decrement in real-time threshold cycles in cDNA obtained from RNA extracts that was treated with DNase prior to reverse transcription, in comparison with the undigested ones [1]. The good outcome of the DNase digestion can be then checked by performing an RT-PCR assay in which the reverse transcriptase is not included.

The following protocols describe two different DNase treatments that can be chosen on the basis of RNA availability.

18.2 DNase Digestion Followed by Phenol/Chloroform Extraction

This protocol allows complete removal of DNase enzyme after DNA digestion [2]. The adjustable volumes of reaction allow digestion of even large amounts

[1]Pseudogenes are DNA non-functional sequences present in the genome that have strong similarities to mRNA but, in general, are unable to be transcribed. The nonfunctionality of the pseudogenes is often caused by the lack of functional promoters or other regulatory elements. Pseudogenes are quite difficult to identify, because their characterization is performed through sequence calculations and alignments rather than biologically proven. A comprehensive list of identified pseudogenes can be found at http://www.pseudogene.org.

I. Dotti (✉) and S. Bonin
Department of Medical, Surgical and Health Sciences,
University of Trieste, Cattinara Hospital,
Strada di Fiume 447, Trieste, Italy

G. Stanta (ed.), *Guidelines for Molecular Analysis in Archive Tissues*,
DOI: 10.1007/978-3-642-17890-0_18, © Springer-Verlag Berlin Heidelberg 2011

of DNA, also when RNA concentration is very low. This protocol is recommended when an RNA solution is not pure (presence of co-extracted contaminants) or when a high RNA quantity should be used in the following analyses. It is not the best choice when the starting amount of RNA is very low (for example, when recovered from small biopsies or after microdissection). In such case, Protocol 18.3 is suggested.

18.2.1 Reagents

Reagents from specific companies are reported here, but reagents of equal quality purchased from other companies may be used.

All reagents should be RNase free[2] and DEPC treated

- *DEPC-treated water (DEPC H_2O)*: add 1 ml of diethylpirocarbonate (DEPC) (0.1% final concentration) to 1 l of sterile water at 37°C overnight under fume hood and then autoclave[3]
- *10 U/μl DNaseI FPLC pure* (cod. 27-0514 GE Healthcare)
- *10× Reaction buffer*: 400 mM Tris-HCl (pH 7.5), 60 mM $MgCl_2$
- *Phenol/H_2O*
- *Chloroform*
- *0.8 M LiCl in absolute EtOH*
- *1 mg/ml Glycogen*
- *Absolute EtOH*

18.2.2 Equipment

- *Adjustable pipettes,*[4] range: 2–20 μl, 20–200 μl, 100–1,000 μl
- *Nuclease-free aerosol-resistant pipette tips*

- *1.5 ml tubes and 0.2 ml tubes* (autoclaved)
- *Thermomixer* (e.g., Eppendorf)

18.2.3 Method

- A 20 μl reaction volume is prepared, adding the following components in a 1.5 ml tube:

10 × Reaction buffer	2 μl
10 U/μl DNaseI	1 μl
Total RNA	2 μg[5]
DEPC H_2O	to 20 μl final[6]

- Incubate the mixture at 37°C for 20′. Spin down the condensate from the tube walls.
- Bring the final volume to 100 μl with DEPC-treated H_2O and add 70 μl phenol/H_2O + 30 μl chloroform.[7] Vortex, keep on ice for 20′ and spin the tube for 20′ at 14,000 rpm in a microcentrifuge at 4°C.
- Carefully transfer the supernatant to a new tube (avoid touching the interphase), and add three volumes of 0.8 M LiCl in absolute EtOH and 5 μl of glycogen solution. Mix the solution by inversion.
- Leave the solution at –20°C for 24 h.
- Spin the tube for 20′ at 14,000 rpm in a microcentrifuge at 4°C; remove the supernatant without disturbing the pellet.
- Wash the pellet with 100 μl ice cold absolute EtOH and spin the tube at 14,000 rpm for 10′ at 4°C.
- Remove the supernatant, air dry the RNA pellet and resuspend it in the proper amount of DEPC-treated H_2O.
- Determine the new RNA concentration photometrically at 260 and 280 nm, adding 1 μl RNA solution to 199 μl of H_2O.[8]

---

[2]Sterile, disposable plasticware should preferably be used because it is RNase free. If general laboratory glassware or plasticware is used, it should be pre-soaked in 0.1% DEPC-treated H_2O for 2 h at 37°C. The DEPC-treated items should be rinsed thoroughly with DEPC H_2O and then autoclaved. Alternatively, glassware could be put in an oven at 250°C for 2 h.

[3]After this treatment the water is essentially free from RNases and can be used to prepare solutions or to rinse any items to use for RNA isolation. DEPC is a carcinogen and should be handled with care under a fume hood.

[4]Clean the pipettes with a disinfectant (e.g., Meliseptol®rapid) and leave them under the UV lamp for at least 10 min. Alternatively it is possible to autoclave the pipette depending on the provider instructions. Pipettes dedicated only to RNA extraction and treatment should be preferably used.

---

[5]If more than 5 μg RNA is digested, increase the DNase units. If a larger amount of RNA is treated, scale up the reaction volume because the viscosity of the solution may increase, preventing DNase activity.

[6]The volume of reaction can be increased up to 100 μl when, for example, the starting RNA is resuspended in large volumes; in such case, the DNase unit concentration has to be maintained by increasing the quantity.

[7]Alternatively, it is possible to purify DNase-treated RNA by the use of silica columns or the direct performance of an on-column DNase digestion (for details, see RNase free DNase set on http://www.qiagen.com/).

[8]See Chap. 16 for details.

- The solution is ready to be used for RT-PCR. Alternatively, RNA can be stored at –80°C until use; it is better if it is divided into aliquots to avoid repeated freezing and thawing.

18.3 DNase Digestion Followed by Heat Inactivation

This protocol can be used when the starting amount of RNA is low and DNase removal by extraction could determine its complete loss [3].

18.3.1 Reagents

Reagents from specific companies are here reported, but reagents of equal quality purchased from other companies may be used.

All reagents should be RNase free[9] and DEPC treated

- *DEPC-treated water (DEPC H_2O)*: Add 1 ml of diethylpirocarbonate (DEPC) (0.1% final concentration) to 1 l of sterile water at 37°C overnight and then autoclave[10]
- *10 U/μl DNaseI FPLC pure* (cod. 27-0514 GE Healthcare)
- *10× Reaction buffer*: 400 mM Tris-HCl (pH 7.5), 60 mM $MgCl_2$
- *25 mM EDTA*[11]

18.3.2 Equipment

- *Adjustable pipettes,*[12] range: 2–20 μl, 20–200 μl, 100–1,000 μl
- *Nuclease-free aerosol-resistant pipette tips*
- *1.5 ml tubes and 0.2 ml tubes* (autoclaved)
- *Thermomixer* (Eppendorf)

18.3.3 Method

- Hereafter, a 20 μl reaction volume is reported, adding the following components in a 1.5 ml tube:

10 × Reaction buffer	2 μl
10 U/μl DNaseI	1 μl
Total RNA	2 μg[13]
DEPC H_2O	to 20 μl final[14]

- Incubate the mixture at 25°C for 15′.[15] Spin the tube to collect the condensate from the tube walls.
- Add 2 μl of 25 mM EDTA to the solution; mix and heat at 65°C for 10′ to inactivate DNase. Immediately, chill the solution on ice and spin the tube.
- Determine the RNA concentration photometrically at 260 and 280 nm, adding 1 μl RNA solution to 199 μl of H_2O.[16] Alternatively, for a more approximate quantification, determine the new RNA concentration simply taking into account the starting RNA concentration and the reaction volume before and after DNase treatment.

---

[9]Sterile, disposable plasticware should be preferably used because it is RNase free. If general laboratory glassware or plasticware is used, it should be pre-soaked in 0.1% DEPC-treated H_2O for 2 h at 37°C. The DEPC-treated items should be rinsed thoroughly with DEPC H_2O and then autoclaved. Alternatively, glassware could be put in an oven at 250°C for at least 2 h.

[10]After this treatment, the water is essentially free from RNases and can be used to prepare solutions or to rinse any items to be used for RNA isolation. DEPC is a carcinogen and should be handled with care under a fume hood.

[11]The EDTA helps to protect RNA at high temperatures.

[12]Clean the pipettes with a disinfectant (e.g., Meliseptol®rapid) and leave them under the UV lamp for at least 10 min. Alternatively it is possible to autoclave the pipette according to the provider instructions. Pipettes dedicated only for RNA extraction and treatment should be preferably used.

[13]If more than 4 μg RNA are digested, increase DNase units. If a larger amount of RNA is digested, scale up the reaction volume as more RNA may increase the viscosity of the solution preventing DNase activity.

[14]The volume of reaction can be increased up to 100 μl when the starting RNA is resuspended in large volumes; in this case also the DNase units must be increased.

[15]This incubation time and temperature has been shown to be sufficient to remove contaminant DNA (manufacturer's protocol suggests 10–20' at 37°C).

[16]For spectrophotometric quantification of RNA after DNase treatment remember to use a DNase solution complete of buffer and EDTA without RNA as a blank.

• The mixture can be used directly for RT-PCR. Alternatively, RNA can be stored at –80°C until use; it is better if it is divided into aliquots.

References

1. Okello JB, Zurek J, Devault AM, Kuch M, Okwi AL, Sewankambo NK, Bimenya GS, Poinar D, Poinar HN (2010) Comparison of methods in the recovery of nucleic acids from archival formalin-fixed paraffin-embedded autopsy tissues. Anal Biochem 400(1):110–117

2. Siracusano S, Niccolini B, Knez R, Tiberio A, Benedetti E, Bonin S, Ciciliato S, Pappagallo GL, Belgrano E, Stanta G (2005) The simultaneous use of telomerase, cytokeratin 20 and CD4 for bladder cancer detection in urine. Eur Urol 47(3):327–333

3. Dotti I, Bonin S, Basili G, Nardon E, Balani A, Siracusano S, Zanconati F, Palmisano S, De Manzini N, Stanta G (2010) Effects of formalin, methacarn, and FineFix fixatives on RNA preservation. Diagn Mol Pathol 19(2):112–122

Reverse Transcription

19

Isabella Dotti, Serena Bonin, and Ermanno Nardon

Contents

I. Dotti (✉) and S. Bonin
Department of Medical, Surgical and Health Sciences,
University of Trieste, Cattinara Hospital,
Strada di Fiume 447, Trieste, Italy

E. Nardon
Department of Medical, Surgical and Health Sciences,
University of Trieste, Trieste, Italy

19.1 Introduction and Purpose

Formalin fixation and paraffin embedding has been the method of choice for the processing and long-term storage of diagnostic tissue samples for a century. A number of studies has shown that such archival material is in principle amenable to the analysis of mRNA and several protocols have been established for the retrieval of mRNA from routinely processed paraffin-embedded tissues [1–3]. Several methods are available to analyze mRNA expression in tissues, including Northern hybridization, subtractive hybridization, RNase protection assay, and cDNA microarrays. However, these methods are significantly limited by the requirement of fresh, unfixed tissues to permit isolation of abundant and high-quality mRNA. Conversely, reverse transcription-polymerase chain reaction (RT-PCR) offers a number of advantages because it allows the use of relatively small amounts of molecules that may be fragmented or degraded, such as those obtained from embedded tissues. RT can be primed with oligo-dT, but this approach is not suitable in formalin-fixed and paraffin-embedded (FFPE)-derived material because of the degradation and loss of polyA tails during fixation. For this reason, random oligonucleotides or specific antisense primers are recommended when analyzing these tissues. Complementary DNA synthesis from RNA is a critical step in gene expression analysis and it is a major source of variability for all RT-PCR assays when these latter need to be performed in a quantitative manner (RT-qPCR). In fact, reverse transcription yield can vary up to 100-fold depending on reverse transcriptase [4], priming strategy [5], and quantity of RNA used [6], and the variation can also be gene dependent [4, 5]. The technical variability is further exacerbated when using RNA from FFPE samples due to intrinsic tissue heterogeneity, to the presence of inhibitory copurified components [7], and to variable degradation of extracted RNA. Standardization of the assays is necessary to allow the analysis of gene expression at the RNA level in standard diagnostic specimens in a clinical setting. The RT-qPCR assay can be performed combining both RT and PCR enzymes in the same tube (single-step RT-PCR system), using an enzyme endowed with both RT and polymerase activity (as the Tth polymerase: one enzyme/one tube system) or separating RT and amplification steps (two-step RT-PCR). Furthermore, commercial RT kits or home-made reagents can be used. We think that higher sensitivity and potential for optimization make the use of the two-enzyme protocol the best choice especially when using home-made reagents [8].

Herein we propose two protocols which differ for the type of reverse transcriptase and for the priming strategy used. The choice of the protocol depends mainly on RNA availability and on the expected expression levels of the target genes.

19.2 Reverse Transcription with MMLV-RT and Random Hexamers

This protocol combines the use of MMLV-RT (Moloney Murine Leukemia Virus Reverse Transcriptase) with "random priming" strategy. Random hexamers prime RT at multiple points along the transcript; this method is nonspecific but yields the most cDNA and is the most useful for transcripts with significant secondary structures. Furthermore, this strategy generates the least bias in the resulting cDNA, because the same reverse transcription reaction can be used for the detection of several genes in the same sample [9]. It is not the method of choice if the mRNA target is present at low levels, as less abundant targets appear to be reverse transcribed less efficiently than more abundant ones: this is probably due to the lower specific efficiency or to aspecific priming of ribosomal RNA [9]. However, when the starting RNA is low, a specific reverse transcriptase with reduced RNase H activity[1] (e.g.,

MMLV-RT RNase H minus, Superscript™) can be used. Moreover, recent results obtained by one of the IMPACTS groups showed a clear improvement of RT yield when using high concentration of random primers (3.35 nmol/reaction) in the reaction with MMLV enzyme [10].

The RT reaction step involving MMLV and random hexamers has been optimized specifically for quantitative studies in FFPE samples. This method improves the detection of scarcely expressed genes in paraffin-embedded tissues and should allow comparable results between fresh and fixed samples [11]. Such protocol is described as follows.

19.2.1 Reagents

Reagents from specific companies are reported here, but reagents of equal quality purchased from other companies may be used.

All reagents should be RNase free or DEPC treated. Refer to Chapter "RNA extraction from FFPE Tissues" for precautions against RNase contamination.

- *DEPC-treated water (DEPC H$_2$O)*: Add 1 ml of diethylpyrocarbonate (DEPC) (0.1% final concentration) to 1 l of sterile water at 37°C overnight and then autoclave[2]
- *200 U/µl MMLV-RT,*[3] supplied with 5× MMLV-RT buffer and 0.1 M DTT (Invitrogen)
- *Mix of 10 mM dNTPs*: Add 80 µl of each 100 mM dNTP (dATP, dCTP, dGTP, and dTTP[4]) (Promega) to 480 µl DEPC H$_2$O and aliquot

---

[1]RNase H is an endonuclease that specifically degrades the RNA in RNA: DNA hybrids. It is commonly used to destroy the RNA template after cDNA synthesis, as well as in procedures such as nuclease protection assays. Since an RNase H activity is also present in the reverse transcriptases, they have been genetically altered to remove this activity, resulting in an increase of full-length cDNA products.

[2]DEPC is a carcinogen and should be handled with care under a fume hood.

[3]When testing mRNA sequences rich in secondary structures, the use of a thermostable reverse transcriptase such as Superscript III or AMVTM is recommended. In such case, the RT step should be carried out at 60°C if using amplicon-specific primer (see Sect. 19.3) or at lower temperatures if using oligo-dT or random primers.

[4]dUTP instead of dTTP may be used both in RT and in PCR reaction to avoid reamplification of carryover PCR products. In such case, an additional incubation of mixes at 50°C for 2′ with the enzyme uracil N-glycosilase is necessary before the transcription procedure. 10′ incubation at 95°C is performed to heat-inactivate the enzyme. Because UNG is not completely deactivated at 95°C, the PCR reaction temperatures should be kept higher than 55°C.

- *100 ng/µl Pd(N)$_6$ random hexamers (Amersham):* Resuspend the lyophilized powder with DEPC H$_2$O to 1 µg/µl stock solution and then make aliquots of 100 ng/µl each
- *40 U/µl RNase inhibitor* [5](Promega)
- *25 mM MgCl$_2$* [6]

19.2.2 Equipment

- *Disinfected adjustable pipettes,*[7] range: 2–20 µl, 20–200 µl, 100–1,000 µl
- *Nuclease-free aerosol-resistant pipette tips*
- *1.5 and 0.2 ml tubes* (autoclaved)
- *Centrifuge* suitable for centrifugation/spinning of 0.2 ml tubes at 13,200 or 14,000 rpm
- *Thermal Cycler*

19.2.3 Method

1. All RT assays should include a positive control (e.g., good-quality RNA from cell lines) and a negative one (DEPC H$_2$O or parallel extraction from a paraffin block without tissues). The amplifiability of genomic DNA (gDNA) in subsequent PCR should be ruled out performing a "no-RT" control, lacking the enzyme, or a control containing only gDNA. Add the following components to a 0.2 ml microcentrifuge tube, to a final volume of 7 µl:

100 ng/µl Random hexamers	2.5–3 µl (250–300 ng final)
Total RNA	1 ng–5 µg[8,9,10]
DEPC H$_2$O	to 7.05 µl

2. Heat the mixture of RNA, random primers, and water in a thermal cycler at 65°C for 10′ and quick chill on ice.
3. Add the following components to the mastermix, to a final volume of 20 µl:

5× MMLV-RT buffer	4 µl (1× final)
10 mM dNTPs	2 µl (1 mM final)
0.1 M DTT	2 µl (10 mM final)
40 U/µl RNase inhibitor	0.1 µl (4 units final)
200 U/µl MMLV-RT	1.25 µl (250 units final)
25 mM MgCl$_2$	3.6 µl (7.5 mM final, considering the amount contained in the buffer)

4. Use the following Thermal Cycler programme:
   - 1×: 25°C for 10′
   - 1×: 37°C for 60′
   - 1×: 70°C for 15′
   - 1×: hold at 4°C
5. The cDNA can now be used as a template for PCR amplification. Store at –20°C until use. Avoid repeated freeze-thawing. The volume of first-strand reaction mixture to be used in the PCR should not exceed 10% of the amplification mixture volume, since larger amounts may result in decreased PCR product.

19.3 Reverse Transcription with AMV-RT and Specific Reverse Primer

It is possible to prime the RT using an antisense oligonucleotide, designed to target a specific mRNA transcript. Since specific primers' melting temperature is higher than that of random oligos, the use of a thermostable enzyme, such as AMV (Avian Myeloblastosis Virus Reverse Transcriptase) or Superscript is advisable.

---

[5]The use of RNase inhibitor is critical for the successful processing of RNA as naked RNA from tissue samples is extremely susceptible to degradation by endogenous ribonucleases (RNases).

[6]Additional MgCl$_2$ may be useful if the RNA has been previously digested by DNase according to the protocol described in the chapter about DNase digestion followed by heat inactivation, because the chelating properties of EDTA, used to inactivate DNase, can reduce free Mg^{2+} concentration

[7]Clean the pipettes with a disinfectant (e.g., Meliseptol®rapid) and leave them under the UV lamp for at least 10 min. Alternatively it is possible to autoclave the pipette depending on the provider instructions.

[8]Do not add more than 5 µg of total RNA per reaction because efficiency of cDNA synthesis can be reduced by higher RNA quantities. Use the same amount of RNA in all reactions.

[9]Total RNA is preferred to polyA RNA when reverse transcription is performed on RNA extracted from FFPE as polyA enrichment requires additional purification steps that cause RNA loss.

[10]When low amounts of RNA are obtained from microdissected or from very small FFPE samples, a fixed volume instead of a fixed quantity can be used for RT reaction.

This protocol uses the AMV reverse transcriptase associated with a specific reverse primer. Target-specific primers synthesize the most specific cDNA and provide the most sensitive method of quantification [12], so it is the method of choice for the detection of scarcely expressed genes. The main disadvantage is that this approach requires separate priming reactions for each target gene, which is wasteful if only limited amounts of RNA are available. The use of more than one specific reverse primer in a single reaction tube (multiplex) is possible, but it requires careful experimental design and optimization of reaction conditions.

19.3.1 Reagents

Reagents from specific companies are reported here, but reagents of equal quality purchased from other companies may be used.

All reagents should be RNase free[11] or DEPC treated

- *DEPC-treated water (DEPC H_2O)*: add 1 ml of diethylpyrocarbonate (DEPC) (0.1% final concentration) to 1 l of sterile water at 37°C overnight and then autoclave[12]
- *10 U/μl AMV-RT*, supplied with *5× AMV-RT buffer* (Promega)
- *Mix of 10 mM dNTPs* (Promega): add 80 μl of each 100 mM dNTP (dATP, dCTP, dGTP, and dTTP)[13] to 480 μl DEPC H_2O and aliquot
- *30 pmol/ul specific reverse primer*[14]: resuspend the lyophilized powder with DEPC H_2O to 300 pmol/μl

stock solution and then make diluted aliquots of 30 pmol/μl each
- *40 U/μl RNase inhibitor*[15] (Promega)

19.3.2 Equipment

- *Disinfected adjustable pipettes,*[16] range: 2–20 μl, 20–200 μl, 100–1,000 μl
- *Nuclease-free aerosol-resistant pipette tips*
- *1.5 and 0.2 ml tubes* (autoclaved)
- *Centrifuge* suitable for centrifugation/spinning of 0.2 ml tubes at 13,200 or 14,000 rpm
- *Thermal Cycler*

19.3.3 Method

1. All RT assays should include a positive control (e.g., good-quality RNA) and a negative one (DEPC H_2O). The amplifiability of gDNA in subsequent PCR should be ruled out performing a "no-RT" control, lacking the enzyme, or a control containing only gDNA.
   Add the following components to a 0.2 ml microcentrifuge tube, to a final volume of 10 microliters:

5× AMV-RT buffer	2 μl (1× final)
10 mM dNTPs	1 μl (1 mM final)
30 pmol/μl Specific reverse primer	0.5 μl (15 pmol final)
10 U/μl AMV-RT	0.2 μl (2 units final)
40 U/μl RNase inhibitor	0.1 μl (4 units final)
DEPC H_2O	to 10 μl
Total RNA	1 ng–2 μg[17,18,19]

---

[11]Refer to chapter dedicated to RNA extraction from FFPE for precautions against RNases contamination.

[12]DEPC is a carcinogen and should be handled with care under a fume hood.

[13]dUTP instead of dTTP may be used both in RT and in PCR reaction to avoid reamplification of carryover PCR products. In such case, an additional incubation of mixes at 50°C for 2′ with the enzyme uracil N-glycosilase is necessary before the transcription procedure. 10′ incubation at 95°C is performed to heat-inactivate the enzyme. Because UNG is not completely deactivated at 95°C, the PCR reaction temperatures should be kept higher than 55°C.

[14]As a general rule, primers should be between 15 and 25 bases long to maximize specificity, with a G/C content of around 50%. Avoid primers with secondary structures or with sequence complementarities at the 3′ ends that could form dimers. Specific software can be used for the design of primers for both endpoint and real-time PCR (e.g., Primer3, http://frodo.wi.mit.edu/primer3/, or IDTDNA, http://eu.idtdna.com/scitools/scitools.aspx).

[15]The use of RNase inhibitor is critical for the successful processing of RNA, as naked RNA from tissue samples is extremely susceptible to degradation by endogenous ribonucleases (RNases).

[16]Clean the pipettes with a disinfectant (e.g., Meliseptol®rapid) and leave them under the UV lamp for at least 10 min. Alternatively it is possible to autoclave the pipette depending on the provider instructions.

[17]Do not add more than 2 μg of total RNA per reaction because efficiency of cDNA synthesis can be reduced by higher RNA quantities. Use the same amount of RNA in all reactions.

[18]Total RNA is preferred to polyA RNA when reverse transcription is performed on RNA extracted from FFPE tissues as polyA enrichment requires additional purification steps that cause RNA loss and because polyA is degraded in these tissues.

[19]When low amounts of RNA are obtained from microdissected or from very small FFPE samples, a fixed volume instead of a fixed quantity can be used for the RT reaction.

Table 19.1 Problems and solutions with the RT reaction

Problem	Possible reason	Solution
No amplification	Low RNA or RNA degradation	Increase RNA content; check RNA integrity; use an RT enzyme with reduced RNase H activity
	Secondary structure	Incubate RNA at 65°C for 10′ before RT. Perform RT at a higher temperature and/or use a thermostable enzyme
	Wrong primer design	Change primers
Aspecific amplification	Too high primer concentration	Reduce primer concentration
	Use of oligo-dT or random primers	Use specific primer
Lower amplification than expected	Inhibitory effect of cDNA solution on PCR reaction	Dilute cDNA solution before PCR amplification
		Perform a phenol-chloroform extraction of RT mixture
	Presence of RT inhibitors in the RNA solution	Improve RNA extraction method
Higher amplification than expected	Use of random hexamers in RT reaction	Use specific primer in the RT reaction
Positive no-RT control	DNA contamination	Perform DNase treatment, use intron-spanning primers, avoid target genes with pseudogenes

2. Use the following Thermal Cycler programme:
   - 1×: 42°C for 60′
   - 1×: hold at 4°C
3. The cDNA can now be used as a template for amplification in PCR. The whole reverse transcription product can be used, provided that amplification reaction is performed in 50 µl in order to avoid inhibition of PCR. In such conditions, a PCR buffer without MgCl$_2$ is required.

19.4 Troubleshooting

Possible troubleshooting in the reverse transcription step is detected after PCR amplification (in this context, an optimized PCR assay is assumed). Refer to Table 19.1 for the most common problems and solutions concerning the cDNA synthesis step.

References

1. Lehmann U, Kreipe H (2001) Real-time PCR analysis of DNA and RNA extracted from formalin-fixed and paraffin-embedded biopsies. Methods 25(4):409–418
2. Stanta G, Bonin S (1998) RNA quantitative analysis from fixed and paraffin-embedded tissues: membrane hybridization and capillary electrophoresis. Biotechniques 24(2):271–276
3. Stanta G, Bonin S, Perin R (1998) RNA extraction from formalin-fixed and paraffin-embedded tissues. Methods Mol Biol 86:23–26. doi:10.1385/0-89603-494-1:23
4. Stahlberg A, Kubista M, Pfaffl M (2004) Comparison of reverse transcriptases in gene expression analysis. Clin Chem 50(9):1678–1680
5. Stahlberg A, Hakansson J, Xian X, Semb H, Kubista M (2004) Properties of the reverse transcription reaction in mRNA quantification. Clin Chem 50(3):509–515
6. Chandler DP, Wagnon CA, Bolton H Jr (1998) Reverse transcriptase (RT) inhibition of PCR at low concentrations of template and its implications for quantitative RT-PCR. Appl Environ Microbiol 64(2):669–677
7. Nolan T, Hands RE, Bustin SA (2006) Quantification of mRNA using real-time RT-PCR. Nat Protoc 1(3):1559–1582
8. Bustin SA (2000) Absolute quantification of mRNA using real-time reverse transcription polymerase chain reaction assays. J Mol Endocrinol 25(2):169–193
9. Bustin SA, Nolan T (2004) Pitfalls of quantitative real-time reverse-transcription polymerase chain reaction. J Biomol Tech 15(3):155–166
10. Nardon E, Donada M, Bonin S, Dotti I, Stanta G (2009) Higher random oligo concentration improves reverse transcription yield of cDNA from bioptic tissues and quantitative RT-PCR reliability. Exp Mol Pathol 87(2): 146–151
11. Godfrey TE, Kim SH, Chavira M, Ruff DW, Warren RS, Gray JW, Jensen RH (2000) Quantitative mRNA expression analysis from formalin-fixed, paraffin-embedded tissues using 5′ nuclease quantitative reverse transcription-polymerase chain reaction. J Mol Diagn 2(2):84–91
12. Lekanne Deprez RH, Fijnvandraat AC, Ruijter JM, Moorman AF (2002) Sensitivity and accuracy of quantitative real-time polymerase chain reaction using sybr green I depends on cDNA synthesis conditions. Anal Biochem 307(1):63–69

Part V

Nucleic Acid Amplification

General Protocol for End-Point PCR

Serena Bonin and Isabella Dotti

Contents

20.1 Introduction and Purpose

This protocol provides a general workflow for PCR amplification of DNA or cDNA obtained from formalin-fixed and paraffin-embedded samples (FFPE) [1, 2]. The protocol is dedicated not to quantitative, but to qualitative, PCR analysis. Prerequisites for successful PCR include the design of optimized primer pairs, the use of appropriate primer concentrations, and optimization of the PCR conditions [3].

PCR products can be detected by using different tools, ranging from gel electrophoresis to capillary electrophoresis and Southern blot, depending on the analysis requisites.

PCR setup should be performed in a separate area from PCR analysis to ensure that reagents used for PCR do not become contaminated with PCR products (carry-over). Similarly, pipettes used for analysis of PCR products should never be used to set up PCR.

The major problem with FFPE tissues is the extent of degradation of the extracted nucleic acids. In routinely fixed samples, the available fragments of DNA are usually about 300 bases long, while for RNA, (refer to this value if the starting material is cDNA) they are about 100 bases long.[1] In any case, we recommend the designing of amplicons[2] no longer than 200–250 bases for DNA and about 60–80 bases for

---

[1]Shorter amplicons are recommended in extracts from autopsy tissues, because of the higher degradation level. For DNA, 90 bases amplicons and for cDNA 70 bases fragments are detectable, on average, in autopsy tissues.

[2]Check homology of the sequences by BLAST analysis: http://www.ncbi.nlm.nih.gov/BLAST/.

S. Bonin (✉) and I. Dotti
Department of Medical, Surgical and Health Sciences,
University of Trieste, Cattinara Hospital, Strada di Fiume 447,
Trieste, Italy

G. Stanta (ed.), *Guidelines for Molecular Analysis in Archive Tissues*,
DOI: 10.1007/978-3-642-17890-0_20, © Springer-Verlag Berlin Heidelberg 2011

cDNA. Shorter amplicons increase the efficiency of the molecular amplification.

20.2 Protocol

20.2.1 Reagents

- *10× PCR buffer with MgCl$_2$* [3]: We report the standard composition of PCR buffer: 150 mM MgCl$_2$, 500 mM KCl, 100 mM Tris pH 8.3 at 25°C[4]
- *Commercial 100 mM stock solution of each dNTP (pH 8)*
- *dNTPs:* Dilute the dNTP stock solution to prepare a 10 mM solution of each dNTP in sterile water or DEPC-treated water
- *Primers* [5]: Lyophilized primers should be dissolved in a small volume of distilled water or 10 mM Tris pH 8 to make a concentrated stock solution. Prepare small aliquots of working solutions containing 10–30 pmol/µl to avoid repeated thawing and freezing. Store all primer solutions at – 20°C. Primer quality can be checked on a denaturing polyacrylamide gel; a single band should be seen
- *Taq DNA Polymerase*
- *Positive and negative controls*: Inclusion of control reactions in the PCR assay is essential to PCR interpretation. A positive control should be included to check if the used PCR conditions can successfully amplify the target sequence. As PCR is extremely sensitive, requiring only a few copies of target template, a negative control containing no template DNA should always be included to ensure that the solutions used for PCR have not become contaminated with template DNA

20.2.2 Equipment

- *Disinfected*[6] *adjustable pipettes,* range: 2–20 µl, 20–200 µl, 100–1,000 µl
- *Nuclease-free aerosol-resistant pipette tips*1.5, 0.5, and thin walled 0.2 ml tubes[7] (autoclaved)
- *Centrifuge* suitable for centrifugation of 1.5 ml tubes at 13,200 or 14,000 rpm
- *Laminar Flow Cabinet equipped with a UV lamp*
- *Thermal Cycler*: A large number of programmable thermal cyclers marketed by different companies are available for use in PCR. The choice of the commercial instruments depends on the investigator's inclination, budget, etc… The main requisite is that the instrument must have the license for use in PCR. The thermal cycler must be equipped with a heated lid in order to prevent internal contamination during the amplification process

20.2.3 Method

Preparation of the reaction mixture should be performed under a laminar flow cabinet. Fresh gloves should be worn.

- In a sterile 0.2 ml tube, add, in the following order, the 10× amplification buffer, dNTPs (10 mM), primers, Taq Polymerase, and finally, the template DNA. The suggested final volume for PCR reaction is 25–50 µl. Each reaction sample must contain the proper amount of target DNA,[8] 50 mM KCl, 10 mM Tris pH 8.3, 1.5–2.0 mM MgCl$_2$, 200 µM of each dNTP, 15–25 pmol of forward primer, 15–25 pmol of reverse primer,[9] and 1–5 units of Taq Polymerase.[10]

---

[3]There are commercial PCR premixed solutions for convenient PCR setup, containing PCR buffer and dNTPs. Usually, Taq Polymerase is provided with its dedicated 10× buffer. Check the composition for the presence of MgCl$_2$. Different commercial PCR buffers are MgCl$_2$ free, but a 25 mM or 50 mM MgCl$_2$ solution is supplied with buffer and enzyme.

[4]Because of the high temperature dependence of the pKa of Tris, the pH of the reaction will drop to about 7.2 at 72°C.

[5]Primer design could be performed using a particular kind of software, please refer to http://molbiol-tools.ca/PCR.htm for on line available software.

[6]Clean the pipettes with alcohol or another disinfectant and leave them under the UV lamp for at least 10 min. Alternatively it is possible to autoclave the pipette depending on the provider instructions.

[7]Micro titer plates could also be used to run PCR.

[8]The amount of target DNA required varies according to the complexity and level of target sequence. Usually, it ranges from 1 pg to 1 µg.

[9]The primer and Mg^{2+} concentration in the PCR buffer and the annealing temperature of the reaction may need to be optimized for each primer pair for a more efficient PCR.

[10]In order to prevent artifacts, a hot start PCR using a thermostable Taq Polymerase (e.g., AmpliTaq Gold by Applied Biosystems) is suggested.

- For every PCR run, include a positive control, a bystander DNA that does not contain the sequence of interest and a blank negative control (H_2O or better the extraction obtained from a paraffin block without tissue).[11]

- Displace the tubes in the thermal cycler and amplify the target DNA using the proper PCR program made up by denaturation, annealing, and elongation steps. For small stretches, the time required in every step of the PCR could be shorter, ranging from 30 s to 1 min, with a final 5–7 min extension step at 72°C. The number of PCR cycles depends on the amount of template DNA in the reaction mix and on the expected yield of the PCR product. For DNA extracted from FFPE, usually at least 35–40 cycles are performed.[12]

- Programme the instrument in order to maintain the amplified samples at 4°C at the end of amplification, before analysis or storing.

- Check the amplification by agarose electrophoresis stained with ethidium bromide,[13] by means of Southern blot or dot blot (using an internal probe) (see Chap. 17, Sect. 17.1 and Appendix A for agarose gel preparation). Store the PCR products at –20°C avoiding repeated thawing and freezing.

20.2.4 Troubleshooting

- If the bands of the desired product are sharp but faint, both in positive controls and samples, inefficient priming or inefficient extension may have occurred. To solve this problem, set up a series of PCRs containing different concentrations of the two primers (e.g., from 5 to 30 pmol each/reaction). Once the optimal primer concentration is found, set up a series of PCRs containing different concentrations of Mg^{2+} (e.g., ranging from 1 to 4.5 mM) to determine the optimal concentration.

- If there are bands in the negative controls, contamination of solutions or plasticware with template DNA may have occurred. In such case, make up new reagents and operate in separate spaces to prepare the PCR solution and to analyze the products of amplification.

- If distinct bands of unexpected molecular weight appear, they are the result of nonspecific priming by one or both primers. To eliminate this problem, decrease the annealing time and/or increase the annealing temperature. Furthermore, it is possible to perform a hot start by utilizing a Taq polymerase requiring thermal activation.

- If a generalized smear of amplified DNA is present, too much template DNA was introduced in the reaction. Try the analysis with a scaling down of the target amount.

- If the amplification is weak or nondetectable, there could be different possible causes: defective reagents, defective thermal cycler, or programming error. If you are setting up the PCR system, the cause could also be a suboptimal extension of annealed primers, an ineffective denaturation, or you may have chosen too long stretches for your PCR analysis.

---

[11]If the thermal cycler is not fitted with a heated lid, overlay the reaction mixture with one drop (about 20 μl) of light mineral oil to prevent evaporation and cross contamination.

[12]Too many PCR cycles could promote aspecific amplifications, and/or PCR product carry-over.

[13]EtBr is a potentially carcinogenic compound. Always wear gloves and use it under a laminar hood. Used EtBr solutions must be collected in containers for chemical waste and discharged according to the local hazardous chemical disposal procedures.

References

1. Gilbert MT, Haselkorn T, Bunce M, Sanchez JJ, Lucas SB, Jewell LD, Van Marck E, Worobey M (2007) The isolation of nucleic acids from fixed, paraffin-embedded tissues-which methods are useful when? PLoS ONE 2(6):e537
2. Lehmann U, Kreipe H (2001) Real-time PCR analysis of DNA and RNA extracted from formalin-fixed and paraffin-embedded biopsies. Methods 25(4):409–418
3. Sambrook J, Russel DW (2001) Molecular cloning. A laboratory manual, 3rd edn. Cold Spring Harbour Laboratory Press, New York

Quantitative End-Point PCR

21

Serena Bonin and Isabella Dotti

Contents

21.1 Introduction and Purpose

In quantitative PCR the amount of PCR product should reflect the number of target sequences in the original specimen. Absolute quantitative PCR is not possible in formalin-fixed paraffin-embedded (FFPE) tissues because of the variables that could affect the efficiency of the amplification. These variables include the type of tissue, length of fixation, interval before fixation, age of the paraffin block, size of the section, and the amount of extracted DNA or the amount of synthesized cDNA added to the PCR mixture. All these factors, and perhaps more, conspire to confound a proportional linkage between the final amplification signal and the original amount of target in an absolute quantification. The only method to quantify PCR products in FFPE material is a relative quantification [1–5]. For a precise comparison, the results must be standardized against the quantity of detected housekeeping genes.

Although currently real-time PCR is the golden standard to perform quantitative PCR or RT-PCR, an end-point approach is also possible. In this context PCR products can be quantified using different tools, ranging from southern blot and dot blot to PCR ELISA and capillary electrophoresis. In any case, the PCR product must be associated to a value (Counts Per minute – CPM – , Digital Light Units – DLU – or others) in order to perform the quantitative analysis. In all these cases, the amplicons are detected using a probe internal to the PCR sequence. Alternatively, without any PCR assays, quantitative analyses of RNA in FFPE has been developed by Panomics by the use of branched DNA technology (for details visit www.panomics.com).

S. Bonin (✉) and I. Dotti
Department of Medical, Surgical and Health Sciences,
University of Trieste, Cattinara Hospital,
Strada di Fiume 447, Trieste, Italy

G. Stanta (ed.), *Guidelines for Molecular Analysis in Archive Tissues*,
DOI: 10.1007/978-3-642-17890-0_21, © Springer-Verlag Berlin Heidelberg 2011

In end-point quantitative PCR, to quantify a DNA or a cDNA target sequence, three oligonucleotides are synthesized, two sense and one antisense oriented. The first sense and the antisense oligonucleotides are used for the PCR amplification, while the second sense oligonucleotide[1] will be used as a probe, to detect only specific amplification products. When RNA sequences have to be detected, intron-spanning primers are recommended in order to avoid amplification of contaminating genomic DNA, although this strategy is not useful when the target mRNA presents pseudogenes at the DNA level (a comprehensive database of identified pseudogenes is published at http://www.pseudogene.org/).

Set up of the proper amount of Mg^{2+}, forward and reverse primers, dNTPs, and Taq Polymerase for PCR is essential to obtain quantitative results.

The amplification conditions must be determined carefully. These conditions are related to the amount of starting material (DNA or cDNA) and the number of amplification cycles. Usually, amplifications performed in nucleic acids from FFPE tissues require more PCR cycles in comparison with fresh tissues because of the presence of inhibitors and degraded starting material. On average, the suggested number of cycles is 40–45.

A linear relationship exists between the log of the quantity of cDNA (or DNA) and the log of the specific amplified gene product being examined (Fig. 21.1).

This relationship is valid in a specific range, defined by the quantity of starting nucleic acid and the number of cycles of the PCR. These variables, known as linearity conditions, change with target genes, starting material (type of samples, fresh or FFPE tissues), but also with PCR primer sets. These conditions, for linear amplification, must be determined empirically. To find the linearity conditions, first of all a constant amount of nucleic acid (DNA or cDNA), usually about 100 ng, is amplified varying the number of cycles, for example, from 30 to 50, with an increment of five cycles for each variation. The proper number of cycles is determined by the middle point in the calibration curve that plots the log of PCR products vs. the number of cycles far from plateau conditions. The second step to determine the linearity condition consists in the amplification with an already selected number of cycles and with increasing quantities of nucleic acid. For example, when analyzing gene expression, the suggested range is 5–500 ng of cDNA by twofold dilutions

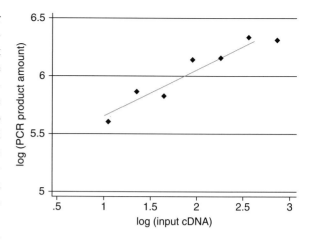

Fig. 21.1 Linearity conditions for MPR1 gene expression. The plot represents the linearity conditions between the log of the PCR products of MRP1 (as log of DLU – Digital Light Unit – of the dot blot hybridization) and the log of the starting amount of cDNA (range 700–11 ng of cDNA). PCR was run for 45 cycles and amplified a sequence of 96 bases corresponding to a fragment of MRP1 gene. The linearity is preserved between 11 and 350 ng of starting cDNA

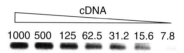

Fig. 21.2 Southern blot representing the linearity conditions of EGFr mRNA using six 1:2 dilutions of cDNA, range 1,000–7.8 ng (see Chap. 23 for the detailed Dot-Blot procedure)

(Fig. 21.2). Again the optimal conditions are chosen for the quantity related to the middle point in the calibration curve avoiding plateau. Once these conditions are defined, it is possible to analyze a large case study using the previously chosen linearity conditions at a lower cost than real-time PCR.

The determination of the linearity conditions must be performed for all the target genes and for the reference genes (usually a housekeeping gene in the case of cDNA analysis[2] and a DNA reference sequence in case of DNA analysis). At this point, it is possible to run the PCR assays for the case study and determine for each sample and each gene the amount of PCR product according to the used detection system (CPM, DLU, area value for capillary electrophoresis...).

---

[1]The probe must not overlap with PCR primers. Its sequence has to be internal to the amplification product.

[2]The selection of the proper reference gene is a critical step in the RT-PCR assay because the inappropriate selection of the reference gene could create unreal quantitative results. A possible suggestion is to analyze the expression of some five candidate housekeeping genes in about ten cases and select for the case study one to three of them, characterized by lower variability among samples (see Sect. 25.5 in Chap. 25).

21.2 Precautions

PCR setup should be performed in a separate area from PCR analysis to ensure that reagents used for PCR do not become contaminated with PCR products. Similarly, pipettes used for analysis of PCR products should never be used for setting up PCR.

The major problem with formalin-fixed and paraffin-embedded tissues is the extent of degradation of the extracted nucleic acids. Usually in routinely fixed samples the available length of DNA fragments is about 300 bases, while for RNA fragments no longer than 100 bases should be used. When amplification of around 60-base long fragments is performed, efficiency in FFPE samples can be similar to that obtained in fresh frozen tissues [6]. In any case, we recommend the designing of amplicons[3] no longer than 200 bases for DNA from bioptic specimens and about 60–80 bases for cDNA. For quantitative analysis the amplifications must be performed in linearity conditions, where there is a linear relationship between the log of the initial quantity of DNA or cDNA and the log of the quantity of amplified product. The quantitative results are then standardized with reference housekeeping genes. This comparison must be performed in every sample in order to normalize the degradation level of the nucleic acids. Inclusion of control reactions is essential for monitoring the success of the PCR reaction. A positive control should be included to check if the used PCR conditions can successfully amplify the target sequence. As PCR is extremely sensitive, a negative control containing no template DNA should always be included to ensure that the solutions used for PCR reaction are not contaminated with the template DNA from previous homologous amplifications.

21.3 Protocol

The following protocol provides a general method for quantitative PCR analysis of DNA or cDNA obtained from formalin-fixed and paraffin–embedded samples. The aim of this chapter is to provide a standard workflow for quantitative PCR when real-time instruments are not used. An example of end point PCR related to

the analysis of MRP1 gene in FFPE liver tissues is reported below. In this example, β-Actin has been used as a housekeeping reference gene [7].

21.3.1 Reagents

- *10× PCR buffer with MgCl2.*[4] We report the standard composition of PCR buffer: 15 mM $MgCl_2$, 500 mM KCl, 100 mM Tris pH 8.3 at 25°C.[5] This composition can be modified for specific purposes[6]
- *Commercial 100 mM stock solution of each dNTP (pH 8).* Dilute the dNTP stock solution to prepare 10 mM solution of each dNTP in sterile water or DEPC-treated water
- *Primers*[7]: Lyophilized primers should be dissolved in a small volume of distilled water or 10 mM Tris pH 8 to make a concentrated stock solution. Prepare small aliquots of working solutions containing 10–30 pmol/μl to avoid repeated thawing and freezing. Store all primer solutions at − 20°C. Primer quality can be checked on a denaturing polyacrylamide gel; a single band should be seen. Aliquots of 20 μM forward and reverse primers are used for the amplification. See Table 21.1 for the primer sequences used in the reported example
- *Taq DNA Polymerase*
- *Positive and negative controls*: Negative controls could be water or extracts from a paraffin block, tissues not included, or DNA without the specific target sequence

---

[3]Check homology of the sequences by BLAST analysis: http://www.ncbi.nlm.nih.gov/BLAST/.

[4]There are commercial PCR premixed solutions for convenient PCR setup, containing PCR buffer and dNTPs. Usually, Taq Polymerase is provided with its dedicated 10× buffer. Check the composition for the presence of $MgCl_2$. Different commercial PCR buffers are $MgCl_2$ free, but a 25 mM or 50 mM $MgCl_2$ solution is supplied with buffer and enzyme. $MgCl_2$ content could be incremented in cases of multiplex PCR or decremented in cases of high fidelity PCR.

[5]Because of the temperature dependence of the pKa of Tris, the pH of the reaction will drop to about 7.2 at 72°C.

[6]PCR reaction mixture could contain other components; among them, some cosolvents (for example DMSO and formamide) are used in cases of higher GC composition and BSA is used to reduce the amount of aspecific products.

[7]Primer design could be performed using specific software, please refer to http://molbiol-tools.ca/PCR.htm for the online available software.

Table 21.1 Primer sequences of MRP1 target gene and β-Actin reference gene

Label	Sequence
MRP1 up	TGT TCT CGG AAA CCA TCC A
MRP1 dw	CAA TCA ACC CTG TGA TCC A
MRP1 probe	ACC CTA ATC CCT GCC CAG AGT CCA
β-Act up	AAG GCC AAC CGC GAG AAG ATG A
β-Act dw	TGG ATA GCA ACG TAC ATG GCT G
β-Act probe	CCC AGA TCA TGT TTG AGA CCT TCA ACA CCC

21.3.2 Equipment

- *Disinfected[8] adjustable pipettes*, range: 2–20 μl, 20–200 μl, 100–1,000 μl
- *Nuclease-free aerosol-resistant pipette tips*
- *1.5, 0.5, and 0.2 ml tubes[9]* (autoclaved)
- *Centrifuge* suitable for centrifugation of 1.5 ml tubes at 13,200 or 14,000 rpm
- *Spectrophotometer or Nanodrop*
- *Thermocycler*: A large number of programmable thermal cyclers marked by different companies are available for use in PCR. The choice among the commercial instruments depends on the investigator's inclination, budget… The main requisite is that the instrument must have the license for use in PCR. The thermal cycler must be equipped with a heated lid in order to prevent contamination. For repeated amplification tests it is better to use the same thermocycler, because the small differences between instruments could give different amplification efficiencies
- *A system to quantify the PCR products* (dot-blot, southern-blot, capillary electrophoresis, PCR ELISA…)

21.3.3 Method

- To 5 μl of cDNA solution (corresponding to 50 ng cDNA), add 45 μl of PCR mastermix containing 1x

PCR buffer with $MgCl_2$,[10] 15 pmol of forward and reverse primers for MRP1, 200 μM of each dNTP, and 1.2 units of Taq polymerase. For every PCR run, include a diluted positive control, a bystander DNA that does not contain the sequence of interest (cDNA of a tissue negative for MRP1) and a blank control (H_2O).[11]

- After the initial denaturation of 3 min at 95°C, carry out thermal cycling for: five amplification cycles of 95°C/1 min; 59°C/1 min; and 72°C/1 min, followed by 40 cycles of 95°C/30 s; 59°C/30 s; and 72°C/30 s. Program the instrument in order to maintain the amplified samples at 4°C before storing.

- At the same time, perform the amplification for β-Actin using 10 ng of cDNA. After the initial denaturation of 3 min at 95°C, carry out thermal cycling for: five amplification cycles of 95°C/1 min; 55°C/1°min; and 72°C/1 min, followed by 35 cycles of 95°C/30 s; 55°C/30 s; and 72°C/30 s. Program the instrument in order to maintain the amplified samples at 4°C before storing.

- Transfer the amplification products onto a nylon membrane by dot blotting, as already reported in the dedicated protocol (see Chap. 23).

- Hybridize the membrane using the internal oligonucleotides as probes. Follow the hybridization procedures described in the Nested-PCR protocol for DIG-dUTP labeling (see Chap. 22, Sect. 22.2.3).

- Alternatively, it is possible to quantify the PCR products for β-Actin and MRP1 using capillary electrophoresis. Depending on the Capillary Electrophoresis instrument, set-up the separation conditions (capillary temperature, injection condition, kV for separation[12]) in order to separate different cDNA templates.[13] The amount of the different

---

[8]Clean the pipettes with a disinfectant (e.g. Meliseptol®rapid) and leave them under the UV lamp for at least 10 min. Alternatively it is possible to autoclave the pipette depending on the provider instructions.

[9]Microtiter plates could also be used to run PCR.

[10]If the RT step is performed with specific primers and AMV reverse transcriptase, the content of $MgCl_2$ in the PCR reaction should be checked because of the high concentration of $MgCl_2$ in the AMV RT working buffer. When the RT with AMV is made in 10 μl final volume and is used entirely for the successive PCR reaction, we recommend using PCR buffer without $MgCl_2$.

[11]If the thermal cycler is not fitted with a heated lid, overlay the reaction mixture with one drop (about 20 μl) of light mineral oil to prevent evaporation and cross contamination.

[12]For our analyses we used a coated capillary (100 μm id and 47 cm) filled with TBE buffer containing a replaceable linear polyacrylamide sieving matrix.

[13]As the PCR fragments amplified for MRP1 and β-Actin, for example, could differ in amplicon length, several amplification products could be coinjected and simultaneously analyzed in the same capillary.

PCR products is related to the area of the peak of the electropherogram for each analyzed sample.

- To standardize the data ("value$_{std}$"), calculate the mean value[14] of the reference genes[15] for every sample included in the case study. Then use this value to calculate the mean value of the reference genes of all the examined cases. This mean value is then divided by the value of the same reference gene for every sample. This result represents the normalization factor ("NF") that has to be multiplied by the value of the target gene for every sample of the case study ("value$_{sample}$").[16]
- The formula is the following:

$$Value_{std} = NF \times value_{sample}$$

21.3.4 Troubleshooting

Sometimes membrane hybridization could generate imperfections that result in incorrect quantification. In such case, there are different possibilities. In case of radiolabeled systems, we suggest washing the membrane even over the weekend at 37°C. In case of nonradioactive systems, we suggest repetition of hybridization, after membrane stripping, or repetition of the blot and the use of more stringent conditions for hybridization. In any case, to confirm the quantitative data we suggest performing dot-blot or ELISA plate in duplicate.

References

1. Lehmann U, Kreipe H (2001) Real-time PCR analysis of DNA and RNA extracted from formalin-fixed and paraffin-embedded biopsies. Methods 25(4):409–418
2. Sambrook J, Russel DW (2001) Molecular cloning. A laboratory manual, 3rd edn. Cold Spring Harbour Laboratory Press, New York
3. Stanta G, Bonin S (1998) RNA quantitative analysis from fixed and paraffin-embedded tissues: membrane hybridization and capillary electrophoresis. Biotechniques 24(2): 271–276
4. Stanta G, Bonin S, Lugli M (2001) Quantitative RT-PCR from fixed paraffin-embedded tissues by capillary electrophoresis. Methods Mol Biol 163:253–258
5. Stanta G, Bonin S, Utrera R (1998) RNA quantitative analysis from fixed and paraffin-embedded tissues. Methods Mol Biol 86:113–119
6. Specht K, Richter T, Muller U, Walch A, Werner M, Hofler H (2001) Quantitative gene expression analysis in microdissected archival formalin-fixed and paraffin-embedded tumor tissue. Am J Pathol 158(2):419–429
7. Bonin S, Pascolo L, Croce LS, Stanta G, Tiribelli C (2002) Gene expression of ABC proteins in hepatocellular carcinoma, perineoplastic tissue, and liver diseases. Mol Med 8(6): 318–325

---

[14]In case of capillary electrophoresis, the values related to the quantity of each gene for each sample are peak areas, while for membrane hybridization, they could be DLU (Digital light unit) in cases of chemiluminescence, CPM or DLU for radiolabeled systems and also colorimetric values for PCR-ELISA.

[15]Only if more than one reference gene has been used.

[16]The selection of small target sequences in a range of 60–80 bp enables the detection of fragmented and degraded RNA. As shown by Specht et al. [6], mRNA levels in routinely prepared formalin-fixed, paraffin-embedded tissues are comparable to matched frozen specimens when the target sequences are around 70 bases-long.

Nested-PCR

22

Serena Bonin and Isabella Dotti

Contents

S. Bonin (✉) and I. Dotti
Department of Medical, Surgical and Health Sciences,
University of Trieste, Cattinara Hospital,
Strada di Fiume 447, Trieste, Italy

22.1 Introduction and Purpose

This section aims to provide the general guidelines with a specific protocol, as an example of how to apply nested-PCR to DNA or cDNA obtained from formalin-fixed and paraffin-embedded (FFPE) samples [1–3]. For general requisites for PCR reaction, please refer to the related section in Chap. 20. The goal of this document is to provide a standard workflow for nested amplification, along with a non-isotopical system for PCR product detection. Nested-PCR is related to the reamplification of a first product of PCR by using a second set of primers within the first amplification product. The utilized primers should be different in both reactions. This methodology increases both the sensitivity and specificity of the assay. On the other hand, the risk of contamination increases significantly because of the greater amount of amplification products and working steps involved. In the traditional workflow of nested-PCR, the primers for the second-round PCR were added directly after the first step PCR. Nowadays, only a small amount (a few micro liters) of the first-round PCR product is amplified in a second PCR reaction. Alternatively, to avoid the first-round product contamination, it is also possible to perform a single-tube two-round PCR in single closed reaction tubes [4]. In addition, single-tube two-round PCR could be performed by mixing immediately the primers for the first- and the second-round PCR if they show marked differences in their annealing temperatures. The selected high annealing temperature of the first-round primers results in high specificity in the first-round PCR. This approach has also been described by the use of real-time PCR [5].

The time required for the whole procedure is 3 days.

The extremely high sensitivity of the nested-PCR system could often result in contaminations from previous reactions, then resulting in false-positive analyses. To check the validity of the assay, we strongly recommend using negative controls and taking wide routine precautions. Positive controls are also recommended in the nested-PCR, but to avoid contamination, we suggest using highly diluted controls. We recommend making at least two different sample aliquots and storing them in a laboratory room outside the PCR space. Furthermore, reagent preparation and extractions should be performed in a separate flow-laminar cabinet. First and second nested-PCR rounds should be performed in different areas, e.g. in two different laminar flow hoods, which should be distinct from the extraction ones. In these different spaces, we also suggest employing different supplies, pipette tips, eppendorf tubes, and any common reagent such as deionized water and PCR buffer. Finally, the detection process should be carried out in a dedicated laboratory. Operators are encouraged to wear disposable gloves in each PCR space, changing them frequently, especially when moving from one dedicated area to others.

For its high sensitivity, nested-PCR systems are commonly used in the detection and genotyping of pathogenic viruses and bacteria. Hereafter, we are going to describe a system of nested-PCR to detect borrelial DNA, the causative agent for Lyme disease, in DNA from FFPE specimens. The method presented here has been previously described by Wienecke et al. [6].

22.2 Protocol

22.2.1 Reagents

Reagents from specific companies are here reported, but reagents of equal quality purchased from other companies may be used.

- *10× PCR buffer with MgCl₂*[1] *(stock solution):* 150 mM $MgCl_2$, 500 mM KCl, 100 mM Tris pH 8.3 at 25°C[2]

Table 22.1 Primer sequences

Label	Sequence (5′–3′)
Bb1	AAACGAAGATACTCGATCTTGTAATTGC
Bb2	TTGCAGAATTTGATAAAGTTGG
Bb3	TCTGTAATTGCAGAAACACCT
Bb4	GAGTATGCTATTGATGAATTATTG
Bb-Hyb	TTGAATTAAATTTTGGCTT

- *Commercial 100 mM stock solution of each dNTP (pH 8)*
- *dNTPs:* Dilute the dNTPs stock solution to prepare a 10-mM solution of each dNTP in sterile water or DEPC-treated water
- *Primers*[3]: Oligonucleotide primers and the method here described for Borrelia detection were designed by Wienecke et al. [6]. The outer primer pair (Bb1 and Bb2) flanks a 171-bp fragment, while the inner primer pair (Bb3 and Bb4) gives a 92-bp product. An internal hybridization control (Bb-hyb) is used. Primer sequences are reported in Table 22.1
- Lyophilized primers should be dissolved in a small volume of distilled water or 10-mM Tris pH 8 to make a concentrated stock solution. Prepare small aliquots of working solutions containing 10–30 pmol/μl to avoid repeated thawing and freezing. Store all primer solutions at − 20°C. Primer quality can be checked on a denaturing polyacrylamide gel: a single band should be seen
- *Taq DNA Polymerase*
- *Agarose powder electrophoresis grade*
- *Agarose gel loading buffer (6×):* 100-mM EDTA pH 8, 0.25% (w/v) Bromo phenol blue (BMP), 50% (v/v) glycerol in H_2O
- *Positively charged nylon membrane* for nucleic acid transfer (e.g., Hybond-N+, GE Healthcare)
- *5× TBE (stock solution):* 225-mM Tris base, 225-mM Boric acid, 5-mM EDTA pH 8
- *Ethidium Bromide, 10 mg/ml (stock solution)*[4] Stock solution of EtBr should be stored at 4°C in a dark bottle
- *0.5-M NaOH, 1.5-M NaCl solution*

---

[1]There are commercial PCR premixed solutions for convenient PCR setup, containing PCR buffer and dNTPs. Taq Polymerase is usually provided with its dedicated 10× buffer. Check the composition for the presence of $MgCl_2$: Different commercial PCR buffers are $MgCl_2$-free, but a 25-mM or 50-mM $MgCl_2$ solution is supplied with buffer and enzyme.

[2]Because of the high temperature dependence of the pKa of Tris, the pH of the reaction will drop to about 7.2 at 72°C.

[3]Primer design could be performed using a specific software; please refer to http://molbiol-tools.ca/PCR.htm for online available software.

[4]EtBr is a potentially carcinogenic compound. Always wear gloves and work under a chemical hood. Used EtBr solutions must be collected in containers for chemical waste and discharged according to the local hazardous chemical disposal procedures.

- *20× SSC:* 3-M NaCl, 0.3-M Na$_3$ citrate·2H$_2$O. Adjust to pH 7 with HCl 1 M
- *N-laurylsarcosine* (30% stock solution-Fluka)
- *Poly dATP*
- *10% SDS (w/v) (stock solution)*[5]
- *DIG Oligonucleotide Tailing Kit, second Gen.* (Roche, cat. No. 03 353 583 910)
- *DIG Nucleic Acid Detection Kit* (Roche, cat. No. 11 175 041 910)
- *3-MM Paper*

22.2.2 Equipment

- *Disinfected*[6] *adjustable pipettes*, range: 2–20 μl, 20–200 μl, 100–1,000 μl
- *Nuclease-free aerosol-resistant pipette tips*
- *1.5-, 0.5- and thin-walled 0.2-ml tubes*[7] (autoclaved)
- *Centrifuge* suitable for centrifugation of 1.5-ml tubes at 13,200 or 14,000 rpm
- Laminar Flow
- Cabinet equipped with a UV lamp
- *Thermal cycler:* a large number of programmable thermal cyclers marketed by different companies are available for use in PCR. The choice among the commercial instruments depends on the investigator's inclination, budget, etc. The main requisite is that the instrument must have the licence for use in PCR. For diagnostic purpose, the thermal cycler must be equipped with a heated lid in order to prevent contamination.
- *Horizontal electrophoresis devices*
- *UV transilluminator*
- *Hybridization oven*
- *UV Crosslinker*

22.2.3 Method

- In a sterile 0.2-ml tube, mix 10 μl of extracted DNA and 40 μl of PCR reaction mixture for a final volume of 50 μl. Each reaction contains 10-mM Tris-HCl pH 8.3, 50-mM KCl, 1.5-mM MgCl$_2$, 200-μM each dNTP, one unit of Taq DNA Polymerase, 25 pmol each of the outer primers Bb1 and Bb2. For every PCR run, include a diluted positive control, a bystander DNA that does not contain the sequence of interest (DNA of another bacterial pathogen), a blank control (H$_2$O).[8]
- Displace the tubes in the thermal cycler. After the initial denaturation for 5 min at 95°C, run 30 amplification cycles as follows: 95°C/1 min; 50°C/1 min ; and 72°C/1 min. Program the instrument in order to maintain the amplified samples at 4°C before storing.
- Samples and controls derived from the first PCR run are submitted to the nested step.[9] In a sterile 0.2-ml tube, mix 2 μl of first-round PCR product and 23-μl of PCR reaction mixture for a final volume of 25 μl. Each reaction contains 10-mM Tris-HCl pH 8.3, 50-mM KCl, 1.5-mM MgCl$_2$, 200-μM each dNTPs, one unit of Taq DNA Polymerase, 25 pmol each of the inner primers Bb3 and Bb4.
- Displace the tubes in the thermal cycler and amplify the target DNA using the previous PCR programme. Set the instrument in order to maintain the amplified samples at 4°C before storing.
- Analyze 10 μl of the final products by nondenaturing 2.5% agarose gel electrophoresis. In a 100-ml pirex bottle, weigh 1.25 g of agarose powder (electrophoresis grade). Add 50 ml of 1× TBE buffer. Melt the powder in the microwave oven, avoiding loss of material. Cool the gel solution at about 50°C, and add 3 μl of EtBr stock solution. Pour the melted agarose solution into the gel chamber, and let the gel thicken.
- Load 10 μl of amplification product using 2 μl of loading buffer 6×. Separate the PCR products at 75 V in 1× TBE buffer until the dye reaches ¾ of the gel length.
- Check the gel at the UV light.
- Denature the separated PCR products by soaking the gel for two times in the 0.5-M NaOH, 1.5-M NaCl solution for 15 min at room temperature with gently swirling.

---

[5]Weigh the proper amount of SDS powder under a fume hood, because it is harmful.

[6]Clean the pipettes with a disinfectant (e.g., Meliseptol®rapid) and leave them under the UV lamp for almost 10 min. Alternatively it is possible to autoclave the pipette depending on the provider instructions.

[7]Microtiter plates could also be used to run PCR.

---

[8]If the thermal cycler is not fitted with a heated lid, overlay the reaction mixture with one drop (about 20 μl) of light mineral oil to prevent evaporation and cross-contamination.

[9]Special precautions should be taken in this step for the possibility of carry-over (false-positive results due to contamination among different tubes). It is recommended to work in a flow hood with samples cooled in ice and with anti-aerosol pipette tips.

- Prepare the following sandwich for Southern transfer (bottom-up): three sheets of paper 3 MM presoaked in 0.5-M NaOH and 1.5-M NaCl, gel upside down, positively charged membrane over the gel surface avoiding bubble formation,[10] one sheet of paper 3 MM wet with deionized water, three dry sheets of paper 3 MM, some paper towels, finally a glass with a weight on the top. Allow the DNA to transfer onto the membrane for at least 6 h.[11]

- Wash the membrane in 2× SSC for 5′, put the membrane between two sheets of paper to dry it, and then store it at room temperature until it is used for hybridization.

- Before hybridization, crosslink blotted nucleic acids to the membrane at the UV crosslinker twice with face up (each crosslinking takes about 1′ and 18″ and goes from 0.25 to 0 J).

- Perform probe hybridization in 5× SSC, 0.1%-N-laurylsarcosine, 0.02%-SDS, 5-μg/ml poly-dATP, 10-ng/ml of detection oligonucleotide Bb-Hyb, which has been tailed with digoxigenin-dUTP, according to the protocol of the manufacturer (Roche, Dig Oligonucleotide Tailing Kit). Incubate overnight in a hybridization oven at 41°C, swirling gently.

- Wash the membrane three times with 0.1× SSC, 0.1%-SDS, at 28°C. Detect the system according to the DIG nucleic acid detection kit that uses an enzyme immunoassay and enzyme-catalyzed colour reaction with NBT/BCIP, contained in the DIG Nucleic Acid Detection Kit (Roche).

22.2.4 Troubleshooting

- If the bands of the desired product are sharp but faint both in positive controls and samples, inefficient priming or inefficient extension may have occurred. To solve this problem, set up a series of PCRs containing different concentrations of the two primers. Find the optimal primer concentration and then set up a series of PCRs containing different concentrations of Mg^{2+} to determine the optimal concentration (see Chap. 20, Sect. 20.2.4).

- If there are bands in the negative controls, contamination of solutions or plastic ware with template DNA may have occurred. In this case, make up new reagents and operate in separate spaces to run PCR and to analyze the products of amplification.

- If distinct bands of unexpected molecular weight are present, they are the result of nonspecific priming by one or both primers. To solve this problem, decrease the annealing time and/or increase the annealing temperature.

- If a generalized smear of amplified DNA is present, too much template DNA was introduced.

- If the amplification is weak or non-detectable, there could be different possible causes: defective reagents, defective thermal cycler, or programming errors. If you are setting up the PCR system, the cause could also be a suboptimal extension of annealed primers, an ineffective denaturation, or the stretches of DNA you have chosen to amplify may be too long.

- For troubles in hybridization detection, refer to the DIG nucleic acid detection kit data sheet instructions (http://www.roche-applied-science.com/pack-insert/1175041a.pdf).

---

[10]Bubbles can be removed by rolling a Pasteur pipette on the surface.

[11]Alternatively to Southern blot, it is possible to transfer directly the nested PCR products onto the membrane by dot-blot apparatus without electrophoretic separation (See Chap. 23 for details). In this case, after the dot-blot transfer, go directly to the UV cross linking of the membrane and follow the protocol for membrane hybridization.

References

1. Cseke L, Kaufman P, Podila G, Tsai C (2004) Handbook of molecular and cellular methods in biology and medicine, 2nd edn. CRC, Boca Raton
2. Sambrook J, Russel DW (2001) Molecular cloning. A laboratory manual, 3rd edn. Cold Spring Harbour Laboratory Press, New York
3. Tarrago-Asensio D, Avellon-Calvo A (2005) Nested PCR and multiplex nested PCR. In: Fuchs J, Podda M (eds) Encyclopedia of diagnostic genomics and proteomics. Marcel Dekker, New York, pp 906–910
4. Olmos A, Cambra M, Esteban O, Gorris MT, Terrada E (1999) New device and method for capture, reverse transcription and nested PCR in a single closed-tube. Nucleic Acids Res 27(6):1564–1565
5. Berg J, Nagl V, Muhlbauer G, Stekel H (2000) Rapid-cycle PCR in temporarily compartmentalized capillaries: two-round PCR in a single capillary prevents product carry-over. Biotechniques 29(4):680–684
6. Wienecke R, Neubert U, Volkenandt M (1993) Molecular detection of borrelia burgdorferi in formalin-fixed, paraffin-embedded lesions of Lyme disease. J Cutan Pathol 20(5):385–388

General Protocol for Dot-Blot

23

Serena Bonin and Isabella Dotti

Contents

S. Bonin (✉) and I. Dotti
Department of Medical, Surgical and Health Sciences,
University of Trieste, Cattinara Hospital,
Strada di Fiume 447, Trieste, Italy

23.1 Introduction and Purpose

This protocol provides the general workflow to simply entrap nucleic acids (in this case DNA) onto a Nylon membrane [1–3]. Dot-blot is generally a simple, fast and sensitive technique that enables to transfer a known amount of sample onto an inherent support, such as a Nylon membrane. The quantity of the specific target is then determined by probe hybridization.

Denatured samples (i.e., PCR products or crude extracts) are directly applied to the membrane with the dot-blot apparatus, without electrophoretic separation. Since the specificity of the probe for the target sequence cannot be determined by this method, generally the specificity of the hybridization signal is preliminarily tested in a Southern blot experiment by running the target sequence on a polyacrilamide gel (see Chap. 22).

The time required for the whole procedure is half an hour.

23.2 Protocol

23.2.1 Reagents

- *Nylon membrane*
- *20X SSC*: 3 M NaCl, 0.3 M Na$_3$ citrate·2H$_2$O. Adjust pH to seven with 1 M HCl
- *Dye for Dot-Blot*: 0.25% bromophenol blue, 2.5% ficoll in water. Store at 4°C

G. Stanta (ed.), *Guidelines for Molecular Analysis in Archive Tissues*,
DOI: 10.1007/978-3-642-17890-0_23, © Springer-Verlag Berlin Heidelberg 2011

23.2.2 Equipment

- *Disinfected*[1] *adjustable pipettes*, range: 2–20 μl, 20–200 μl
- *Pipette tips*
- *1.5 ml tubes* (autoclaved)
- *Centrifuge* suitable for centrifugation of 1.5 ml tubes at 13,200 or 14,000 rpm
- *Dot-blot apparatus*
- *Vacuum system*
- *UV Crosslinker*

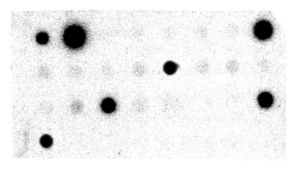

Fig. 23.1 Autoradiography of a dot blot membrane hybridized with a radiolabeled probe

23.2.3 Method

- For each sample to be blotted, transfer 20 μl of PCR product or DNA in a 1.5 ml tube.
- Denature at 95°C for 10 min[2] and chill on ice.
- Spin down briefly to remove the condensate from the tube walls.
- For 20 μl of PCR product add 30 μl of 20X SSC supplemented with dye.[3]
- In the meantime equilibrate the membrane for 5–10 min in 10X SSC.
- Load the membrane on the dot-blot apparatus, switch on the vacuum pump and load the samples on the membrane.[4]
- When the transfer is finished, remember first of all to detach the vacuum alimentation and then to switch off the pump, otherwise you could have a reflux in the apparatus.
- Mark the membrane using a needle or scissors (for example, you can cut every time the same corner of the membrane).
- Dry the membrane in oven at max 65°C for 15–20′.
- Crosslink two times face up in the UV cross linker (each cross linking takes about 1′ and 18″ and goes from 0.25 to 0 J).
- For the hybridization procedure refer to Chap. 22 (Fig. 23.1)

23.2.4 Troubleshooting

- If the dots are irregularly shaped, check the vacuum system and repeat the transfer.

---

[1]Clean the pipette with alcohol or another disinfectant and leave them under the UV lamp for at least 10 min. Alternatively it is possible to autoclave the pipette depending on the provider instructions.

[2]It is possible to use a thermoblock, the Thermomixer or the thermocycler. If the latter has been chosen, load the samples directly in the PCR tubes.

[3]Add approximately 500 μl of dye for dot blot in 20 ml of SSC 20x. The use of this dye gives the possibility to evaluate the shape of the dot and consequently the transfer.

[4]Sometimes it could be convenient to use the dot-blot apparatus without the upper part (the cover) to check the transfer of the samples. In this way, when the vacuum is applied, the membrane adheres directly to the wells.

References

1. Cseke L, Kaufman P, Podila G, Tsai C (2004) Handbook of molecular and cellular methods in biology and medicine, 2nd edn. CRC, Boca Raton
2. Sambrook J, Russel DW (2001) Molecular cloning. A laboratory manual, 3rd edn. Cold Spring Harbour Laboratory Press, New York
3. Stanta G, Bonin S, Utrera R (1998) RNA quantitative analysis from fixed and paraffin-embedded tissues. Methods Mol Biol 86:113–119

PCR-ELISA

24

Christiane Schewe, Wilko Weichert, and Manfred Dietel

Contents

24.1 Introduction and Purpose

This chapter deals with PCR analysis coupled to an ELISA-based detection system (PCR-ELISA). This type of assay employs probes containing sequences complementary to an amplified target sequence.

PCR-ELISA is advantageous to detect specific DNA and to differentiate among several PCR products. One of the possible applications is to discriminate between products amplified with one set of primers (consensus primers), but differing in the specific sequence spanned by these primer probes. The capture probes for ELISA are designed for the discrimination of the target product from unwanted products that would otherwise mimic a positive result. Furthermore, coupling an ELISA analysis to traditional PCR ensures the specificity of the sequence of an amplified PCR product. As an example, the PCR-ELISA assay to detect *Mycobacterium tuberculosis* complex in formalin-fixed paraffin-embedded (FFPE) tissues is described in detail [1].

24.2 Steps and Information

Nucleic acids are extracted from FFPE specimens either following the respective protocol (see Chap. 7 and other specific chapters dedicated to DNA extraction) or applying a commercial kit for extraction.

In general, five to ten 3-μm sections are cut from a tissue block and collected in a 1.5-ml vial. After workup of each sample, the microtome has to be cleaned and the blade has to be changed. Paraffin is removed by a single extraction with xylene and centrifugation. The pellet is then washed once with ethanol, the supernatant is removed, and the pellet is

C. Schewe (✉), W. Weichert, and M. Dietel
Institute of Pathology, Charité University Hospital,
Berlin, Germany

G. Stanta (ed.), *Guidelines for Molecular Analysis in Archive Tissues*,
DOI: 10.1007/978-3-642-17890-0_24, © Springer-Verlag Berlin Heidelberg 2011

air-dried. In parallel, blank extraction controls containing no mycobacterial material are processed. These controls include extractions containing all reagents and other additives except for tissue, as well as extractions of tissues ascertained to be negative for mycobacterial DNA.

The air-dried tissue pellets are processed using QIAamp DNA Mini tissue kit, (www.qiagen.com). The tissue is suspended in lysis buffer, and heated for 20 min at 95°C. After cooling down to room temperature, the subsequent steps are carried out according to the manufacturer's instructions with the exception that elution of DNA is performed with 60–100-µl buffer.

The proposed DIG-PCR/ELISA method is based on two processes: (1) amplification of template DNA combined with digoxigenin (DIG) labeling, (2) hybridization of the amplified DIG-labeled products to a biotin-labeled oligonucleotide probe specific to the target with colorimetric detection of the probe-bound products.

The time required for the whole procedure is 1 day.

24.3 Protocol

24.3.1 Reagents

Note: Reagents from specific companies are here reported, but reagents of equal quality purchased from other companies may be used.

- *2XReddyMix™ PCR Master Mix (1.5 mM MgCl$_2$)* (Thermo Scientific, ABgene, Epsom, Surrey) refer to manufacturer instructions.
- *PCR DIG Labeling Mix* (Roche, Cat. No. 11 585 550 910).
- *PCR ELISA (DIG Detection)* (Roche, Cat. No. 11 965 409 910).
- *Primer (GenBank accession No. M29899)*: The target is a 123-bp segment within the IS6110 region of *M. tuberculosis* complex genome, which indicates a highly conserved repetitive sequence of the primers IS1 and IS2 selected for PCR reaction. [2–6]
- For internal hybridization control, probe IS3, which is unique for *M. tuberculosis* complex, is used. Sequences for human β-globin gene are used for the control PCR. Primer and probe sequences are listed in Table 24.1. The PCR master mix reagent containing additionally digoxigenine-11-dUTP allows amplification and DIG-labeling in PCR in one step.

Table 24.1 Primer and probe sequences for *M. tuberculosis* and β-globin gene (reference gene)

Label	Sequence 5′→3′
IS6110 element:	*M. tuberculosis* complex PCR:
IS1: sense (BMT 003)	CCT GCG AGC GTA GGC GTC GG
IS2: antisense (BMT 002)	CTC GTC CAG CGC CGC TTC GG
IS3: probe	Biotin-CAT AGG TGA GGT CTG CTA CC′
β-globin gene:	Control PCR:
PC04	CAA CTT CAT CCA CGT TCA CC
GH20	GAA GAG CCA AGG ACA GGT AC

- Lyophilized primers should be dissolved in small volumes of distilled water or 10-mM Tris pH 8 to achieve a stock solution. Small aliquots of working solutions containing 10–30-pmol/µl are recommended to avoid repeated freezing and thawing. Primer and probe solutions are stored at –20°C. Primer quality can be checked on a denaturing polyacrylamide gel; a single band would indicate specificity. The final concentrations of forward and reverse primers are 100 µM each.
- *Agarose powder/low melting Agarose powder, electrophoresis grade.*
- *DNA ladder (100 bp).*
- *10× TAE (stock solution):* 400 mM Tris base, 225 mM acetic acid, 10 mM EDTA pH 8.0.
- *Ethidium bromide (EtBr), 10 mg/ml (stock solution)*[1] Stock solution of EtBr should be stored at 4°C in a dark bottle.

24.3.2 Equipment

- *Disinfected*[2] *adjustable pipettes*, range: 2–20 µl, 20–200 µl, 100–1,000 µl
- *Nuclease-free aerosol-resistant pipette tips*

[1]EtBr is a potentially carcinogenic compound. Always wear gloves. Used EtBr solutions must be collected in containers for chemical waste and discharged according with the local hazardous chemical disposal procedures.

[2]Clean the pipette with alcohol or another disinfectant and leave them under the UV lamp for at least 10 min. Alternatively, it is possible to autoclave the pipette depending on the provider's instructions for the different pipettes.

- *1.5-, 0.5-, and 0.2-ml tubes*[3] (autoclaved)
- *Centrifuge* suitable for centrifugation of 1.5-ml tubes at 13,200 or 14,000 rpm
- *SpectroPhotometer or Nanodrop*
- *Thermocycler:* a large number of programmable thermal cyclers by different companies are available for use in PCR. The choice among the commercial instruments depends on the investigator's inclination, budget, personnel, and experience. The main prerequisite is that the instrument must be licensed for use in PCR. For diagnostic purpose, the thermal cycler must be equipped with a heated lid in order to prevent contamination
- *ELISA reader:* Wavelength 405 nm; reference filter 492 nm

24.3.3 Method

- To standardize the conditions and to estimate the amount of extracted DNA, measure the concentration of nucleic acid spectrophotometrically, and calculate the OD ratio (260 nm/280 nm) to obtain information about the purity of the template. In our experience, amounts of nucleic acids of approx. 100 ng for the control gene and up to 400 ng for the detection of *Mycobacteria* yield the best results.
- To examine whether an amplifiable template is present in the samples, assess the quality of the prepared nucleic acid by control PCR amplifying an appropriate segment of a human housekeeping gene (for example, β-globin gene, PC04/ GH20 primers, 268 bp).
- Detect the PCR products for β-globin gene by standard electrophoresis using a 10-μl PCR aliquot and a 3.3-% agarose/low melting agarose gel (0.8–2.5%) containing ethidium bromide (see Chap. 17).
- Mix 2.5-μl of the nucleic acid preparation and 22,5-μl of the working master mix (12.5-μl Reddy mix, 1.5-μl DIG-labeling mix [120 μM], 1-μl primer mix for *M. tuberculosis* [2.5 μM] and 7.5-μl nuclease-free water).
- Perform the PCR in a thermocycler with the following program: 5 min 94°C, 40 cycles (1 min 94°C, 2 min 68°C, 1.5 min 72°C), 7 min 72°C, and the last

step holds 4°C. The PCR products are now ready for detection by ELISA.
- For ELISA detection, a 2.5-μl aliquot of the denatured digoxigenin-labeled PCR products is hybridized with the biotin-labeled probe in 100-μl hybridization buffer onto a well of a streptavidin-coated microwell plate according to the manufacturer's instruction (Roche, PCR ELISA DIG Detection Kit). Unattached DNA strands are removed by washing. The DIG-containing hybrid can be detected by binding of an anti-DIG peroxidase-conjugated antibody and photometrically quantified based on the substrate color reaction at 405 nm as described in the instruction manual. Read absorbance by means of an ELISA reader when the green color becomes visible or after a fixed time (e.g. 30 min). Correct absorbance for blank.

24.3.4 Troubleshooting

Problems like unexpected color development, questionable readings, drift of values, and poor precision are described together with possible causes and recommendations in the Instruction Manual "PCR DIG labeling and PCR ELISA (DIG Detection)" of Roche. For example:

- Weak or no signal:
  - Check PCR and ELISA reaction and integrity of positive controls.
  - Check protocol (incubation times/temperatures, buffer conditions, concentration of primary antibody or antigen).
  - Check substrate reagent for storage conditions and biological contaminations, use fresh prepared reagents.
- High background signal:
  - Prolong washing procedure.
  - Modulate concentrations of antibody.

References

1. Schewe C, Goldmann T, Grosser M, Zink A, Schluns K, Pahl S, Ulrichs T, Kaufmann SH, Nerlich A, Baretton GB, Dietel M, Vollmer E, Petersen I (2005) Inter-laboratory validation of PCR-based detection of mycobactcrium

---

[3]Microtiter plates could also be used to run PCR.

tuberculosis in formalin-fixed, paraffin-embedded tissues. Virchows Arch 447(3):573–585

2. Eisenach KD, Cave MD, Bates JH, Crawford JT (1990) Polymerase chain reaction amplification of a repetitive DNA sequence specific for mycobacterium tuberculosis. J Infect Dis 161(5):977–981

3. Mangiapan G, Vokurka M, Schouls L, Cadranel J, Lecossier D, van Embden J, Hance AJ (1996) Sequence capture-PCR improves detection of mycobacterial DNA in clinical specimens. J Clin Microbiol 34(5):1209–1215

4. Thierry D, Brisson-Noel A, Vincent-Levy-Frebault V, Nguyen S, Guesdon JL, Gicquel B (1990) Characterization of a mycobacterium tuberculosis insertion sequence, IS6110, and its application in diagnosis. J Clin Microbiol 28(12): 2668–2673

5. Thierry D, Cave MD, Eisenach KD, Crawford JT, Bates JH, Gicquel B, Guesdon JL (1990) IS6110, an IS-like element of mycobacterium tuberculosis complex. Nucleic Acids Res 18(1):188

6. Walker DA, Taylor IK, Mitchell DM, Shaw RJ (1992) Comparison of polymerase chain reaction amplification of two mycobacterial DNA sequences, IS6110 and the 65kda antigen gene, in the diagnosis of tuberculosis. Thorax 47(9): 690–694

Quantitative Real-Time RT-PCR

25

Isabella Dotti, Ermanno Nardon, Danae Pracella, and Serena Bonin

Contents

25.1 Introduction and Purpose

Real-time PCR has become an essential tool for qualitative and quantitative detection of both DNA and RNA targets and is increasingly being utilized in novel clinical diagnostic assays [1].

The targeting of DNA allows detection of infectious pathogens in tissues, single nucleotide polymorphisms (SNPs), and chromosome aberrations (translocations, inversions, gene amplifications/deletions) or point mutations, which are acquired during the progression of many tumours and may significantly affect their aggressiveness. When combined with reverse transcription (RT), real-time PCR can be a valuable analytical tool for the quantitative detection of mRNA targets (quantitative reverse transcription - PCR, qRT-PCR), both in fresh and archival clinical samples. A number of studies have shown that it is possible to measure mRNA levels using formalin-fixed paraffin-embedded (FFPE) tissues as a source of RNA, despite the frequent degradation of RNA to fragments shorter than 200 bases [2, 3]. It has been shown that, when amplicons of less than 100 bp are selected, amplification efficiency between fresh frozen and FFPE tissues is very similar [4]. Benefits of this procedure as compared to conventional methods include its sensitivity, its large dynamic range, and the potential for high throughput as well as accurate quantification. Besides, it is a homogeneous method that combines both amplification and analysis at the same time, with no need for gels, isotopes, or sample manipulation. Its potential is particularly promising at the clinical level for cancer patients, both as a prognostic assay and as a tool for predicting response to therapy. Among the applications of real-time PCR, High Resolution Melting (HRM) technology has recently been introduced as a rapid and robust analysis tool for the detection of DNA methylation and mutations. The method is based on the use of faster real-time

I. Dotti (✉), E. Nardon, D. Pracella, and S. Bonin
Department of Medical, Surgical and Health Sciences,
University of Trieste, Cattinara Hospital,
Strada di Fiume 447, Trieste, Italy

G. Stanta (ed.), *Guidelines for Molecular Analysis in Archive Tissues*,
DOI: 10.1007/978-3-642-17890-0_25, © Springer-Verlag Berlin Heidelberg 2011

PCR instruments, combined with an accurate HRM system. This technique has been recently applied to DNA extracts from FFPE for mutation detection [5–7]. However, currently the above-mentioned tool does not have a diagnostic application in FFPE tissues.

Although real-time qRT-PCR is widely used to quantify biologically relevant changes in mRNA expression levels, there are still a number of biological and technical problems associated with its use (tissue heterogeneity, inherent variability of RNA quality, the choice of the extraction protocol, the presence of copurified inhibitors, the use of different mRNA priming strategies, different reverse transcription yields, PCR efficiency…) [8], and achieving an accurate and reproducible quantification needs standardization. This situation emphasizes the urgent need for the establishment of best practice guidelines for this technology, particularly in the context of its adaptation as a clinical diagnostic assay. Efforts have been already made in this direction by introducing a set of guidelines (the MIQE guidelines) Minimum Information for publication of Quantitative real-time PCR Experiments for the proper handling of qPCR experiments [9]. This work provides a list of minimum criteria that should be considered in a qPCR experiment to ensure its relevance, accuracy, correct interpretation, and reproducibility.

The following protocols, which are intended for gene expression analysis, differ for the chemistry used for amplification signal detection. SYBR Green chemistry exploits the property of SYBR Green dye to increase greatly its fluorescence upon binding to double-stranded DNA. Fluorescence is monitored at each cycle after product extension and reflects the amount of generated PCR product. Nevertheless, this dye detects every double-stranded DNA (dsDNA), including primer dimers and other nonspecific products. Therefore, it is good practice to control for product specificity by performing a melting curve analysis of the amplified product after completing the PCR. SYBR Green-based chemistry is not the best choice when the target sequence concentration is too low (<1,000 copies) since nonspecific amplification and primer dimer products become more pronounced. The use of this dsDNA-binding dye is the most cost-effective chemistry for initial investigations and for primer optimization steps.

TaqMan chemistry exploits the 5′-nuclease activity of the DNA polymerase to hydrolyze a dual-labeled hybridization probe during the elongation step of PCR. Hydrolysis causes uncoupling of fluorescent labels and hence, fluorescence emission. This assay is preferable when target concentration is very low (<1,000 copies/sample) or degraded. The use of a specific probe increases the sensitivity and specificity of the assay by reducing the amplification of nonspecific products and the formation of primer dimers. For this reason it represents the best choice for gene expression analysis in FFPE tissues. The protocol described in this chapter makes use of a derivative of TaqMan probes, called Minor Groove Binder (MGB) probes. MGB probes are shorter than conventional TaqMan probes (12–15 nucleotides) and have a chemical compound attached to their 3′ ends, which binds to the minor groove of the DNA. The MGB moiety gives greater stability to the hybridized probe, which raises its melting temperature during amplification and therefore, its specificity. Other remarkable features of the MGB probes are lower background fluorescence and sensitivity comparable or greater than conventional probes. The protocol is intended for gene expression analysis and combines Moloney Murine Leukemia Virus Reverse Transcriptase (MMLV-RT)+random primers-based reverse transcription with amplification using a premixed solution of primers and MGB probe (Custom TaqMan Gene Expression Assay by Applied Biosystems).

Sections dedicated to assay optimization, data analysis, and quantification methods have also been included in this chapter.

25.2 qRT-PCR Using the SYBR Green Chemistry

25.2.1 Reagents

Note: Reagents from specific companies are reported here, but reagents of equal quality purchased from other companies may be used.

- *Sterile water*
- *RealMasterMix SYBR ROX, 2.5× reaction buffer*[1] (10 mM Magnesium Acetate, 1 mM dNTPs with

---

[1]It is specially formulated to adjust the Mg^{2+} concentration, eliminating the need of optimizing it.

[2]Unlike other HotStart polymerase formulations that block the enzyme activity only prior to the first high temperature step, the inhibitor of HotMaster polymerase is not denatured or inactivated at higher temperatures, allowing control of nonspecific binding during the whole PCR reaction.

[3]ROX dye is used as an internal passive reference to normalize non-PCR-related background fluorescence.

dUTPs, 0.05 unit/µl HotMaster Taq DNA polymerase,[2] PCR buffer, enhancers, stabilizers, ROX dye[3] (5 PRIME)

- *SYBR solution, 20×* (5 PRIME)
- *Specific forward and reverse primers, 30 pmol/µl*[4,5]: Resuspend the lyophilized powder to 300 pmol/µl stock solution in 1x TE buffer and then make diluted aliquots of 30 pmol/µl each
- *cDNA solution*: Optimal quantity is determined by the standard curve (see also Sect. 25. 4. 2)

25.2.2 Equipment

- *Adjustable pipettes,*[6] range: 2–20 µl, 20–200 µl, 100–1,000 µl
- *Filter barrier pipette tips*
- *PCR tubes*
- *PCR microplates/strips* (e.g. Axygen Biosciences)[7]
- *PCR strip caps* suitable for real-time PCR
- *Centrifuge* suitable for centrifugation of 1.5 ml tubes
- *Real-Time Thermal Cycler* (e.g. Mastercycler® ep Realplex by Eppendorf)[8]

25.2.3 Method[9]

All reactions should be performed at least in duplicate. If a threshold cycle $C_t > 35$ cycles is expected, run samples in triplicate.[10] In this protocol a total reaction volume of 35 µl is prepared for each sample;

this mastermix is then split in triplicates of 10 µl each. All assays should include a positive template control (a PCR product, or a linearized plasmid containing the target sequence) to check for consistency of reaction and a negative control (water control) to check for nonspecific signals arising from primer dimers or template contamination.

- Add the following components to a microcentrifuge tube:

2.5x RealMastermix buffer	14 µl
20x SYBR solution	1.75 microliters
30 pmol/microliter specific forward primer	Optimized[11]
30 pmol/microliter specific reverse primer	Optimized
cDNA solution	up to 3.5 microliter[12,13]
Sterile water	to 35 µl

- Spin the tube to collect the content and dispense the mixture into the plate microwells, in triplicates of 10 µl each. Avoid bubble formation, especially at the bottom. Cover the microwells with the strips.

Refer to the instrument's software instructions for setting up the real-time run. Use this general scheme for the Thermal Cycler programme:

1x: 95°C for 1–2′
40 – 45x: 95°C for 30″; 55°C[14] for 30″; 72°C for 30″; 80°C for 20″[15]
1x: 72°C for 10′
1x: melting curve 95°C for 15″, 60°C for 15″, ramp from 60°C to 95°C in 20′, 95°C for 15′ (default conditions)

---

[4]Adding of reverse primer is not necessary when reverse transcription has been performed using AMV combined with specific reverse primer and total RT solution has been used for real-time reaction.

[5]As a general rule, primers should be between 15 and 25 bases long to maximize specificity, with a G/C content of around 50%. Avoid primers with secondary structures or with sequence complementarity at the 3′ ends, which could contribute to primer self-dimerization.

[6]Clean the pipettes with a disinfectant (e.g. Meliseptol®rapid) and leave them under the UV lamp for at least 10 min. Alternatively it is possible to autoclave the pipette depending on the provider instructions.

[7]Depending on the number of samples to be processed in a single experiment, 8-well strips or 96-well plates can be used.

[8]For details, download the brochure "Mastercycler ep *realplex*" from http://www.eppendorf.com website.

[9]For general caveats in setting up a PCR assay, refer to chapter "General Protocol for End-Point PCR."

[10]See Sect. 25.5.

---

[11]Final primer concentrations should be optimized by a preliminary experiment to reduce nonspecific product amplifications and primer-dimer formation (see paragraph "Optimal Primer Concentration" in this chapter). Three hundred nM is usually found as proper concentration.

[12]The volume of the first-strand reaction mixture to be used in the PCR should not exceed 10% of the amplification mixture volume, since larger amounts may result in decreased PCR product.

[13]In this protocol cDNA synthesized with either MMLV + random primers or with AMV + specific primer can be used.

[14]Normally annealing is performed at 5°C below the lower of the PCR primers melting temperatures.

[15]A fourth step can be added for the fluorescence reading when nonspecific products are amplified at temperatures lower than those of the target sequence. The reading temperature is empirically determined and is set between the nonspecific and the specific peaks (usually very close to the amplicon melting).

25.3 qRT-PCR Using the TaqMan Chemistry

25.3.1 Reagents

Note: Reagents from specific companies are reported here, but reagents of equal quality purchased from other companies may be used.

- *Sterile water*
- *JumpStart Taq ReadyMix* for Quantitative PCR (Sigma, cod. D7440), including the *2× JumpStart ReadyMix Taq buffer* (20 mM Tris–HCl, pH 8.3, 100 mM KCl, 3 mM MgCl$_2$ 0.002% gelatin, 0.4 mM of each dNTP, stabilizers, 0.06 unit/µl Taq DNA polymerase, JumpStart Taq antibody[16]), 100× reference dye (ROX),[17] and 25 mM MgCl$_2$ (Sigma)
- *20× Custom TaqMan Gene Expression Assay*,[18] containing an optimized mix of forward and reverse primer and a MGB probe labelled with FAM dye[19] (Applied Biosystems)[20]
- *cDNA solution*: Optimal quantity is determined by the standard curve (see also Sect. 25.4.2)

25.3.2 Equipment

- *Adjustable pipettes*,[21] range: 2–20 µl, 20–200 µl, 100–1,000 µl
- *Filter barrier pipette tips*

---

[16]At room temperature the antibody keeps the polymerase inactive, preventing nonspecific amplifications; when the temperature is raised above 70°C in the first step of cycling process, the complex dissociates and the polymerase becomes fully active.

[17]ROX dye is used as an internal passive reference to normalize non-PCR-related background.

[18]Instead of this optimized ready-to-use mix, the labelled probe and primers can be designed by the operator using the "Primer Express" software by Applied Biosystems and then ordered separately. In such case, primer concentration of 300 nM and probe concentration of 200 nM can be tested as initial conditions.

[19]Fluorescent dyes are light-sensitive, keep solutions protected from light.

[20]Alternatively, primer pairs and TaqMan MGB probe can be purchased separately. In this case, probe concentration is usually 200 nM final while primer concentration is 300 nM. Primer concentration can also be tested as suggested in paragraph "Optimal Primer Concentration" in this chapter.

[21]Clean the pipettes with a disinfectant (e.g. Meliseptol®rapid) and leave them under the UV lamp for at least 10 min. Alternatively it is possible to autoclave the pipette depending on the provider instructions.

- *PCR tubes*
- *Centrifuge* suitable for centrifugation of 1.5 ml tubes
- *PCR microplates/strips* (e.g. Axygen Biosciences)[22]
- *PCR strip caps* suitable for real-time PCR
- *Real-Time Thermal Cycler* (e.g. Mastercycler® ep Realplex by Eppendorf)[23]

25.3.3 Method[24]

All reactions should be performed at least in duplicate. If a $C_t > 35$ cycles is expected, run samples in triplicate.[25] In this protocol a total reaction volume of 35 µl is prepared for each sample; this mastermix is then split in triplicates of 10 µl each. All assays should include a positive template control (a PCR product, or a linearized plasmid containing the target sequence) to check for consistency of reaction and a negative control (water control) to check for nonspecific signals arising from primer dimers or template contamination. Add the following components to a microcentrifuge tube:

2x JumpStart ReadyMix Taq	17.5 µl
20x Primer/probe mix	1.75 µl
100x ROX	0.35 µl
25 mM MgCl$_2$	Depending on the assay[26]
cDNA solution	up to 3.5 µl[27]
Sterile water	to 35 µl

- Spin the tube to collect the content and then dispense the mixture on the plate in triplicates of 10 µl each, avoiding bubble formation while pipetting, especially at the bottom. Cover microwells with the strips.

---

[22]Depending on the number of samples to be processed in a single experiment, 8-well strips or 96-well plates can be used.

[23]For details, download the brochure "Mastercycler ep *realplex*" from http://www.eppendorf.com website.

[24]For general caveats in setting up a PCR assay, refer to chapter "General protocol for end-point PCR."

[25]See Sect. 25.5.

[26]A preliminary experiment must be performed for each Gene Expression Assay to optimize MgCl$_2$ concentration.

[27]The volume of first-strand reaction mixture to be used in the PCR should not exceed 10% of the amplification mixture volume, since larger amounts may result in decreased PCR product.

- Refer to instrument's software instructions for setting up the real-time run. Use the following Thermal Cycler programme (recommended by Applied Biosystems) when Custom TaqMan Gene Expression Assays are used:

    1x: 95°C for 10′
    40 – 45x: 95°C for 15″; 60°C for 1′[28]

25.4 General Guidelines for Optimization of the qRT-PCR Assay

Whenever a new mRNA target sequence has to be introduced in an experiment, the following steps should be considered to optimize the qRT-PCR assay.

25.4.1 PCR Optimization

The target sample used for optimization can be a linearized plasmid containing the sequence of interest, a PCR product, or a cDNA. Choose an amount of template deemed adequate to yield a consistent, reliable amplification (i.e. an expected C_t between 20 and 30 cycles). If cDNA synthesized from total RNA is to be used, 0.5–1 µl of the reverse transcription mixture are usually sufficient to detect even low-expressed mRNA. It is advisable to use a good-quality RNA from a cell line known to express the transcript. In these preliminary experiments SYBR Green chemistry is recommended because useful information can be derived from the melting curve analysis.

25.4.1.1 Optimal Primer Concentration

Test the amplification of a fixed amount of template in presence of different primer concentrations (concentrations should range from 100 to 500 nM for each primer). Different forward/reverse primer concentration ratios should also be tested (from 0.2 to 5). Referring to Sect. 25.5, look at melting curves and amplification plots and select the primer combination

producing minimal primer dimers, giving the lowest Ct value, the highest end point fluorescence (Rn, defined as the magnitude of the signal generated by the given set of PCR conditions), and having no signal in the NTC (no-template-control).

25.4.1.2 Optimal Annealing Temperature

Perform the amplification of a fixed amount of template by testing 10–12 different candidate annealing temperatures, ranging from the Tm of the primers to 10°C below.[29] Follow the steps outlined in Sect. 25.5. Look at the melting curves and at the amplification plots, and select the temperature that gives the lowest C_t signal, the minimal nonspecific products, and minimal primer-dimer formation.

25.4.2 Efficiency, Sensitivity, and Reproducibility Determination

Once the amplification conditions have been optimized, determination of sensitivity, linear range, efficiency, and reproducibility of the assay is carried out by the method of the standard curve, plotting the log amount of target against the corresponding C_t obtained after PCR amplification. The target sample to be used for curve construction can be a linearized plasmid containing the transcript of interest, a PCR product, or a cDNA. When using a linearized plasmid as template, prepare a stock of the target sample of around 10^8 copies/µl and test the amplification of at least five ten-fold serial dilutions of the stock solution. Results obtained with "pure" template should be confirmed in FFPE samples.

- Prepare a 10-fold serial dilution of the template using a minimum of five samples covering the range between 10^7 and 10^2 copies. Although it is possible to quantify a dynamic range of 10 or 11 logs on most systems, in the laboratory practice this is not really necessary. Consider 2–3 replicates for every dilution.
- Add each dilution to a PCR tube and prepare a PCR mastermix according to one of the two above-described protocols, using the optimized primer concentration and the optimal annealing temperature.

---

[28]As the polymerase cleaves the probe only while it remains hybridized to the target DNA, a combined annealing and polymerisation step can be used in the TaqMan system. Fluorescence data are collected during this step.

[29]It is advisable to perform this assay with a thermal cycler including a temperature gradient block.

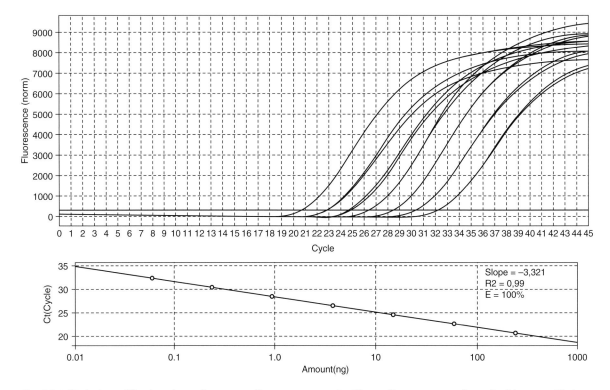

Fig. 25.1 Typical amplification plot and corresponding standard curve generated by serial four-fold dilutions using cDNA as a template. In this example, cDNA was obtained by reverse transcribing a pool of total RNA isolated from five FFPE samples. Two replicates were run for each of the seven dilutions, and a range between 250 and 20 pg of cDNA was covered. By the standard curve the coefficient of determination (R^2) and the efficiency (E) are extrapolated

- Plot Ct values against the log of target concentration and determine the slope and the coefficient of determination (R^2) values of the regression line. This latter parameter is connected with both the reproducibility (standard deviation) of the replicates and the linear range of target detection. It can be said that the target is detected by the assay in a reasonably true linear fashion if R^2 is above 0.98. Interassay reproducibility should be checked by repeating the experiment two-four different times. Efficiency of amplification is calculated from the slope value of the curve [30] using the following formula [10]:

$$E = 10^{(-1/\text{slope})} \qquad (25.1)$$

An E value of 2 (slope of −3.32) corresponds to an efficiency of 100%.

- Confirm the efficiency and reproducibility[31] on a RNA mixture obtained from of a small selection of FFPE samples that is expected to represent the behaviour of the whole case study under investigation. The use of the same chemistry is mandatory whenever the same target gene has to be detected and/or quantified. Prepare a 4–5 fold serial dilution of the cDNA template using a minimum of five dilutions covering the range between 10 pg and 200–400 ng (the choice of the range depends on the supposed expression levels of the target genes) (Fig. 25.1).

- The dynamic range of the standard curve obtained from the FFPE sample mix can be used to determine the optimal cDNA concentration that should be employed for the subsequent expression analysis of the whole case study (it usually corresponds to the median cDNA concentration within the linearity interval of the curve).

---

[30]Normally, slope, R^2, and efficiency are automatically determined by the software connected with the real-time Thermalcycler.

[31]Amplification efficiencies of the target gene should be equal in samples and in standard if an absolute quantification is planned.

25.4.3 Detection of Inhibitory Components

The sensitivity and kinetics of a qRT-PCR assay can be significantly reduced by the presence of inhibitory agents present in the biological material. These components may be reagents used during nucleic acid extraction or copurified contaminants from the biological sample (bile salts, urea, heme, heparin...).

Various methods can be used to assess the presence of inhibitors within biological samples. One of the approaches is the standard curve plot (the so-called "inhibition plot"). In this case, after reverse transcription, cDNA is serially diluted and amplified by real-time PCR. Data are then plotted in a standard calibration curve. A slope suggesting a PCR efficiency greater than 100% may be considered an indicator of PCR inhibition (see also Sect. 25.8). This method can be difficult to perform when a very small amount of RNA is available (for example, when RNA is isolated from microdissected sections). Recently, the SPUD method has been introduced as a valid substitute for the "inhibition plot" [11]. This procedure identifies potential inhibitors of the reverse transcription or PCR steps by recording the C_ts values of an artificial amplicon (SPUD-A) when amplified alone or together with the sample RNA. The presence of an inhibitor is identified as a shift to a higher C_t value in the reaction where both SPUD and RNA sample have been included.

25.5 General Guidelines for Data Analysis

In order to reduce subjectivity of the data analysis and to produce accurate and reproducible results, the assay must be properly evaluated. For this purpose the introduction of a standard curve in the assay is recommended. The data analysis takes into account some essential parameters (Fig. 25.2). Such parameters can be assessed by using the specific software provided with any real-time instrument.

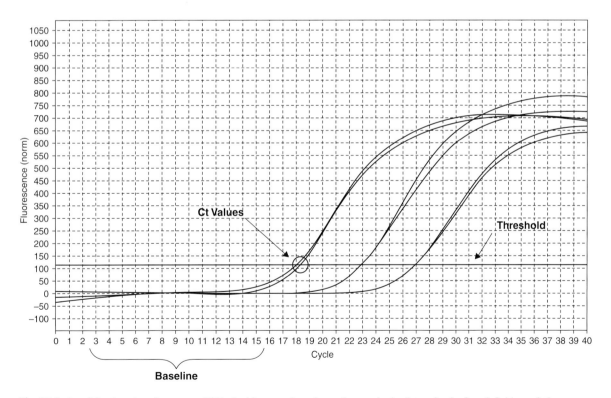

Fig. 25.2 Amplification plot of a target mRNA. In this example RNA has been isolated from three FFPE samples run in duplicate (here SD is <0.5). Ct value relative to the target expression levels in each sample is determined after definition of the proper baseline and threshold. Higher the Ct value, lower the number of copies relative to the target mRNA expressed in the sample

- Baseline: During amplification, the fluorescence is influenced by some factors that determine the behaviour of the curve. During the early PCR cycles, the background signal is evident and no apparent specific product is amplified. This initial "noise" is used to determine the baseline fluorescence. By default, it is usually computed by the system software averaging the background fluorescence between cycles 3 and 15. Since high background levels result in lower sensitivity, this parameter can be changed by the user, depending on the expected expression levels of the considered genes: for example, a low abundant gene requires a wider baseline. NTC (no template control) background fluorescence levels can be used to better define this parameter.

- Threshold: It represents a significant point above the calculated baseline. It is calculated automatically by the software after baseline determination and is placed in the exponential phase of the amplification. Keep the threshold settings constant when the evaluation of the same target expression is to be compared between samples.

- C_t value: It is determined after definition of baseline fluorescence and threshold. It represents the fraction of cycle at which the fluorescence crosses the threshold. Use this value for the following quantitative analysis.

- Replicates: The standard deviation (SD) of the C_t values should be no more than 0.5. At low C_t number the tolerance is lower than at high C_ts. Above 35 cycles the variability is indeed greater and quantification may be unreliable.

- ΔC_t between NTC (no template control) and sample: Check that a minimum difference of five C_t is present between any NTC signal and the sample data. If not, contamination by PCR products is possible.

- Slope: In evaluating a standard curve, ensure that the slope is not higher than −3.321 because this would implicate an unreliable efficiency (over 1) according to the above-described formula. A correct determination of efficiency is important for the choice of the most proper quantification method (see Sect. 25.7). The value of regression coefficient R^2 should be > 0.98, confirming a true linear relationship and a good reproducibility of replicates.

- Dynamic range: Define the dynamic range of the curve, as at < 10 or > 34–35 C_t values, amplification may not show a linear behaviour and the inclusion of these points in the standard curve could impair the efficiency of the calculation. Ensure that all data

collected for the unknown samples are within the dynamic range defined by the standard curve. Especially for FFPE samples, C_t values between 20 and 30 are considered the most reliable range.

- Melting curve: If using SYBR Green chemistry, check the melting curve profiles for fluorescent nucleic acid binding dye detection and ensure that products are specific. If an additional peak at a higher temperature is present, this is probably due to the amplification of longer sequences (e.g. from contaminant genomic DNA containing an intron); lower peaks usually correspond to primer-dimer formation (Fig. 25.3). Their signal can be removed by adding a fourth step in the PCR programme (see footnote 15). To exclude carry-over contamination in the NTC, check that its melting curve corresponds to nonspecific products.

25.6 Determination of the Most Appropriate Reference Gene/Panel of Genes for Normalization of mRNA Expression Levels

All RT-qPCR assay results are subject to variability, caused by technical as well as biological variations. These include variations in sample-to-sample RNA recovery, integrity, and efficiency of cDNA synthesis. Normalization is a rather problematic area and there is no universally accepted method for data normalization [12]. Nevertheless, it is a major concern especially when analyzing the RNA from archive tissues. Normalization of target gene expression to an endogenous reference gene, whose expression is a supposed invariant in the sample set (usually a housekeeping gene, HKG), is the most common method of internal control and it is currently the preferred option. However, it is essential that reference gene expression in the tissue of interest is carefully analyzed and the minimum variability is determined and reported, as inappropriate normalization can lead to the acquisition of biologically wrong data. A number of reports have recently demonstrated that the classic reference genes (GAPDH, beta-actin, 18S…) are inappropriate under certain experimental conditions due to their variability of expression [13]. For this reason, reference genes should always be validated prior to their use [14]. The current gold standard combines the evaluation of a panel of several candidate reference genes by means of a

Fig. 25.3 Melting curve profiles using the SYBR Green-based chemistry. In this image the coamplification of the target sequence and aspecific products is shown. Melting curves can be visualized as a direct measurement of the intensity of the fluorescence (**a**) or as a derivative of it (**b**). The identity of the reaction products can be retrieved looking at the Tm of the peaks, as reported in (**b**)

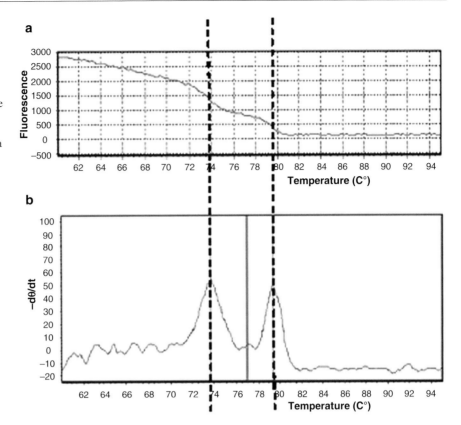

statistical package such as BestKeeper [15], GeNorm [16], or NormFinder [17]. The following procedure is suggested for the determination of the most suitable reference genes:

- In order to narrow the number of candidate genes, check in the literature those genes that have been analyzed and validated in similar archival tissue-based case studies
- According to the statistic software instructions, decide to test the expression of at least 4–5 genes in a set of at least 10, properly selected, FFPE samples
- Set up a standard curve for each candidate reference gene (refer to Sect. 25.4). Use the same chemistry that will be employed for the analysis of the whole case study
- Make sure that all the amplification efficiencies are equal among candidate genes. Define the dynamic range of detection of each gene
- Collect expression data (as C_ts) for each sample. These data should span within the previously defined dynamic ranges. Discard outliers. Submit data to GeNorm or Bestkeeper software analysis,

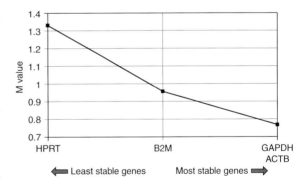

Fig. 25.4 Determination of the most stable reference genes from a set of tested genes as obtained by GeNorm algorithm. In this example, the stability of four candidate genes (ACTB, GAPDH, B2M, and HPRT) has been evaluated. In GeNorm, the stability level of each gene is suggested by the M value, defined as the average pairwise variation V for each gene with all other tested reference genes. The least stable gene (high M value) and the most stable gene (low M value) are visualized in a chart, as reported in the graph above

following the author's instructions. The software will calculate the most stable reference genes including a graphical output in the results (Fig. 25.4).

25.7 Relative Quantification by qRT-PCR

RNA expression levels can be determined by absolute and relative quantification [18]. Absolute quantification is used to determine the number of copies of specific RNA per cell or unit mass of tissue using a calibration curve. Relative quantification is easier to perform than absolute quantification because a calibration curve is not necessary. It is the most commonly used approach to investigate changes in expression levels in biological samples. Relative quantification is normally performed using an internal reference gene or a panel of genes that should always be validated before use in each experimental condition (refer to the previous paragraph). Changes in sample gene expression are then measured based on either an external standard or a reference sample, also known as a calibrator (e.g., the microdissected normal counterpart of a tumoral sample, an untreated sample, or simply an unknown sample from the case study). When using a calibrator, the results are expressed as a target/reference gene ratio. There are numerous mathematical models available to calculate the mean normalized gene expression from relative quantification assay [18]. Among them, the standard curve and the comparative Ct methods are the most commonly used. For large case studies, the comparative Ct method is the best choice but it requires strong validation and optimization of the assay, which is not always so easy to carry out when analyzing RNA from archival material.

25.7.1 Standard Curve Method [19]

The quantity of each experimental sample is first determined using a standard curve and then expressed relative to a single calibrator sample. Gene expression of the calibrator is designated as one-fold, with all experimentally derived quantities reported as an n-fold difference relative to the calibrator. Because the sample quantity is divided by the calibrator quantity, the unit from the standard curve drops out. All that is required of the standards is that their relative dilutions be known. Therefore, a standard curve can be prepared from an RNA pool of FFPE samples from the case study under investigation.

This method is applied when the amplification efficiencies of the reference and the target gene are unequal;

it requires no preparation of exogenous standards and no quantification of calibrator samples. However, it requires the introduction of a standard curve in each experiment, both for endogenous reference and for target genes. The following procedure is suggested:

- Prepare a 4–5 fold serial dilution of the cDNA template (from a pool of RNA of more than one FFPE sample) using a minimum of five dilutions covering the range between 10 pg and 200 ng (the choice of the range depends on the supposed expression levels of the genes) for each gene under investigation
- In parallel, prepare the unknown samples to be quantified. Use an amount of cDNA expected to yield a C_t within the dynamic range of the standard curve
- Use one of the above-described protocols (TaqMan or SYBR Green chemistry) to perform the real-time PCR reactions
- Quantify the input amount for unknown samples extrapolating units from the equation of the standard curve
- Normalize the unknown data against the endogenous reference calculating the ratio between target and reference gene amount for each sample. Normalized data are unitless numbers that can be used to compare the relative amount of target in different samples

25.7.2 Comparative C_t Method [20]

It is a mathematical model that calculates changes in gene expression as relative fold differences between an experimental and calibrator sample. Since this method doesn't require a standard curve, it is useful when assaying a large number of FFPE samples. The following procedure is suggested:

- Before using the comparative C_t method, set up a standard curve to determine the efficiencies of the target and reference genes. To this purpose, prepare a 4–5 fold serial dilution of the cDNA template from a statistically consistent number of FFPE samples using a minimum of five dilutions covering the range between 10 pg and 200 ng (the choice of the range depends on the supposed expression levels of the target genes) for each gene under investigation
- Apply one of the following formulas to calculate changes in target gene expression as fold differences between the experimental and calibrator sample

– If both the efficiencies are approximately equal to 100%:

$$\text{ratio} = 2^{-\Delta\Delta C_t} \qquad (25.2)$$

where $\Delta\Delta C_t = (C_t \text{ of target} - C_t \text{ of reference gene in sample}) - (C_t \text{ of target} - C_t \text{ of reference gene in calibrator})$

– If the efficiencies are not equal:

$$\text{ratio} = \frac{(E_{target})^{\Delta C_t \, target(calibrator-sample)}}{(E_{ref})^{\Delta C_t \, ref(calibrator-sample)}} \qquad (25.3)$$

where $E = 10^{(-1/slope)}$; target = target gene; ref = reference gene

Normalized data are unitless numbers that are used to compare the relative amount of the target gene in different samples relative to the calibrator, for which the fold change in gene expression equals one, by definition.

25.8 Troubleshooting

Troubleshooting that specifically involves the reverse transcription step (as already reported in chapter "Reverse Transcription") can be added to the following list (Table 25.1) as it may affect both endpoint and real-time PCR results.

Table 25.1 Common troubleshooting in real-time PCR reaction

Problem	Possible reason	Solution
High standard deviation in replicates	Inaccurate pipetting	Pipette with care, avoid air bubbles, avoid pipetting less than 2 µl, use the same pipette for the same set of tests
	C_ts are > 35	Use highly concentrated template
Nonspecific amplification	Primer-dimer formation	Change primers, define a different reading step in the in the PCR programme, adjust $MgCl_2$
	DNA contamination	Perform DNase treatment of RNA prior to RT step
No amplification detected	Inappropriate assay design	Consider primers again, annealing temperature, activity of fluorophore, $MgCl_2$ concentration, expiration of the reaction buffer
Efficiency of amplification < 90%	Poor primer and/or assay design	Redesign primers; revise annealing temperature and $MgCl_2$ concentration
Efficiency of amplification >110%	Inhibitory components	Reduce the cDNA input, reduce RNA input in the RT reaction, reduce primer concentration
	Nonlinear amplification conditions	If C_ts are <10, increase dilutions; if C_ts are >35, use higher amounts of template. Exclude outermost points from regression analysis
Low R^2	Inhibitory components	Reduce the cDNA input, reduce RNA input in the RT reaction, reduce primer concentration
	Inaccurate pipetting	Pipette with care, avoid air bubbles, avoid pipetting less than 2 µl, use the same pipette for the same set of tests
	Nonlinear amplification conditions	If C_ts are <10, increase dilutions; if C_ts are >35, use higher amounts of template. Exclude outermost points from regression analysis
Positive NTC control	Carryover contamination	Use new aliquots of all the reagents, treat RNA with Uracil DNA Glycosylase (UDG)
	Primer-dimer formation	Change primers, define a different reading step in the PCR programme, adjust $MgCl_2$

References

1. Bustin SA, Mueller R (2005) Real-time reverse transcription PCR (qRT-PCR) and its potential use in clinical diagnosis. Clin Sci Lond 109(4):365–379

2. Godfrey TE, Kim SH, Chavira M, Ruff DW, Warren RS, Gray JW, Jensen RH (2000) Quantitative mRNA expression analysis from formalin-fixed, paraffin-embedded tissues using 5' nuclease quantitative reverse transcription-polymerase chain reaction. J Mol Diagn 2(2):84–91

3. Stanta G, Bonin S (1998) RNA quantitative analysis from fixed and paraffin-embedded tissues: membrane hybridization and capillary electrophoresis. Biotechniques 24(2): 271–276

4. Specht K, Richter T, Muller U, Walch A, Werner M, Hofler H (2001) Quantitative gene expression analysis in microdissected archival formalin-fixed and paraffin-embedded tumor tissue. Am J Pathol 158(2):419–429

5. Fadhil W, Ibrahem S, Seth R, Ilyas M (2010) Quick-multiplex-consensus (qmc)-PCR followed by high-resolution melting: a simple and robust method for mutation detection in formalin-fixed paraffin-embedded tissue. J Clin Pathol 63(2):134–140

6. Kristensen LS, Wojdacz TK, Thestrup BB, Wiuf C, Hager H, Hansen LL (2009) Quality assessment of DNA derived from up to 30 years old formalin fixed paraffin embedded (FFPE) tissue for PCR-based methylation analysis using SMART-MSP and MS-HRM. BMC Cancer 9:453

7. Ma ES, Wong CL, Law FB, Chan WK, Siu D (2009) Detection of kras mutations in colorectal cancer by high-resolution melting analysis. J Clin Pathol 62(10):886–891

8. Nolan T, Hands RE, Bustin SA (2006) Quantification of mRNA using real-time RT-PCR. Nat Protoc 1(3): 1559–1582

9. Bustin SA, Benes V, Garson JA, Hellemans J, Huggett J, Kubista M, Mueller R, Nolan T, Pfaffl MW, Shipley GL, Vandesompele J, Wittwer CT (2009) The MIQE guidelines: minimum information for publication of quantitative real-time PCR experiments. Clin Chem 55(4):611–622

10. Rasmussen R (2001) Quantification on the lightcycler. In: Meuer S, Wittwer C, Nakagawara K (eds) Rapid cycle real-time PCR, methods and applications. Springer, Heidelberg, pp 21–34

11. Nolan T, Hands RE, Ogunkolade W, Bustin SA (2006) SPUD: a quantitative PCR assay for the detection of inhibitors in nucleic acid preparations. Anal Biochem 351(2):308–310

12. Huggett J, Dheda K, Bustin S, Zumla A (2005) Real-time RT-PCR normalisation; strategies and considerations. Genes Immun 6(4):279–284

13. Dheda K, Huggett JF, Bustin SA, Johnson MA, Rook G, Zumla A (2004) Validation of housekeeping genes for normalizing RNA expression in real-time PCR. Biotechniques 37(1):112–114, 116, 118–119

14. Dheda K, Huggett JF, Chang JS, Kim LU, Bustin SA, Johnson MA, Rook GA, Zumla A (2005) The implications of using an inappropriate reference gene for real-time reverse transcription PCR data normalization. Anal Biochem 344(1):141–143

15. Pfaffl MW, Tichopad A, Prgomet C, Neuvians TP (2004) Determination of stable housekeeping genes, differentially regulated target genes and sample integrity: Bestkeeper–excel-based tool using pair-wise correlations. Biotechnol Lett 26(6):509–515

16. Vandesompele J, De Preter K, Pattyn F, Poppe B, Van Roy N, De Paepe A, Speleman F (2002) Accurate normalization of real-time quantitative RT-PCR data by geometric averaging of multiple internal control genes. Genome Biol 3 (7): RESEARCH0034

17. Andersen CL, Jensen JL, Orntoft TF (2004) Normalization of real-time quantitative reverse transcription-PCR data: a model-based variance estimation approach to identify genes suited for normalization, applied to bladder and colon cancer data sets. Cancer Res 64(15):5245–5250

18. Wong ML, Medrano JF (2005) Real-time PCR for mRNA quantitation. Biotechniques 39(1):75–85

19. Livak KJ (1997) ABI-Prism 7,700 sequence detection system user bulletin no. 2. PE Applied Biosystems

20. Pfaffl MW (2001) A new mathematical model for relative quantification in real-time RT-PCR. Nucleic Acids Res 29(9):e45

DNA Sequencing from PCR Products

26

Giorgio Basili, Serena Bonin, and Isabella Dotti

Contents

26.1 Introduction and Purpose

Direct DNA sequencing can give important information about the role of genetic alterations in the pathogenesis of diseases. DNA mutations are at the origin of many cancer types, such as the germline mutations in BRCA1 and BRCA2 in breast cancer [1] or the activating mutations in the oncogene BRAF in melanoma [2]. Moreover, the detection of mutations can be used to address the choice of the therapy. For example specific tumour-associated somatic mutations in the tyrosine kinase domain of the EGFR can predict sensitivity to tyrosine kinase inhibitors (e.g. Gefitinib), while mutations in K-Ras gene predict response to monoclonal therapy against EGFR (e.g. Cetuximab) [3]. This section provides a procedure and some advice for direct sequencing of DNA extracted from formalin fixed and paraffin embedded (FFPE) tissues. Mutational analysis requires the following main steps: DNA extraction from FFPE tissues (see chapter "DNA Extraction from FFPE Tissues" and other specific chapters dedicated to DNA extraction), PCR amplification, PCR product purification and sequencing.

26.2 Precautions for Detection of Somatic Mutations

Single nucleotide polymorphisms (SNPs) and germline mutations involve all the cells of the organism. Conversely, somatic mutations appear only in tumoral cells. In order to detect the presence of somatic mutations in tumour tissues by DNA sequencing, it is necessary to limit the non-mutated DNA component coming from non-tumoral cells (like tumour infiltrating

G. Basili (✉), S. Bonin, and I. Dotti
Department of Medical, Surgical and Health Sciences,
University of Trieste, Cattinara Hospital,
Strada di Fiume 447, Trieste, Italy

G. Stanta (ed.), *Guidelines for Molecular Analysis in Archive Tissues*,
DOI: 10.1007/978-3-642-17890-0_26, © Springer-Verlag Berlin Heidelberg 2011

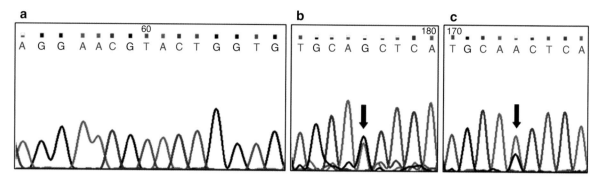

Fig. 26.1 Examples of electropherograms obtained by direct DNA sequencing. (**a**) Electropherogram of a portion of exon 21 of the EGFR gene obtained from a normal tissue without mutations. This is an ideal condition where no background is detected. (**b**) Electropherogram of a portion of exon 20 of the EGFR gene obtained from one tumoral tissue with 100% of cells bearing the sequence alteration in heterozygosis. The alteration is well visible with a double peak despite some background. (**c**) Electropherogram of the portion of exon 20 in another tumoral tissue. Here, although 100% of cells bear the mutation in heterozygosis, the size of the two peaks is not similar and this is a problem when a high background is present

lymphocytes, stromal cells, and endothelial cells). Such component could give electropherograms in which the tumour-associated mutations are not visible because the signal of the non-mutated DNA is prevalent. Usually tumour-associated mutations are present in heterozygosis, resulting in a double peak: one is due to the normal gene and the other one to the mutated gene (Fig. 26.1) [4].

The following limitations and precautions should be considered when using FFPE for DNA mutational analysis:

- Formalin causes a strong DNA degradation [5] and this can result in difficult PCR amplifications that would require higher amounts of starting material.
- PCR amplification on DNA extracted from formalin fixed specimens is usually unsuccessful for fragments longer than 400 bases [6]. Subsequently, the design of short amplicons is essential also in sequence analysis. For a successful DNA amplification, we recommend designing amplicons no longer than 200–250 base pairs. In any case, in establishing the sequencing fragment it must be considered that 20 bases after the primers are not usually correctly read. To bypass this inconvenience we strongly recommend sequencing the PCR product twice using both the forward and reverse primer.
- Sequencing of PCR products obtained by isolating DNA from FFPE can give a number of artifacts four times higher than those detected in DNA from fresh frozen samples[1] [4, 5, 7]. Due to the risk of artifacts, the mutation should be confirmed either by repeating PCR and sequencing or by other means (e.g. use of digestion enzymes capable to discriminate the specific nucleotidic alterations).

In order to overcome these problems, the use of alcoholic fixatives should be taken into consideration as a good alternative to formalin fixation because of their higher preservative properties on nucleic acids (see chapter "Formalin Free Fixatives").

Once these precautions are followed, FFPE can be used as a good source for DNA sequencing. Moreover, FFPE are the ideal material when an enrichment of the tissue sample in specific cell component (e.g.tumoral area) is required because, in comparison to fresh frozen tissues, they allow performing a more precise microdissection and their manipulation is easier.

---

[1]Artifacts are *ex novo* sequence alterations introduced by the Taq polymerase during the PCR reaction that become visible in the electropherogram when the enzyme amplifies a single strand of damaged DNA. This situation appears when few nanograms of DNA template are used (i.e. 10 ng or less) or when DNA template has high degradation levels (such as DNA from formalin-fixed samples). The expected frequency of artifacts can be considered around 1/4,000 in fresh frozen samples and 1/1,000 bp in formalin fixed samples

26.3 Protocol

26.3.1 Reagents

- *10x PCR buffer with MgCl$_2$[2]* (standard composition: 15 mM MgCl$_2$, 500 mM KCl, 100 mM Tris pH 8.3 at 25°C)
- *10 mM dNTPs* prepared from the stock dNTP solution
- *Primers* [3,4]: Lyophilized primers should be dissolved in sterile ddH$_2$O or 10 mM Tris pH 8 to make a concentrated stock solution. Use sterile ddH$_2$O to prepare small aliquots of working solutions containing 10–30 pmol/μl primer to avoid repeated thawing and freezing. Store all primer solutions at –20°C
- *Taq DNA Polymerase*
- *Positive and negative controls*: A positive control (e.g. DNA containing the target sequence) should be included to check that the used PCR conditions successfully amplify the target sequence. Negative controls could be water or extracts from a paraffin block without included tissues
- *DNA purification with silica resin or similar*. The columns with silica membrane, e.g. provided by the "QIAquick Gel Extraction Kit" (QIAGEN), are suitable for this purpose.

- *Low melting agarose*
- *40% Polyacrilamide 29:1*
- *Ethidium Bromide, 100 mg/ml*
- *High degree deionized water (ddH$_2$O)* is recommended for the elution steps. The use of Milli-Q water (Millipore) is suitable for this purpose

26.3.2 Equipment

See Chap. 20 and 21

26.3.3 Method [5]

26.3.3.1 Sample Preparation

- Cool the paraffin blocks in order to properly cut the sections. Cut 5–8 sections of 5 μm thickness from the tissue block, depending on the size of the available sample, discarding the first section and lay the others on glass slides. Stain a section with H&E and use it to evaluate at the microscope if the tissue sample is suitable for microdissection.
- If microdissection is required, mark the area of interest with a pen-marker directly on the coverglass of the H&E stained section. Deparaffinize the corresponding unstained slides (See chapters 4 and 6 for details on microdissection). Fit the unstained sections on the marked H&E slide and scratch away the unnecessary portion of tissue[6] using a scalpel or

---

[2]There are commercial PCR premixed solutions for convenient PCR setup, containing PCR buffer and dNTPs. Usually Taq Polymerase is provided with its dedicated 10x buffer. Check the composition for the presence of MgCl$_2$. Different commercial PCR buffers are MgCl$_2$ free but a 25 mM or 50 mM MgCl$_2$ solution is supplied with buffer and enzyme. MgCl$_2$ content could be incremented in cases of multiplex PCR or decremented in cases of high fidelity PCR

[3]Primer design could be performed using specific software, please refer to http://molbiol-tools.ca/PCR.htm for the on line available software. Since the first 20–40 bases of the electropherograms are not usually reliable, it is necessary to design primers that match about 40–50 bases before the region of interest

[4]Primer design depends on the quality of starting material. For sub-optimal quality of material (for example DNA from FFPE), shorter PCR products are recommended. However, using higher amounts of starting DNA, the probability to have some intact template molecule is higher and products up to 400 bp can be obtained from FFPE or even longer from alcoholic fixatives. When working with highly damaged DNA, we suggest designing primers for amplicons of around 150 base pairs even if it would cover a short region of interest

[5]In the case of cDNA sequencing, see chapters "RNA Extraction from FFPE Tissues", "DNase treatment of RNA" and "Reverse Transcription (RT)", then use the obtained cDNA to proceed with PCR. Use 120 ng of cDNA as template for PCR. If necessary, increase up to 200 ng for difficult amplifications. Primers for cDNA amplification should be intron skipping in order to avoid genomic DNA amplification

[6]In the analysis of tumour associated mutations, the selected region of the tissue slide should not contain more than 20–30% of normal cells in order to obtain a reliable result in the sequencing analysis. In particular, consider the contamination from lymphocytes infiltrating the tumour

the tip of a needle. With the needle collect the tissue of interest in an Eppendorf tube.

- If microdissection is not necessary, cut the whole sections and directly collect them in the Eppendorf tube. Proceed with deparaffinization steps as described in the protocol dedicated to DNA extraction from FFPE.

26.3.3.2 DNA Extraction

- Perform tissue sample digestion following the steps suggested in chapters on DNA extraction from FFPE tissues (Chaps. 7, 8 and 9).
- Resuspend the extracted DNA in a suitable volume (usually 15 μl) of sterile ddH$_2$O.
- Immediately quantify the extracted DNA[7] (see chapter 16, section 16.1).
- Prepare the DNA dilutions in order to use the proper amount of template for the PCR reaction.[8]

26.3.3.3 PCR Procedure

- In a sterile 0.2 ml tube prepare the mastermix including, in the following order, 1X PCR buffer, 200 μM each dNTP, 0.22–0.3 μM of forward and reverse primers,[9] 0.03–0.04 U/μl Taq Polymerase,[10] considering a final PCR reaction volume of 25 μl.[11] The amount of

target DNA in each reaction should range between 5 and few hundreds of nanograms.[12] For every PCR run include a positive and a negative control.

- Place the tubes in the thermal cycler and amplify the target DNA using the proper PCR programme (with the annealing temperature required by the primers in use). In the case of DNA extracted from FFPE, a suitable PCR programme consists of: 5 min denaturation step at 95°C, five amplification cycles made up by denaturation, annealing and elongation steps of 1 min. each, 40–45 amplification cycles with steps of 30 s. each. The indicated times are suitable for amplicons shorter then 600–700 bp. Set the final holding step at 4°C. The PCR products can be stored at 4°C until sequencing.[13]

26.3.3.4 PCR Product Analysis

- Load 2 μl of the PCR products on an 8% polyacrylamide gel to evaluate the amount of PCR product and therefore its suitability for sequencing (to prepare the gel follow the procedure described in Appendix B).
- If the expected band is well detected, proceed to step 26.3.3.6.
- If the presence of primer dimers, aspecific amplification products and low amount of specific products is detected,[14] choose one of the following options:

---

[7]A very accurate quantification of DNA solution is recommended because the PCR for the sequencing analysis requires proper DNA template amounts. When a very little amount of DNA is expected, quantification with a Nanodrop spectrophotometer is recommended because it allows a reliable quantification using only 1 μl of sample

[8]In order to prevent further degradation, whatever the source of the extracted DNA, multiple DNA freezing and thawing should be avoided. For this reason, after quantification, the concentrated DNA sample should be divided into aliquots suitable for the following PCR amplification. The dilutions can be stored at 4°C for a short time (3–4 weeks). The quantification of the concentrated DNA solution should be repeated after each thawing

[9]It is important to reduce the primers to optimal concentration because an excess could result in primer dimers formation and thus interfere with sequencing giving a higher background in the electropherograms

[10]Lowering the amount of Taq polymerase can reduce the presence of aspecific products

---

[11]Amplifying nanograms of target DNA is troublesome with smaller volumes

[12]A good suggestion is to start with 5–20 ng of DNA template (usually the high DNA degradation caused by formalin does not allow amplifying less than 5 ng). If no PCR products are obtained, try to increase the quantity (do not exceed 500 ng; in case design primers for shorter amplicons)

[13]Freeze-thawing of the PCR products causes their degradation and consequently result in electropherograms with high background. For this reason, it is suggested to store the amplified DNA at 4°C. However, if the PCR products have been frozen repeatedly, good sequencing results may be still obtained by using a higher amount of PCR product (50–100 ng) in the sequencing reaction without changing the amount of primers

[14]Do not confuse polyacrylamide artifacts with aspecific products: the former consist in thin and weak bands of high molecular weight (far more than 1 kbp) and do not interfere with the sequencing reaction

- Try to repeat the PCR reaction using different amplification conditions and/or designing more efficient primers ex novo.
- Go to the optional step 26.3.3.5.

26.3.3.5 Optional: PCR Product Purification from Gel

Load all the PCR product on a 1.5–2% low melting agarose gel stained with ethidium bromide (see Appendix A), and run up to one third of the gel (a longer run can reduce the visibility of the band of interest). Excise the band containing the specific PCR product and purify it using a silica resin or equivalent (e.g. the columns provided by the "QIAquick Gel Extraction Kit" - QIAGEN). Elute the DNA in 25–30 µl. Use 5 µl of the DNA solution as template for a second PCR reaction and check again 2 µl of PCR product on an 8% polyacrylamide gel (repeat steps 26.3.3.3. and 26.3.3.4 of this protocol).

26.3.3.6 PCR Product Purification from the Solution

- Once the PCR product is free of aspecific DNA, purify it (all the available volume) by using the specific silica resin or equivalent (e.g. the columns provided by the "QIAquick Gel Extraction Kit" - QIAGEN). This purification step is intended to separate the PCR product from primers, dNTPs and enzyme. After this step, we strongly recommend eluting the DNA with high degree deionized water (ddH_2O) in order to avoid the presence of salts that could interfere with the following sequencing reaction. The use of 25 µl ddH_2O for a complete elution is suggested.
- Concentrate the DNA by partial evaporation to a final volume of about 5 µl, using a thermoblock at 60°C with open tubes.[15]
- Quantify DNA with a spectrophotometer (see chapter 16, section 16.1) and dilute the DNA sample to a concentration suitable for the sequencing reaction (see the following step).

26.3.3.7 Sequencing

There are two possibilities for sequencing:

- Performing the sequencing reaction yourself (in this case the procedure will depend on the sequencing system in use in the laboratory).
- Sending the DNA to a sequencing service. For PCR product sequencing, the sequencing primers correspond to the amplification primers. In this case, prepare two tubes each containing the purified PCR product, then add the forward primer to one tube and the reverse primer to the other one. For the analysis of each sequence both the "forward" lecture and the "reverse" lecture are needed in order to confirm all the detected signals. Follow the sequencing service instruction for the proper quantities to be sent. Usually, to sequence 200 bases 4–8 ng of purified PCR product plus 6.4 pmoles of each primer are enough. In this case, before sending DNA + primers are dehydrated in order to preserve them until the sequencing reaction is performed.[16]

26.3.3.8 Electropherogram Analysis

The results can be visualized as electropherograms using specific software such as "Chromas Lite" (www.technelysium.com.au/chromas_lite.html) easily available on line. The corresponding sequence can be easily exported as a text file. Every sequence alteration visualized in the electropherogram can be considered a real mutation only if detected both in the forward and in the reverse lecture. However, in order to exclude PCR artifacts, even the nucleotidic substitution detected in both directions should be confirmed by repeating PCR amplification and sequence analysis.[17] Nucleotidic alterations in heterozygosis are immediately visible in the electropherogram as double peaks, whereas those in homozygosis can be highlighted by aligning the text file of the sequence with the corresponding consensus

---

[15]Since the DNA is highly diluted, this step is necessary to allow an accurate spectrophotometric DNA quantification

[16]The Dehydration step can be performed at 60°C in a thermoblock with open tube caps

[17]The probability that the artifacts detected in the first round of PCR appear in a second independent PCR amplification is extremely low

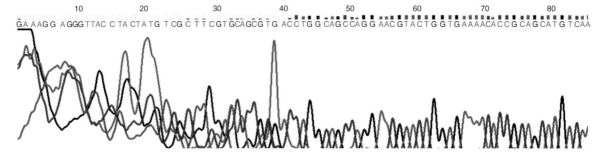

Fig. 26.2 Electropherogram of the first codons of exon 21 of the EGFR gene obtained from a formalin fixed non-tumour sample. In this case the electropherogram is reliable after the first 39–40 nucleotides

Fig. 26.3 Terminal sections of electropherograms from three different specimens (**a**, **b**, **c**). False peaks are evident (*arrows*) when the real peaks at the same position are short, resulting in double peaks

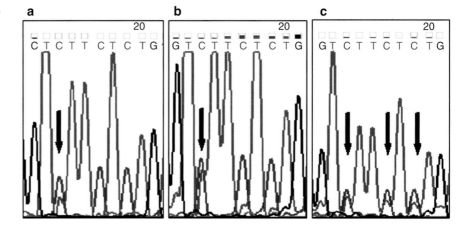

"wild-type" sequence obtained from a database.[18] Consider that usually the first 20–40 nucleotides in the electropherograms are not readable (Fig. 26.2).

Sometimes, recurrent short "background peaks" can appear in the extremities of the electropherogram, reducing the reliability of the sequencing analysis. These background peaks become evident when the peaks of the DNA sequence are short (Fig. 26.3). They can appear again in the same positions when amplification and sequencing are repeated.

26.3.4 Troubleshooting

• The presence of primer dimers and aspecific products will result in low quality electropherograms characterised by high background. In this case, increase the annealing temperature in the PCR programme, reduce the amount of primers used in the PCR reaction or design more efficient primers ex novo.

• If you don't have successful PCR amplifications, try to increase the number of cycles in the PCR programme (55 cycles at most) or, if needed, increase the amount of template up to 200–500 nanograms.

• If you have neither primer dimers nor aspecific products but still you have bad quality electropherograms, be sure that your PCR product has not been frozen too many times or that the sequencing primers and purified PCR products do not contain salts. Alternatively, reduce the amount of Taq polymerase (around 0.03 U/μl).

References

[18]On-line free software, such as "ClustalW" (www.ebi.ac.uk/Tools/clustalw2/index.html) can be used to perform the alignment

1. Gerhardus A, Schleberger H, Schlegelberger B, Gadzicki D (2007) Diagnostic accuracy of methods for the detection of BRCA1 and BRCA2 mutations: a systematic review. Eur J Hum Genet 15(6):619–627

2. Spittle C, Ward MR, Nathanson KL, Gimotty PA, Rappaport E, Brose MS, Medina A, Letrero R, Herlyn M, Edwards RH (2007) Application of a BRAF pyrosequencing assay for mutation detection and copy number analysis in malignant melanoma. J Mol Diagn 9(4):464–471

3. van Krieken JH, Jung A, Kirchner T, Carneiro F, Seruca R, Bosman FT, Quirke P, Flejou JF, Plato Hansen T, de Hertogh G, Jares P, Langner C, Hoefler G, Ligtenberg M, Tiniakos D, Tejpar S, Bevilacqua G, Ensari A (2008) KRAS mutation testing for predicting response to anti-EGFR therapy for colorectal carcinoma: proposal for an European quality assurance program. Virchows Arch 453(5):417–431

4. Lynch TJ, Bell DW, Sordella R, Gurubhagavatula S, Okimoto RA, Brannigan BW, Harris PL, Haserlat SM, Supko JG, Haluska FG, Louis DN, Christiani DC, Settleman J, Haber DA (2004) Activating mutations in the epidermal growth factor receptor underlying responsiveness of non-small-cell lung cancer to gefitinib. N Engl J Med 350(21):2129–2139

5. Srinivasan M, Sedmak D, Jewell S (2002) Effect of fixatives and tissue processing on the content and integrity of nucleic acids. Am J Pathol 161(6):1961–1971

6. Stanta G, Mucelli SP, Petrera F, Bonin S, Bussolati G (2006) A novel fixative improves opportunities of nucleic acids and proteomic analysis in human archive's tissues. Diagn Mol Pathol 15(2):115–123

7. Quach N, Goodman MF, Shibata D (2004) In vitro mutation artifacts after formalin fixation and error prone translesion synthesis during pcr. BMC Clin Pathol 4(1):1

Pyrosequencing Analysis for Detection of KRAS Mutation

27

Gerlinde Winter and Gerald Höfler

Contents

G. Winter and G. Höfler (✉)
Institute of Pathology, Medical University Graz, Graz, Austria

27.1 Introduction and Purpose

KRAS, a Kirsten ras oncogene homolog from the mammalian ras gene family, encodes a protein that is a member of the small GTPase superfamily. The KRAS gene is located in chromosome 12p12.1 (Fig. 27.1).

A single amino acid substitution can be responsible for an activating mutation. The resulting transforming protein is thought to convey a growth advantage for cells from various malignancies, including adenocarcinoma of the lung, ductal adenocarcinoma of the pancreas, and colorectal adenocarcinoma.

The KRAS gene is mutated in 35–45% of tumors in metastatic colorectal cancer (CRC) patients. Studies have shown that KRAS mutation testing can predict which CRC patients will benefit from treatment with epidermal growth factor receptor (EGFR)–inhibiting monoclonal antibodies, such as panitumumab and cetuximab. For this reason, there is a strong demand for KRAS mutation analysis in Europe due to the European Commission's granting of marketing authorization for anti-EGFR therapies for treatment of metastasized colon cancer in patients who carry a non-mutated (wild-type) KRAS gene only.

The therascreen KRAS Pyro Kit (IVD, Ref. 971460, Qiagen, Germany) is a kit validated for pyrosequencing and is used for the quantitative determination of mutations in codon 12, 13, and 61 of the human KRAS gene to select patients for whom treatment with anti-EGFR therapies will be effective [1, 2]. Other commercial kits can be used for the same purposes; the choice of the kit here reported is linked to the direct experience of the authors in the use of Qiagen KRAS kit.

The kit consists of two assays: one for detecting mutations in codon 12 and 13, and the second one for detecting mutations in codon 61 (Fig. 27.2). The two

Fig. 27.1 Chromosomal localization of KRAS (v-ki-ras2 Kirsten rat sarcoma viral oncogene homolog) gene

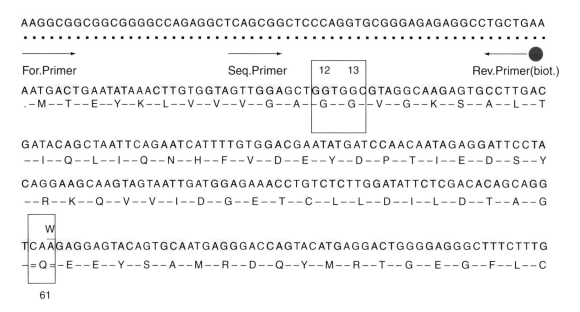

Fig. 27.2 cDNA sequence of the human KRAS gene and position of the oligos included in the PyroMark KRAS Kit. In this figure, the primers detecting mutations in codons 12 and 13 are indicated. KRAS gene sequence has been retrieved from: http://www.ensembl.org/Homo_sapiens/Transcript/Sequence_cDNA?db=core;g=ENSG00000133703;t=ENST00000256078

regions are amplified separately by PCR, and the region of interest is sequenced.[1]

Following PCR using primers targeting codons 12/13 and codon 61, the amplicons are immobilized by the biotinylated primers on Streptavidin Sepharose High-Performance Beads. Single-stranded DNA is prepared, and in the following step, the corresponding sequencing primers anneal to the DNA. The next step is real-time sequencing with a mixture of enzymes, substrates, and nucleotides. The samples are then analyzed with the PyroMark Q24 software.

For the pyrosequencing reaction, sequencing primers are hybridized to the target and the second-strand reaction is carried out in the Pyromark Q24 instrument using an enzyme-mix with DNA polymerase, ATP sulfurylase, luciferase, and apyrase. Adenosine 5′ phosphosulfate and luciferin serve as substrates. To start the

reaction, the enzyme-mix and the substrate-mix are added to the target and the primers that are already in the 24-well plate. For the first step of the reaction, the first dNTP is added and incorporated into the emerging DNA strand if it corresponds to the template strand. In this case, pyrophosphate is released in a quantity that is equimolar to the amount of incorporated nucleotides. For visualization, ATP sulfurylase converts pyrophosphate into ATP in the presence of adenosine 5′ phosphosulfate. ATP is needed for the conversion of luciferin to oxyluciferin that is catalyzed by luciferase and results in the generation of visible light that is proportional to the amount of ATP. This reaction is detected separately in every well by the CCD sensors and recorded as peaks in the Pyrogram™.

Before the next nucleotide can be added, unincorporated nucleotides and ATP must be destroyed by apyrase. Nucleotides are added sequentially according to the dispensation order that is based on the target sequence and the variations expected to build the new strand.

---

[1]For more details, consult the PyroMark KRAS Kit Handbook that can be found at the QIAGEN website (www1.qiagen.com).

The following protocol is intended to be used with the therascreen KRAS Pyro Kit for 24 samples and consists of four main steps: DNA extraction from FFPE, PCR and check-gel, immobilization, and sample preparation. The total time required is two working days: one for DNA extraction, and one for PCR, check-gel, immobilization, and preparation of samples.

27.2 Precautions

- Wear disposable gloves.
- Handle preamplification and postamplification reagents in separate rooms.
- Thaw all components thoroughly at room temperature, and mix them well before use.

27.3 DNA Extraction from FFPE

27.3.1 Reagents

Note: Reagents from specific companies are reported here, but reagents of equal quality purchased from other companies may be used.

Refer to reagents reported in Chap. 7.

27.3.2 Equipment

Refer to equipment listed in Chap. 7.

27.3.3 Method

- Identify the areas of the FFPE samples to be submitted to the analysis (e.g., primary tumor and/or metastasis). For a proper histologic selection, the use of tissue microdissection is recommended (see Chap. 4 and Chap. 6).
- Perform DNA extraction from FFPE material following the protocol described in Chap. 7.

- Determine DNA concentration (e.g., using Nanodrop). Dilute samples in order to test amplification of both 10 ng and 30 ng genomic DNA in 5 µl.[2,3]
- For the PCR reaction, prepare two tubes for each sample when performing duplicate testing as recommended.

27.4 PCR Amplification

27.4.1 Reagents

Note: Reagents from specific companies are reported here, but reagents of equal quality purchased from other companies may be used

- *therascreen KRAS Pyro Kit (24)* (Qiagen Cat. No.: 971460),

  24-µl PCR Primer KRAS 12/13
  24-µl PCR Primer KRAS 61
  100-µl Unmethylated Control DNA, 10ng/µl
  24-µl Seq Primer KRAS 12/13
  24-µl Seq Primer KRAS 61
  850µl PyroMark PCR Master Mix
  2x 1,2ml CoralLoad Concentrate, 10x
- *High purity water*: Pyrophosphate – free (MilliQ 18.2 MΩ × cm or equivalent)

27.4.2 Equipment

- *Disinfected variable pipettes and aerosol-resistant pipette tips* for 10 µl, 20 µl, and 200 µl
- *0.5-ml thin-walled thermo-well tubes* (e.g., Costar Cat. No.: 6530)
- *1.5-ml micro tubes, sterile* (e.g., Sarstedt Cat. No.: 72.692.005)
- *Microcentrifuge* suitable for 0.5 ml (e.g., Costar tubes)

---

[2]According to the manufacturer's instructions, optimal DNA concentration is 10–20 ng in 5 µl. Since 10 ng sometimes results in rather small peaks in the pyrosequencing analysis, we prefer using 10 ng and 30 ng DNA in 5 µl and recommend adapting the amount of DNA according to your own experience.

[3]If the concentration of DNA is less than 10 ng, the amplification reaction might still work. However, it is recommended to check the quantity of the amplification product using gel electrophoresis (2% agarose gel) and to use a control size ladder PCR.

Table 27.1 Amplification Mix

Component	Volume per sample	24(+2) x
PyroMark PCR master mix, 2x	12.5 µl	325 µl
CoralLoad concentrate (10x)	2.5 µl	65 µl
PCR primer KRAS 12/13 or 61	1.0 µl	26 µl
High purity water	4.0 µl	104 µl
Total volume	20.0 µl	520 µl

- *Thermal cycler* (e.g., Perkin-Elmer Gene Amp System 9700)

27.4.3 Method

- Thaw primer solutions and the reaction mix for the PCR in the master mix room.
- Prepare the Amplification Mix in a microtube, depending on the number of reactions needed (consider + two more reactions to compensate for loss of pipetting, see Table 27.1).
- Mix the Amplification Mix thoroughly, and dispense 20 µl into each thermo-well tube.
- For non-template control: 5 µl of high purity water.
- In the DNA preparation room, pipette 5 µl containing the appropriate amount of diluted DNA (10 and 30 ng respectively) into each tube already containing the Amplification Mix and mix by pipetting up and down 2–3 times. Total volume is 25 µl.
- Include the reactions with unmethylated Control DNA as positive control. Spin down mixture.
- Perform the following amplification program:

95°C	15 min	Initial activating step
95°C	20 s	
53°C	30 s	40–42 cycles
72°C	20 s	
72°C	5 min	
4°C	∞	

- After performing the PCR check the amplification products on a 2% agarose gel with slots that contain 6–8 µl of each PCR product (Fig. 27.3).

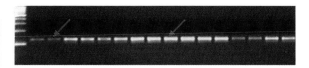

Fig. 27.3 Visualization of the amplification products of KRAS gene on 2% agarose gel. Although the amount of DNA is adjusted by determining the DNA concentration (e.g,. Nanodrop), amplification can result in different amounts of PCR product, because of differences in DNA quality. The results from the check-gel are used to adjust the amount of PCR product for the following step: e.g., 17 µl for weak (red arrow) bands to 10 µl for strong (blue arrow) bands

27.5 Immobilization of PCR Products to Streptavidin Sepharose High-Performance Beads

27.5.1 Reagents

Note: Reagents from specific companies are reported here, but reagents of equal quality purchased from other companies may be used.

- *Streptavidin Sepharose™High Performance* (GE Healthcare Cat. No.: 17-5113-01)
- *PyroMark Binding Buffer* (Qiagen Cat. No.: 979306 included in the Box 2 of the therascreen KRAS Pyro Kit) (4°C)
- *High purity water* (MilliQ 18.2 MΩ × cm or equivalent)

27.5.2 Equipment

- *Plate mixer Monoshake X826.1* (e.g., Carl Roth) orbital shaking up to 1,400 rpm[4]
- *24-well PCR plate or 96-well plate (cuttable) and strip caps*

---

[4]Orbital shaking is recommended to coat the Sepharose beads to their full capacity and to avoid sedimentation of the beads.

27.5.3 Method

- Gently shake the bottle containing Streptavidin Sepharose High-Performance Beads until it is a homogeneous solution.
- Prepare the following Mastermix (volumes are chosen according to the intensity of the gel bands):

	Low intensity	High intensity
Streptavidin Sepharose High-Performance Beads:	3 µl	3 µl
Pyromark binding buffer:	40 µl	40 µl
High purity water:	20 µl	27 µl
PCR product:	17 µl	10 µl
Total volume:	80 µl	80 µl

- Add 63 µl, respectively 70 µl, of the master mix to wells of a 24-well plate (cut 3 rows of a 96 well plate).
- Add 17 µl, respectively 10 µl, of biotinylated PCR product (according to the intensity of the bands on the gel), for a total volume of 80 µl/well.
- Seal the plate and agitate constantly at room temperature for 10–15 min, 1,400 rpm on an orbital shaker.
- During the previous step, prepare the vacuum workstation.
- Fill the troughs with solutions indicated to line according to the scheme:

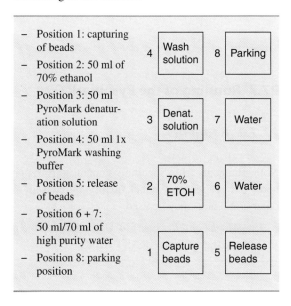

- – Position 1: capturing of beads
- – Position 2: 50 ml of 70% ethanol
- – Position 3: 50 ml PyroMark denaturation solution
- – Position 4: 50 ml 1x PyroMark washing buffer
- – Position 5: release of beads
- – Position 6 + 7: 50 ml/70 ml of high purity water
- – Position 8: parking position

27.6 Sample Preparation (Separation of the DNA Strands and Release of the Samples in Q24 Plate)

27.6.1 Reagents

Note: Reagents from specific companies are reported here, but reagents of equal quality purchased from other companies may be used)

- *70% Ethanol*
- All following reagents are included in the therascreen KRAS Pyro Kit
- *PyroMark Denaturation Solution*[5] *(4°C)*
- *PyroMark Wash Buffer, concentrate (10x) (4°C)*
- *PyroMark Validation Oligo (-20°C)*
- *PyroMark Control Oligo (-20°C)*
- *Seq Primer KRAS 12/13 or 61 from the Kit* *(-20°C)*
- *PyroMark Annealing Buffer (4°C)*

27.6.2 Equipment

- *PyroMark Q24 MDx vacuum workstation* (Qiagen Cat. No.: 9001515/9001517)
- *PyroMark Vacuum Prep Filter Probes* (100) (Qiagen Cat. No.: 979010)
- *PyroMark Q24 Plate* (Qiagen Cat. No.: 979301)
- *Heating block* at 80°C

27.6.3 Method

This protocol is used for preparation of single-stranded DNA and annealing of the sequencing primer to the template prior to pyrosequencing analysis on the Pyromark Q24 MDx.

27.6.3.1 Sequencing Primer

- Fill a Q24 plate with 0.3 µM sequencing primer (codon 12/13 or codon 61) in 30 µl Annealing

---

[5]This reagent should be handled with caution because it contains NaOH.

Buffer included in the Box 2 of the therascreen KRAS Pyro Kit in each well (0,96 µl Sequencing primer (10 µM) + 29,04 µl Annealing Buffer).

- For example, for a Q 24 plate, prepare a Mastermix for 26 samples:

Sequencing primer (10 µM)	24,96 µl
Annealing buffer	755,04 µl
Total volume	780 µl (30 µl/well)

27.6.3.2 Vacuum Prepstation

- Dip the filter probes for a few seconds into water to get small drops on the tops of the filter probes.
- Switch on the vacuum, and check if all filters are working, indicated by the disappearance of drops.
- Wash and rehydrate the filter probes by flushing them with approximately 70 ml water.

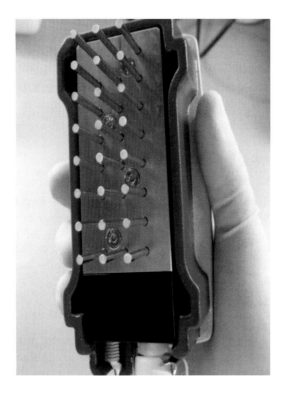

- Capturing of beads must take place immediately following agitation, because the beads sediment quickly.
- Leave the filter probes in the well plate for 15 s to capture all beads. Position 1.
- Plate must be in the proper orientation!
- Leave the vacuum switch on to do the following steps:
  - 10 s – position 2 – 70% Ethanol
  - 10 s – position 3 – Denaturation Solution
  - 15 s – position 4 – Washing Buffer
- Move to beyond 90° vertical for a few seconds after each step to avoid backdrain!
- Switch off the vacuum for 5 s before releasing the beads into the sequencing primer/Annealing Buffer (otherwise the sequencing primer solution can be exhausted into the filters) and shake the tool from side to side – position 5.
- Heat the Q24 plate with beads and primers at 80°C for 2 min using a prewarmed Q24 plate holder.
- Remove the plate from the holder and let the samples cool to room temperature for at least 10 min, otherwise enzyme and substrate for the sequencing reaction will be damaged.
- Clean the filter probes by agitating the filter probes for 10 s in water on position 6 and then leave 70 ml water from position 7 flush through the filter probes (Vacuum on).
- Final step for the filter probes is washing with Ethanol and drying. Then park the tool in the parking position until the next run.

27.7 Running of the Pyromark Q24

27.7.1 Reagents

Note: Reagents from specific companies are reported here, but reagents of equal quality purchased from other companies may be used.

- *PyroMark Gold Reagents* (4°C) included in the Box 2 of the therascreen KRAS Pyro Kit

27.7.2 Equipment

- *PyroMark Q24 MDx* (Qiagen Cat. No.: 9001513)
- *PyroMark Q24 MDx Software* (Qiagen Cat. No.: 9019063)
- *PyroMark Q24 Cartridge* (Qiagen Cat. No.: 979302)

27.7.3 Method

This protocol describes the run set up with the PyroMark software, the loading of the cartridge with the PyroMark Gold Q24 Reagents, and the procedure for the analysis of the run.

27.7.3.1 Run Setup

- Open the PyroMark software and create an Assay Setup for the KRAS assay (Fig. 27.4).
- Create a new run file:

- Enter run parameters
  - Run name – it is given when the run is saved, e.g., 20100106_KRAS_X.
  - Instrument method – instructions are supplied with the product.
  - Plate ID, Bar code, and Reagent ID – are optional, but we recommend entering the reagent ID to avoid any unexpected problems.
- Set up a plate by adding the corresponding assay files for both codon 12/13 and codon 61.
- Enter sample ID (name, number of tissue) and note the concentration (ng, dilution…).
- Open the Prerun information (list of required volumes of enzymes, substrate-mix, and nucleotides) from the Tools menu and print.
- Close the run file, and copy it to the USB memory stick.

27.7.3.2 Loading of Pyromark Q24 Gold Reagents into the Pyromark Q24 Cartridge

- Load the Cartridge with the appropriate amount of enzyme, substrate, and nucleotides according to the

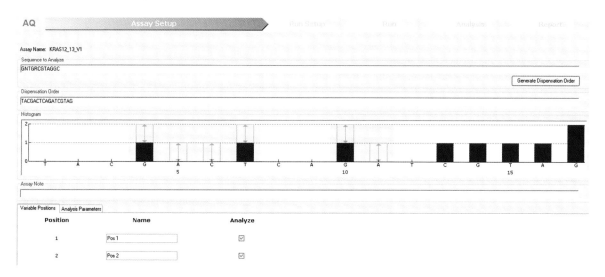

Fig. 27.4 Features of the assay setup screen opened with the PyroMark software

Prerun information. The label of the cartridge must be facing the front.

- There are some limitations for the cartridge –an intact surface is guaranteed for 30 runs. Watch the water jet at the rinsing step (step 6) – the water jet must be precisely straight down into the wells.
- Open the instrument and the cartridge gate and insert the filled cartridge with label facing out. Take care that the cartridge is inserted correctly, then close the gate.
- Open the plate holding frame and place the Q24 well plate to the heating block. Close the frame and the lid.
- Insert the USB stick with the run file – select run – press o.k. – select run file – start run.
- When the run has finished – after about 22 min for KRAS – the run file should have been saved on the stick. Press – close.
- Open the instrument, discard the plate, and remove the cartridge (wash with water and press some drops through the thin capillaries).
- Alternatively, if you plan to use the cartridge again immediately, refill with enzyme, substrate, and nucleotides, respectively, without washing (important!).

27.7.3.3 Analysis

- Insert the USB stick with the processed run into the USB port of the computer.
- Open the run file in AQ mode of the PyroMark Q24 Software and analyze the run either by analyzing all wells or analyzing the selected wells.
  - Well information: Assay name, sample ID, notes, and any analysis warnings
  - Pyrogram: The results and the quality assessment are displayed above the variable position
- Quality assessments are displayed by quality bars in the overview and background color of the analysis results:

    ☐ Passed

    ☐ Check (if the single peak height is lower than 30 RLU, but detectable results)

    ■ Failed (if single Peak height is lower than 15 RLU)

(a) For reliable results, a single peak height above 30 is recommended!

(b) The software analyzes only the region of interest and looks for the percentage of bases other than the wild-type sequence (see Run Setup).

(c) Nucleotide 35 is checked first since it is the most frequently mutated.

(d) Reanalyze all samples with "no mutation detected in nucleotide 35 (GGT_GGC)" as well as all samples that received CHECK or FAILED quality assessment with the sequence to analyze targeting mutations at nucleotide 34 (GGT_GGC)!

(e) Samples that are suspicious to contain a mutation with a peak height near or at the limit of detection should be reanalyzed in duplicate and compared to a normal sample also analyzed in duplicate!

27.7.3.4 Results

(a) The most frequent mutations in the KRAS gene are found at nucleotide 35 GGT (the second base in codon 12) (Fig. 27.5).

(b) If a sample contains a mutation in nucleotide 34 (first base of codon 12) (Fig. 27.6), the sequence to analyze must be changed to analyze the mutations at this position. To do so, go to analysis setup and change sequence to analyze from GNT-GRCGTAGGC to NGTGRCGTAGGC.
Press apply button and click – to all or to selected samples

(c) Base 2 of codon 12 can be also affected by a less frequent mutation (Fig. 27.7).
The results 1% A and 1% C in the figures above should be considered as background and result from the injection of the nucleotides.

27.7.4 Troubleshooting

- If there is no product visible on the check-gel, further analysis is not recommended because it is very unlikely that peaks of sufficient height are obtained (for a valid sequencing result, a threshold of 15 RLU is necessary). Possible solution: Repeat PCR and try to increase yield.

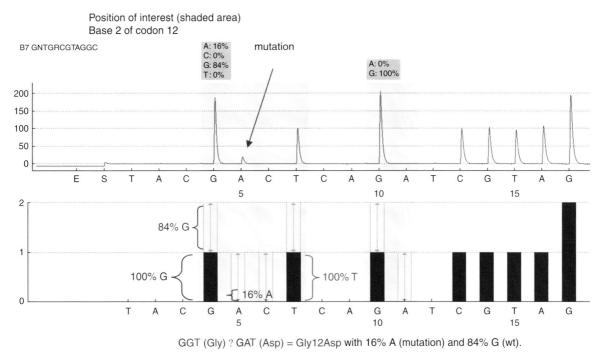

Fig. 27.5 Graphical representation of KRAS mutation in the second base of codon 12, as visualized by the PyroMark software

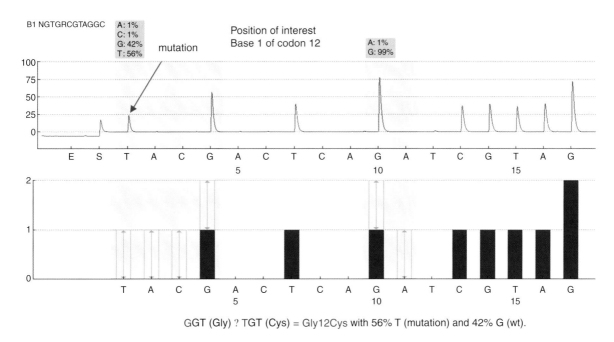

Fig. 27.6 Graphical representation of KRAS mutation in the first base of codon 12, as visualized by the PyroMark software

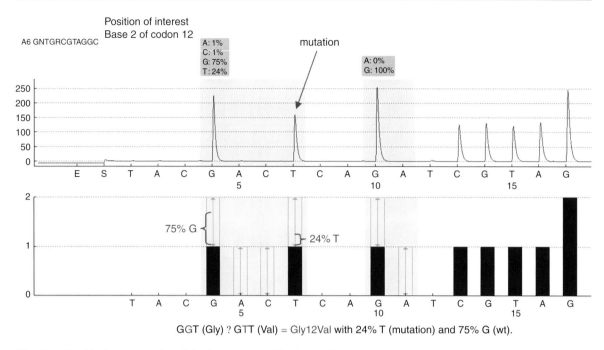

GGT (Gly) ? GTT (Val) = Gly12Val with 24% T (mutation) and 75% G (wt).

Fig. 27.7 Graphical representation of the less common KRAS mutation in the second base of codon 12, as visualized by the PyroMark software

- If the PCR product is clearly visible on the check-gel but sequencing peaks are too small, it is possible that not enough PCR products have been bound to the beads or beads have been lost during cleanup steps. Possible solutions: Carefully suspend beads in the master mix, use orbital shaker at 1,400 rpm to coat beads evenly. Quickly transfer the mix to the preptool to avoid sedimentation of the beads.
- Problems performing the sequencing reaction. Possible solution: heat plate containing beads coated with single-strand amplicon and sequencing primers for 2 min (not longer!) at 80°C. Avoid evaporation of the buffer since the salt concentration for the subsequent sequencing reaction would be too high. Cool at least for 10 min prior to adding the enzymes to avoid denaturation.

After dissolving the lyophilized enzyme-mix and reaction substrate, store at −20°C up to 2 weeks, not longer.

References

1. Allegra CJ, Jessup JM, Somerfield MR, Hamilton SR, Hammond EH, Hayes DF, McAllister PK, Morton RF, Schilsky RL (2009) American society of clinical oncology provisional clinical opinion: testing for KRAS gene mutations in patients with metastatic colorectal carcinoma to predict response to anti-epidermal growth factor receptor monoclonal antibody therapy. J Clin Oncol 27(12): 2091–2096
2. Ciardiello F, Tortora G (2008) EGFR antagonists in cancer treatment. N Engl J Med 358(11):1160–1174

Microsatellite Instability (MSI) Detection in DNA from FFPE Tissues

28

Damjan Glavač and Ermanno Nardon

Contents

28.1 Introduction and Purpose

Colorectal cancer (CRC) is usually classified as arising from at least two distinct mutational pathways, involving chromosomal instability (CIN) or microsatellite instability (MSI) [1–3]. CIN is the most common genetic alteration, occurring in 75–80% of CRC [4, 5], while MSI occurs in approximately 15%, including those arising in the Hereditary Non-Polyposis CRC familiar syndrome (HNPCC, 1–6% of total CRC) [6–9]. MSI results from the absence of a functional mismatch repair (MMR) enzyme, typically hMLH1 or hMSH2 [10], and is characterized by size alteration in the length of a microsatellite allele due to either insertion or deletion of repeated units.

MSI tumours have generally been reported as presenting a series of molecular defects and clinicopathological features distinct from CIN ones. Remarkably, MSI CRCs seem to display a less aggressive behaviour and better prognosis [6, 8, 11, 12]. Moreover, MMR impairment confers a higher resistance to alkylating agents and to antimetabolic chemotherapics [13–15].

An MMR defect can be detected by means of immunohistochemistry, but this approach could be inadequate if a nonfunctional protein is expressed, or difficult to perform in formalin-fixed paraffin-embedded (FFPE) tissues, because of epitope degradation [16]. Microsatellite instability is a phenotypic indicator of defective DNA mismatch repair system and, to date, the detection of this phenomenon is regarded as the standard screening method prior to mutation analysis. In 1997, participants in a National Cancer Institute workshop developed the international criteria (Bethesda guidelines) for microsatellite instability testing [17], identifying a reference panel of five microsatellite

D. Glavač (✉)
Department of Molecular Genetics, Institute of Pathology,
University of Ljubljana, Ljubljana, Slovenia

E. Nardon
Department of Medical, Surgical and Health Sciences,
University of Trieste, Trieste, Italy

G. Stanta (ed.), *Guidelines for Molecular Analysis in Archive Tissues*,
DOI: 10.1007/978-3-642-17890-0_28, © Springer-Verlag Berlin Heidelberg 2011

markers (BAT25, BAT26, D5S346, D2S123, and D17S250), which had previously been proposed and validated by a German multicenter study group [18, 19]. With regard to MSI status, tumours can be classified into three groups. The first one is characterized by MSI-high (MSI-H, showing instability in ≥40% of tested markers); the second one by MSI-low (MSI-L, showing instability in <40% of tested markers); and the third one by no instability (MSS, showing microsatellite stability). The NCI guidelines also suggest testing a second panel of alternative markers if only one marker of the first reference panel tests unstable, to rule out low-level MSI (MSI-L). In this case, the result is interpreted as MSI-H if at least 30% of all markers show instabilities; otherwise it is regarded as MSI-L [17, 20].

The Bethesda panel provides a uniform set of markers and criteria in MSI analysis. This panel, however, has some limitations because of the three dinucleotide markers used [20]: (a) each dinucleotide repeat generally shows instability in only 60–80% of MSI-H tumors [21], (b) their highly polymorphic nature requires the analysis of tumour and corresponding germline DNA (not always available), (c) misclassification of MSI can occur if samples from two individuals are mixed up [22].

In order to improve the existing panel, more mononucleotide markers have been tested for germline polymorphisms and several of them showed quasimonomorphic nature [23, 24]. Suraweera et al. proposed a set of five quasimonomorphic mononucleotide microsatellite markers (BAT-25, BAT-26, NR-21, NR-22, and NR-24) analyzed together in pentaplex PCR with nearly 100% sensitivity and specificity, eliminating the need for corresponding germline DNA. Tumour is defined as MSI-H when at least three out of five mononucleotides show instability [21, 25, 26]. Bethesda guidelines were revised in 2002 and accepted the aforementioned panel of markers as an alternative to previously used panels [20].

Recent evidence has shown that the sensitivity and specificity of the mononucleotide repeats panel remain substantially unaffected whether a cutoff threshold of ≥2 or ≥3 unstable markers is used to detect an MMR defect. In addition, the ≥2 markers threshold would provide a slightly higher sensitivity for a MSH6 defect, which usually escapes detection when using the classic Bethesda panel [27].

Herein three technical approaches to MSI analysis (MSA) are proposed:

- A basic method involving singleplex amplification of the Bethesda panel microsatellite sequences [17] and subsequent PAGE run and silver stain detection of PCR products;
- A multiplex PCR amplification of the five mononucleotide markers of the alternative panel [25, 26] coupled with DHPLC (Denaturing High-Performance Liquid Chromatography) analysis of PCR products;
- A multiplex PCR amplification of the five mononucleotide markers of the alternative panel [25, 26] coupled with capillary electrophoresis.

Since DNA from FFPE could be quite variable in quality, a control PCR for the assessment of amplifiability and integrity of sample DNA is also suggested.

All these three methods involve a common step of DNA extraction.

The interlaboratory reproducibility of MSI testing with the five mononucleotide markers has been assessed recently by the institutions involved in the IMPACTS project. The overall concordance at single locus level was 97.7%, comparable to that of the NCI panel (95.0%). A 100% agreement was reported at the patient-sample level both using the mononucleotide and the NCI markers. With these latter, however, concordance lowered to 85.7% if considering the MSI-L phenotype [28].

28.2 DNA Extraction from FFPE for Microsatellite Analysis

The following protocol is an adaptation of a previously published method of total DNA extraction [29] and is currently in use in one institution of the IMPACTS group (other protocols that can give good-quality DNA can be found in Chaps. 7 and 8; alternatively, one can avail of a commercial kit). For MSA, DNA from both the tumoral and the normal tissue of the same patient are required. If both normal and tumour tissues are present in the same paraffin block, it is necessary to perform microdissection (manual or laser capture, see Chaps. 4 and 6). The sections of tumor tissue should contain more than 50% of neoplastic cells [6], in order to avoid false negatives. The time required for the whole procedure is 3 days.

28.2.1 Reagents

Note: Reagents from specific companies are reported here, but reagents of equal quality purchased from other companies may be used.

- *Xylene* (Fluka or Sigma-Aldrich)
- *100% and 70% Ethanol* (Fluka or Sigma-Aldrich)
- *Lysis stock solution 10×*: 500 mM Tris-HCl pH 8, EDTA 10 mM, 5% Tween 20 (Sigma)
- *Proteinase K* stock solution, 20 mg/ml (Sigma P2303): dissolve Proteinase K in 50% sterile glycerol[1] in DNA-grade water. Store at −20°C
- *Phenol*, equilibrated with 10 mM Tris HCl, pH 8 (Sigma P4557)
- *Chloroform* (Sigma)
- *DNA precipitating solution*: 100% EtOH, 3 M Sodium Acetate, pH 7
- *1 mg/ml Glycogen* in H_2O (Sigma) used as precipitation carrier
- *10× TE buffer*: 100 mM Tris HCl, pH 8, 10 mM EDTA

28.2.2 Equipment

- *Adjustable pipettes*, range: 2–20 μl, 20–200 μl, 100–1,000 μl
- *Nuclease-free aerosol-resistant pipette tips*
- *1.5 ml nuclease-free autoclaved tubes*
- *Microtome*, with new blades
- *Chemical hood*
- *Tabletop refrigerated centrifuge* suitable for centrifugation of 1.5 ml tubes
- *Thermomixer*
- *Spectrophotometer*

28.2.3 Method

Wear gloves when isolating and handling DNA to minimize the activity of endogenous nucleases.

- Using a clean microtome blade,[2] cut 3–10 sections[3] of about 5 μm thickness. Discard the first section and displace the other ones in a clean 1.5 ml tube.[4]
- Add 1 ml of xylene,[5] vortex for 10 s. and then incubate the tube at room temperature (RT) for approximately 5 min.
- Spin the tube for 10 min at maximum speed (14,000 rpm) in a microcentrifuge and then carefully remove and discard the supernatant.[6] Repeat wash with a fresh aliquot of xylene.
- Wash the pellet by adding 1 ml of 100% Ethanol. Flick the tubes to dislodge the pellet and then vortex the tubes for 10 s. Incubate at RT for approximately 5 min.
- Spin the tube for 10 min at maximum speed (14,000 rpm) in a microcentrifuge and then carefully remove and discard the supernatant.
- Repeat washes twice, first using 100% ethanol and then 70% ethanol.
- After removing 70% ethanol, allow the tissue pellet to air dry at 37°C for 15–30 min.
- Complete the digestion solution by mixing 100 μl of 10× lysis solution, 25 μl of 20 mg/ml Proteinase K (final conc. 0.5 mg/ml), and 875 μl of sterile water per 1 ml of solution. Keep in ice until use.
- Add 200–400 μl of complete digestion solution to the pellet, depending on the pellet size. The digestion buffer must cover the tissue completely. Pipette the solution until the pellet is resuspended.
- Incubate for at least 16 h in a thermomixer at 55°C, shaking moderately.[7]

---

[1]The solubilization of proteinase K in 50% sterile glycerol avoids the freezing effect at −20°C, thus maintaining an optimal enzymatic activity.

[2]Clean the microtome with xylene. Cool the paraffin blocks at −20°C or on dry ice in aluminum foil to obtain thin sections.

[3]The number may vary, according to tissue type and area.

[4]Use some sections from a paraffin block without any tissue for the negative controls.

[5]When working with xylene, avoid breathing fumes; it is better to perform the deparaffinization step under a chemical hood.

[6]Discard the supernatant using a micropipette or a glass Pasteur pipette. Disposal of chemicals should be done in keeping with your laboratory safety rules.

[7]If low amounts of DNA are obtained, digestion can be prolonged up to 72 h by adding new Proteinase K solution every 24 h.

- Spin down the tubes to collect the condensate from the walls before opening. Add one volume of a 1:1 mixture of phenol-chloroform[8] per volume of digestion solution. Shake vigorously.
- Leave in ice for 10 min.
- Spin the tube for 10 min at maximum speed (14,000 rpm) at 4°C in a tabletop refrigerated microcentrifuge.
- Transfer the upper, aqueous phase to a new 1.5 ml tube; do not touch the interphase disc, which contains most of the protein residues.
- Add 5 μl of 1 mg/ml glycogen and 3 volumes of DNA precipitating solution per volume of recovered aqueous phase.
- Allow precipitation overnight at −20°C.
- Spin the tube for 20 min at maximum speed (14,000 rpm) at 4°C. A small pellet should form at the bottom of the tube.
- Discard the supernatant. Do not touch the pellet.
- Add 100 μl of 70% ice-cold ethanol, without resuspending the pellet.
- Spin the tube for 5 min at maximum speed (14,000 rpm) at 4°C.
- Discard the supernatant. Leave the pellet to air dry at 37°C for 15–30 min.
- Resuspend the pellet in 22 μl of 1× TE buffer.
- Determine the DNA concentration spectrophotometrically at 260 and 280 nm.[9] Absorbance ratio between readings at 260 and 280 nm should be ≥1.6, which is indicative of an acceptable level of contamination from proteins (see also Chap. 16).
- Store the DNA at −20°C until use.

28.2.4 Troubleshooting

- If neither a pellet nor a "smear" on the tube wall is visible after precipitation, add 5 μl of glycogen solution and again leave overnight at −20°C.

28.3 MSA: Basic Method

Microsatellite marker amplifications are performed as singleplex PCR reactions using DNA from both tumoral and normal tissue of the same patient. In accordance to the recommendations by the NCI [17] the validated five microsatellite panel is tested first. If only one marker shows instability, or if only the dinucleotide repeats are mutated, a second panel of five additional markers (BAT40, D10S197, D13S153, MYCL1, and D18S58) [17, 20, 30] is checked.[10] This latter panel is currently adopted by the institutions of the IMPACTS group in a diagnostic setting. See Tables 28.2 and 28.3 for primer details.

PCR products are separated on an acrylamide gel and visualized by means of silver staining.

Since DNA extracted from FFPE can be variably degraded and may contain PCR inhibitors, we suggest performing a preliminary quality control to test if sample DNA is suitable for MSA and to determine the optimal quantity for amplification. For this purpose, a 251 bp fragment of the β-globin gene is amplified. Since β-globin gene is present in all the cells (it never undergoes deletions) and is not polymorphic, it is a suitable target for the control PCR [31]. The size of the amplified fragment (251 bp) reflects the length of the largest PCR product that can be obtained.

28.3.1 PCR Reaction

28.3.1.1 Reagents

Note: Reagents from specific companies are reported here, but reagents of equal quality purchased from other companies may be used.

- *10X PCR buffer II without MgCl₂*: we report the standard composition of PCR buffer II: 500 mM KCl, 100 mM Tris pH 8.3 at 25°C
- *25 mM MgCl₂ solution* (Applied Biosystems)

---

[8]Phenol and chloroform are toxic by inhalation; work under a chemical hood.

[9]The concentration of dsDNA expressed in μg/μL is obtained as follows: [DNA] = A_{260} x dilution factor x 50 x 10^{-3}.

---

[10]The primer sequences for D2S123 and D13S153 have been modified with respect to those previously published, in order to reduce the length of the amplified product.

- *dNTPs stock solution pH8, 100 mM each* (Amersham): dilute the dNTP stock solution in sterile water to prepare a 10 mM solution of each dNTP
- *Primers (Sigma)*: Lyophilized primers should be dissolved in a small volume of distilled water or 10 mM Tris pH 8 to make a concentrated (e.g. 300 pmol/μl) stock solution. Prepare small aliquots of working solutions containing 30 pmol/μl to avoid repeated thawing and freezing. Store all primer solutions at −20°C. See Tables 28.1–28.3 for primer details
- *AmpliTaq Gold* (Applied Biosystems, N8080247) 5 U/μl

Table 28.1 Control PCR primers

Marker name	Genomic position	Sequences (5′–3′)	$T°m$	Product (bp)
β-globin	11p15.5	F: ACACAACTGTGTTCACTAGC	53.4	251
		R: GAAAATAGACCAATAGGCAG	53.9	

F forward primer, *R* reverse primer

Table 28.2 MSA primers, first panel

Marker name	Genomic position	Sequences (5′–3′)	$T°m$	Range (bp)
D2S123	2p16-2p16	F: ACATTGCTGGAAGTTCTGGC	62.6	121–141
		R: CCTTTCTGACTTGGATACCA	57.5	
D5S346	5q21-5q22	F: ACTCACTCTAGTGATAAATCGGG	58.7	96–122
		R: AGCAGATAAGACAGTATTACTAGTT	52.1	
D17S250	17q11.2-17q12	F: GGAAGAATCAAATAGACAAT	50.6	151–169
		R: GCTGGCCATATATATATTTAAACC	57.2	
BAT25	4q12-4q12	F: TCGCCTCCAAGAATGTAAGT	59.7	124
		R: TCTGCATTTTAACTATGGCTC	57.0	
BAT26	2p16.3-2p16.3	F: TGACTACTTTTGACTTCAGCC	57.0	121
		R: AACCATTCAACATTTTTAACCC	59.0	

See footnote 10
F forward primer, *R* reverse primer

Table 28.3 MSA primers, second panel

Marker name	Genomic position	Sequences (5′–3′)	$T°m$	Range (bp)
BAT40	1p13.1-1p13.1	F: ATTAACTTCCTACACCACAAC	54.0	123
		R: GTAGAGCAAGACCACCTTG	56.7	
MYCL1	1p32-1p32	F: TGGCGAGACTCCATCAAAG	56.7	140–209
		R: CCTTTTAAGCTGCAACAATTTC	54.7	
D18S58	18q22.3-18q23	F: GCTCCCGGCTGGTTTT	54.3	144–160
		R: GCAGGAAATCGCAGGAACTT	57.3	
D13S153	13q14.1-13q14.36	F: TAACCGACTCCTGTTTCTCC	56.8	162–186
		R: AAGGTCTAAGCCCTCGAGTT	57.0	
D10S197	10p12-10p12	F: ACCACTGCACTTCAGGTGAC	59.4	161–173
		R: GTGATACTGTCCTCAGGTCTCC	62.1	

See footnote 10
F forward primer, *R* reverse primer

28.3.1.2 Equipment

- *Adjustable pipettes,* range: 2–20 µl, 20–200 µl, 100–1,000 µl
- *Nuclease-free aerosol-resistant pipette tips*
- *0.2 and 1.5 ml nuclease-free microtubes*
- *Sterile laminar flow hood*
- *Tabletop refrigerated centrifuge*: suitable for centrifugation of 0.2 ml tubes
- *Thermal cycler*
- *Electrophoresis unit for agarose gel*
- *UV transillumination unit*

28.3.1.3 Method

Controls

Negative control for β-globin and MSA: see section 28.2.3.

Positive control for β-globin and MSA amplifications: DNA from normal human lymphocytes, 50 ng/µl.[11]

Quality Control PCR

- Prepare the following dilutions of sample DNA in 1× TE: 400 ng/µl, 100 ng/µl, and 25 ng/µl.
- Prepare the PCR master mix in a sterile laminar floodhood.
- PCR is performed in a final volume of 50 µl, containing:
    - 5 µl 10X PCR Buffer II........... 1X final
    - 3 µl MgCl$_2$, solution, 25 mM....1.5 mM final
    - 1 µl dNTP, 10 mM each.......... 0.2 mM final
    - 0.5 µl β-globin forward primer, 30 pmol/µl.............................. 0.3 pmol/µl final
    - 0.5 µl β-globin reverse primer, 30 pmol/µl.............................. 0.3 pmol/µl final
    - 0.25 µl AmpliTaq Gold, 5 U/µl......................................0.025 U/µl final
    - 1 µl of diluted sample DNA
    - or
    - 1 µl of undiluted negative control-or

- 1 µl of positive control.
- H$_2$O to volume
- Overlay the reaction mixture with 20 µl of mineral oil.
- Thermal cycling: 94°C 10"+5 × (94°C 60", 52°C 60", 72°C 60")+35 × (94°C 30" 52°C 30" 72°C 30")+72°C 5'.
- Gel visualization: mix 10 µl of PCR product with 2 µl of 6× loading buffer[12]; load on a 2% agarose gel prepared with 1× TBE containing 0.5 µg/ml ethidium bromide.[13] Include a 100 bp marker ladder (e.g., Amersham 27400701). Run at 80 V constant until bromophenol blue reaches 1/2 of the gel. Inspect under a UV source. A single band should be visible in the sample and in the positive control lanes.
- Sample dilution yielding the strongest amplification signal will be used in MSA. Should two dilutions yield equal signals, we suggest using the one containing the highest DNA concentration. Indeed, PCR artifacts may occur when too low DNA quantities from FFPE are used [32].

PCR for First or Second Microsatellite Panel

- Operate in a sterile laminar floodhood. Prepare a different master mix for each microsatellite marker (singleplex).
- PCR is performed in a final volume of 50 µl, containing:
    - 5 µl 10X PCR Buffer II........... 1X final
    - 3 µl MgCl$_2$ solution, 25 mM... 1.5 mM final
    - 1 µl dNTP, 10 mM each.......... 0.2 mM final
    - 0.5 µl forward primer, 30 pmol/µl..............................0.3 pmol/µl final
    - 0.5 µl reverse primer, 30 pmol/µl..............................0.3 pmol/µl final
    - 0.25 µl AmpliTaq Gold, 5U/µl......................................0.025 U/µl final

---

[11]It is recommended that a high-quality DNA, e.g. a DNA extracted from a cell line, be used.

[12]6× loading buffer: 0.25% bromophenol blue, 0.25% xylene cyanol, 30% glycerol in H$_2$O.

[13]Ethidium bromide is a potentially carcinogenic compound. Always wear gloves. Used EtBr solutions must be collected in containers for chemical waste and discharged according to the local hazardous chemical disposal procedures.

- 1 µL of diluted sample DNA
- or
- 1 µL of undiluted negative control-or
- 1 µL of positive control for amplification
- H$_2$O to volume
- Overlay reaction mixture with 20 µl of mineral oil.
- Thermal cycling: 94°C 10'+5 × (94°C 60", 55°C 60", 72°C 60")+35 × (94°C 30" 55°C 30" 72°C 30") + 72°C 5'.
- Gel visualization: mix 10 µl of PCR product with 2 µl of 6× loading buffer[14]; load on a 2% agarose gel prepared with TBE 1×, containing 0.5 µg/ml ethidium bromide. Include a 100 bp marker ladder (e.g., Amersham 27400701). Run at 80 V constant until bromophenol blue reaches 1/2 of the gel. Inspect under a UV source. A single band should be visible in the sample and in the positive control lanes.

28.3.1.4 Troubleshooting

- No sample dilution yields a visible band. DNA could be degraded to an extent that does not allow amplification of a 251 bp fragment. Repeat amplification performing five more cycles. Run the sample in duplicate/triplicate. If only one of two/three replicas amplifies, stochastic amplification is suspected due to very low concentration or highly degraded DNA and results should be cautiously evaluated. Sample re-extraction is recommended.

28.3.2 Silver Stain Detection and Interpretation of the Results

28.3.2.1 Reagents

Note: Reagents from specific companies are reported here, but reagents of equal quality purchased from other companies may be used.

- *40% Acrylamide-bis acrylamide (19:1) solution*[15] (Sigma)
- *10% Ammonium Persulfate* (Sigma) in H$_2$O

- *TEMED   (N,N,N′,N′-Tetramethylethylenediamine)* (Sigma)
- *TBE buffer, 10X* (1 M Tris-HCl, 1 M Boric Acid, 20 mM EDTA, pH 8.3)
- *100 bp marker ladder* (Amersham 27400701)
- *10% Ethanol in H$_2$O*. Prepare at the moment
- *1% Nitric Acid in H$_2$O*. Prepare at the moment. Keep in the dark
- *12 mM Silver Nitrate* (Sigma) in H$_2$O. Prepare at the moment. Keep in the dark
- *0.28 M Sodium carbonate anhydrous in H$_2$O*. Prepare at the moment
- *Formalin* (Sigma 47629). Prepare at the moment. Keep in the dark
- *10% Acetic acid in H$_2$O*. Prepare at the moment

28.3.3 Equipment

- *Adjustable pipettes,* range: 2–20 µl, 20–200 µl, 100–1,000 µl
- *Nuclease-free aerosol-resistant pipette tips*
- *Electrophoresis unit for acrylamide gel, sequencing size*
- *Polyethylene tray, acid resistant*
- *White light transilluminator*

28.3.4 Method

28.3.4.1 Silver Stain Detection

Perform all the steps in a different room from the one dedicated to DNA extraction or amplification. Wear gloves. Glass plates should be at least 30 cm long; spacers and combs should be 0.8–1 mm thick.

- Prepare 100 ml 8% acrylamide gel by mixing 10 ml of 10X TBE buffer, 20 ml of Acrylamide-bis acrylamide 40% solution, 69.5 ml of distilled H$_2$O, 500 µl of 10% Ammonium Persulfate, and 60 µl of TEMED. Allow polymerization to occur for at least 1 h.
- Mix the MSA-PCR amplification product[16] with 2 µl of 6× loading buffer; add 1X TBE to a final

---

[14]See Footnote 12.

[15]Toxic and carcinogenic. Avoid contact with eyes and skin. Avoid breathing fumes.

[16]The choice of amplification product volume (usually 1–5 µl) relies on band signal intensity at visual inspection of the agarose gel.

Fig. 28.1 MSA of (**a**) BAT25 mononucle-otide and (**b**) D5S346 dinucleotide markers visualized by silver nitrate staining. N_1 and T_1 are, respectively, the normal and tumour counterpart of an MSS sample, N_2 and T_2 are the normal and tumour counterpart of an MSI sample. The MSS sample features equal-sized allele bands. The T_2 MSI sample shows a single prominent allele band, lower than its normal counterpart in the mononucleotide marker, and shows additional, different-sized bands in the dinucleotide marker

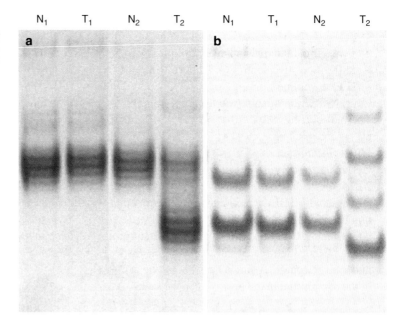

volume of 12 µl. Flush the wells of the gel with 1X TBE before loading the samples. Normal and tumour samples from the same patient should be loaded in contiguous lanes to facilitate comparison. Include the 100 bp marker ladder.

- Perform the electrophoretic run in 1X TBE at 250 V constant for 16 h (overnight) at RT. Xylene cyanol should be at the bottom of the gel.[17]
- Carefully transfer the gel from the glass plates to the polyethylene tray.
- Perform the fixation in 10% ethanol for 10′.
- Wash the gel twice in demineralized H_2O.
- Perform the oxidation with 1% nitric acid for 3–4 min under shaking.
- Wash the gel twice in demineralized H_2O.
- Leave the gel in 12 mM silver nitrate for 20′.
- Wash the gel twice in demineralized H_2O
- Perform the reduction with 0.28 M Sodium carbonate containing 0.05% (v/v) formalin solution until the appearance of the bands.
- Remove the reducing solution; block the reduction with 10% acetic acid for 2′.

---

[17]As impedance may vary according to elecrophoretic system, buffer volume, etc., voltage and run time should be adapted.

- Wash the gel twice in demineralized H_2O.
- Transfer the gel over the white light transilluminator for inspection.

28.3.4.2 Interpretation of the Results

Microsatellite alleles might show very complex band patterns after PAGE separation and silver staining. In fact, artifactual fragments larger than true alleles are usually detected in nondenaturing conditions. This may be due to the Taq polymerase slippage during PCR, to recombination of heteroduplex amplicons, or to secondary DNA structures, etc. [33]. This can make allele sizing sometimes cumbersome.

- *First microsatellite panel*: a sample is scored as MSS if the allele bands are at the same height in both normal and tumour tissue. A sample is scored as MSI if the tumour tissue presents extra bands, or a shift in band height, with respect to normal tissue (Fig. 28.1). If a sample shows MSI in two or more of the five markers, it is considered MSI-H. If only one marker tests unstable, the second microsatellite panel is tested.
- *Second microsatellite panel*: a sample is scored as MSI-L if no microsatellite marker shows instability or only 1 marker from this panel tests unstable. A sample is scored as MSI-H if it shows instability in two or more markers from this panel [17].

28.3.4.3 Troubleshooting

- One or more markers from the first panel fails to amplify: if the sample tests unstable in two or more loci, it is considered MSI-H, regardless of PCR failure [17]. If only one marker or no marker shows instability, repeat amplification performing five more cycles and/or using a different quantity of DNA. Run sample in duplicate/triplicate. If only one of two/three replicas amplifies, stochastic amplification is suspected and results should be cautiously evaluated.
- Band pattern of one or more markers is smeared and uninterpretable because of too intense silver staining: adjust the quantity of amplified product to be used.
- Band pattern of one or more markers is smeared and uninterpretable, but silver staining is normal: poor DNA quality and/or too many PCR cycles. Same as point 1.
- (Rare) Normal sample bands are of unexpected size (out of allele range): this may be due to PCR carry-over contamination from an MSI sample.
- (Rare) Both normal and tumour tissues show the same band pattern, but of unexpected size (out of allele range): poor microdissection of the normal sample before DNA extraction could be the cause.

28.4 MSA: DHPLC Method

Microsatellite marker amplifications are performed as multiplex PCR reactions on DNA from tumour tissue. In accordance with the revised recommendations by the NCI, [20] the validated, five mononucleotide microsatellite markers are tested [25, 26].

PCR products are separated on a DHPLC column and visualized by means of UV detection.

28.4.1 PCR Reaction

28.4.1.1 Reagents

Note: Reagents from specific companies are reported here, but reagents of equal quality purchased from other companies may be used.

- *10X PCR HotMaster^{TM} Buffer with 25 mM MgCl$_2$ (5 Prime GmbH)*
- *Commercial 25 mM stock solution of each dNTP (Applied Biosystems): dilute the dNTP stock solution to prepare 2.5 mM solution of each dNTP in sterile water*
- *Primers*: lyophilized primers should be dissolved in a small volume of distilled water or 10 mM Tris pH 8 to make a concentrated stock solution. Prepare small aliquots of working solutions containing 5 pmol/μl to avoid repeated thawing and freezing. Store all primer solutions at −20°C. See Table 28.4 for primer details
- *HotMaster^{TM} DNA Taq polymerase (5 Prime GmbH), 5 U/μl*

28.4.1.2 Equipment

- *Adjustable pipette;* range: 2–20 μl, 20–200 μl, 100–1,000 μl
- *Nuclease-free aerosol-resistant pipette tips*

Table 28.4 MSA primers for DHPLC analysis

Marker name	Genomic position	Sequence (5′–3′)	$T°m$	Size (bp)
BAT 26	2p16.3-2p16.3	F: CTGCGGTAATCAAGTTTTTAG R: AACCATTCAACATTTTTAACCC	60.1 63.1	183
BAT25	4q12-4q12	F: TACCAGGTGGCAAAGGGCA R: TCTGCATTTTAACTATGGCTC	63.1 59.8	153
NR 24	2q11.2-2q11.2	F: GCTGAATTTTACCTCCTGAC R: ATTGTGCCATTGCATTCCAA	58.1 66.0	131
NR 21	14q11.2-14q11.2	F: GAGTCGCTGGCACAGTTCTA R: CTGGTCACTCGCGTTTACAA	62.4 63.5	109
NR 27	11q22-11q22	F: AACCATGCTTGCAAACCACT R: CGATAATACTAGCAATGACC	64.5 54.4	87

F forward primer, *R* reverse primer

- *0.2 and 1.5 ml nuclease-free microtubes*
- *Sterile laminar flow hood*
- *Tabletop refrigerated centrifuge* suitable for centrifugation of 0.2 ml tubes
- *Thermal cycler*
- *Electrophoresis unit for agarose gel*
- *UV transillumination unit*

28.4.2 Method

Operate in sterile laminar flowhood. Prepare a multiplex master mix for five microsatellite markers.

- PCR is performed in a final volume of 20 µl, containing:
  - 4 µl 10X PCR HotMasterTM Buffer...................................... 2 X final
  - 2.4 µl dNTP, 2.5 mM each...... 0.3 mM final
  - 0.4 µl BAT 26 forward primer, 5 pmol/µl................................. 0.1 pmol/µl final
  - 0.4 µl BAT 26 reverse primer, 5 pmol/µl................................. 0.1 pmol/µl final
  - 0.4 µl BAT 25 forward primer, 5 pmol/µl................................. 0.1 pmol/µl final
  - 0.4 µl BAT 25 reverse primer, 5 pmol/µl................................. 0.1 pmol/µl final
  - 0.4 µl NR 24 forward primer, 5 pmol/µl................................. 0.1 pmol/µl final
  - 0.4 µl NR 24 reverse primer, 5 pmol/µl................................. 0.1 pmol/µl final
  - 0.4 µl NR 21 forward primer, 5 pmol/µl................................. 0.1 pmol/µl final
  - 0.4 µl NR 21 reverse primer, 5 pmol/µl................................. 0.1 pmol/µl final
  - 0.4 µl NR 27 forward primer, 5 pmol/µl................................. 0.1 pmol/µl final
  - 0.4 µl NR 27 reverse primer, 5 pmol/µl................................. 0.1 pmol/µl final
  - 0.6 µl HotMasterTM DNA Taq polymerase, 5 U/µl...................................... 0.15 U/µl final
  - 3 µl of 10 ng/µl DNA sample
  - or
  - 1 µl of undiluted negative control
  - 1 µl of positive control for amplification.
  - H_2O to volume
- Thermal cycling: 94°C 2" + 35 × (94°C 20", 55°C 10", 65°C 30") + 65°C 7'.

- Gel visualization: mix 10 µl of PCR product with 2 µl of 6× loading buffer[18]; load on a 2% agarose gel prepared with TBE 1×, containing 0.5 µg/ml ethidium bromide. Include a 100 bp marker ladder (e.g., Amersham 27400701). Run at 80 V constant until bromophenol blue reaches 1/2 of the gel. Inspect under a UV source. Five bands should be visible in the sample and in the positive control lanes.

28.4.3 DHPLC Detection and Interpretation

28.4.3.1 Reagents

- *Buffer A* (0.1 M TEAA solution) (e.g., Transgenomic Inc., Omaha, NE, USA)
- *Buffer B* (0.1 M TEAA containing 25% acetonitrile solution) (e.g., Transgenomic)
- *Solution D* (75% acetonitrile solution) (e.g., Transgenomic)
- *Syringe wash solution* (4% acetonitrile solution) (e.g., Transgenomic)

28.4.3.2 Equipment

- *Adjustable pipettes*, range : 2–20 µl, 20–200 µl, 100–1,000 µl
- *Nuclease-free aerosol-resistant pipette tips*
- *WAVE system* (e.g., Transgenomic)
- *DNASep cartridge* (e.g., Transgenomic)
- *WAVEMAKER™ software* (e.g., Transgenomic)

28.4.3.3 Method

DHPLC Detection

PCR products are automatically injected into DNASep cartridge (Transgenomic Inc., Omaha, NE, USA) by WAVE system and eluted at a flow rate of 0.9 ml/min through a linear gradient of acetonitrile containing 0.1 M triethylammonium acetate (TEAA).

---

[18]See Footnote 12.

Elution gradient of Buffer A (0.1 M TEAA solution) and Buffer B (0.1 M TEAA containing 25% acetonitrile solution) is automatically adjusted using WAVEMAKER™ software (Transgenomic Inc., Omaha, NE, USA). Oven temperature is set to 50°C as recommended for double-stranded DNA sizing. Run is performed using multiple fragments double-stranded DNA sizing application, which allows simultaneous analysis of five microsatellites. Gradient conditions are set for elution of 100 base-pair products in 4.5 min. The sizes of the largest and the smallest PCR product are set at 77 and 189 bp, respectively. UV detection of DNA fragments eluted from the column is performed at 260 nm wavelength.

Interpretation of the Results

DHPLC elution profile of each sample contains five peaks representing five microsatellite markers separated according to the size of the PCR products (Fig. 28.2a). Microsatellite instability is detected by the presence of different-sized alleles and is visible as additional peaks in the vicinity of the peak representing normal-sized alleles (Fig. 28.2b). When additional peaks are present in at least three of five tested microsatellite markers, the sample is scored as MSI-high, when additional peaks are present in less than three markers the sample is scored as MSI-low, and when at five markers no additional peaks are detected, the sample is classified as MSS.

28.4.3.4 Troubleshooting

- Less than five peaks are visible on elution profile after separation:
  - This usually occurs due to inappropriate elution gradient of buffers, which causes too fast elution of the shortest fragment from the column. To solve this problem, it is necessary to adjust the elution gradient of buffers A and B by WAVEMAKER™ software;
  - Multiplex PCR failed to amplify all five fragments. Solution: optimize PCR conditions.
- The signal is too low:
  - Low quality of PCR products. Solution: increase the injection time of the sample;
  - DNASep cartridge is dirty. Solution: perform clean procedures according to the manufacturer's protocol; if the detection is not improved, change cartridge;
  - The power of the UV lamp is too low. Solution: change the UV lamp.

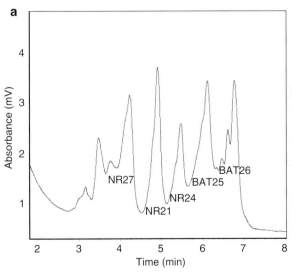
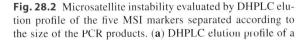
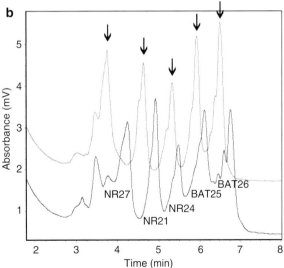

Fig. 28.2 Microsatellite instability evaluated by DHPLC elution profile of the five MSI markers separated according to the size of the PCR products. (**a**) DHPLC elution profile of a normal-sized allele; (**b**) Allele with microsatellite instability that is visible as the presence of additional peaks (*arrows*) in proximity to the peaks representing normal-sized alleles

28.5 MSA: Capillary Electrophoresis Method

Microsatellite marker amplifications are performed as multiplex PCR reactions on DNA from tumour tissue. In accordance with the revised recommendations by the NCI, [20] the validated, five mononucleotide microsatellite markers are tested.

PCR products are separated by capillary electrophoresis and visualized by means of fluorescent laser detection.

28.5.1 PCR Reaction

28.5.1.1 Reagents

Note: Reagents from specific companies are reported here, but reagents of equal quality purchased from other companies may be used.

- *10X PCR HotMaster™ Buffer* with 25 mM MgCl₂
- *Commercial 25 mM stock solution of each dNTP* (Applied Biosystems): dilute the dNTP stock solution to prepare 2.5 mM solution of each dNTP in sterile water
- *Primers*: Lyophilized primers should be dissolved in a small volume of distilled water or 10 mM Tris pH 8 to make a concentrated stock solution. Prepare small aliquots of working solutions containing 5 pmol/μl to avoid repeated thawing and freezing. Store all primer solutions at −20°C (Table 28.5).
- *HotMaster™ DNA Taq polymerase* (5 Prime GmbH), 5 U/μl

28.5.1.2 Equipment

- *Adjustable pipettes*; range: 2–20 μl, 20–200 μl, 100–1,000 μl
- *Nuclease-free aerosol-resistant pipette tips*
- *0.2 and 1.5 ml nuclease-free microtubes*
- *Sterile laminar flow hood*
- *Tabletop refrigerated centrifuge* suitable for centrifugation of 0.2 ml tubes
- *Thermal cycler*
- *Electrophoresis unit for agarose gel*
- *UV transillumination unit*

28.5.1.3 Method

Amplifications and gel visualization are performed in the same way as in section 28.4.2.

28.5.2 Capillary Electrophoresis Detection

28.5.2.1 Reagents

Note: Reagents from specific companies are reported here, but reagents of equal quality purchased from other companies may be used.

- *Hi-Di™ formamide* (Applied Biosystems)
- *Performance Optimized Polymer 4™ (POP4)* (Applied Biosystems)
- *EDTA Buffer (10X)* (Applied Biosystems)
- *GenScan™-500 Rox™ Size Standard* (Applied Biosystems)

Table 28.5 MSA primers for capillary electrophoresis analysis

Marker name	Genomic position	Sequence (5′–3′)	$T°m$	Size (bp)
BAT 26	2p16.3-2p16.3	F: FAM-CTGCGGTAATCAAGTTTTTAG	60.1	183
		R: AACCATTCAACATTTTTAACCC	63.1	
BAT25	4q12-4q12	F: HEX-TACCAGGTGGCAAAGGGCA	63.1	153
		R: TCTGCATTTTAACTATGGCTC	59.8	
NR 24	2q11.2-2q11.2	F: NED-GCTGAATTTTACCTCCTGAC	58.1	131
		R: ATTGTGCCATTGCATTCCAA	66.0	
NR 21	14q11.2-14q11.2	F: FAM-GAGTCGCTGGCACAGTTCTA	62.4	109
		R: CTGGTCACTCGCGTTTACAA	63.5	
NR 27	11q22-11q22	F: HEX-AACCATGCTTGCAAACCACT	64.5	87
		R: CGATAATACTAGCAATGACC	54.4	

F forward primer, *R* reverse primer

28.5.2.2 Equipment

- *Adjustable pipettes*; range: 2–20 μl, 20–200 μl, 100–1,000 μl
- *Nuclease-free aerosol-resistant pipette tips*
- *ABI PRISM 310 Genetic Analyzer* (Applied Biosystems)
- *Capillaries (5–47 cm × 50 μm)* (Applied Biosystems)
- *GeneScan™ Software* (Applied Biosystems)

28.5.2.3 Method

Capillary Electrophoresis Detection

After multiplex amplification of five microsatellite markers with fluorescently labeled primers, the samples are diluted in double distilled water at the ratio of 1:15. 3 μl of diluted PCR product is mixed with 8 μl Hi-Di™ formamide and 0.3 μl GenScan™-500 Rox™ Size Standard. The mixture is heated for 2 min at 95°C and placed on ice until ready to be placed in the instrument. The samples are placed in ABI PRISM 310 Genetic Analyzer and automatically analyzed.

The detection is performed by the laser excitation of fluorescent labels attached to PCR products.

Interpretation of the Results

Analysis of raw data is performed with GeneScan™ Software using Dye filter set D. The obtained electropherograms show five peaks representing five microsatellite markers separated according to the size of the PCR products. Microsatellite instability is detected by the presence of different-sized alleles and is visible as additional peaks in the vicinity of the peak representing normal-sized alleles (Fig. 28.3). Evaluation of MSI status is the same as in the case of DHPLC.

28.5.2.4 Troubleshooting

- The detected signal is too low to evaluate the results. Solution: increase the injection time or decrease the dilution of the PCR products.
- The detected signals go off-scale. Solution: decrease the injection time or increase the dilution of the PCR products.

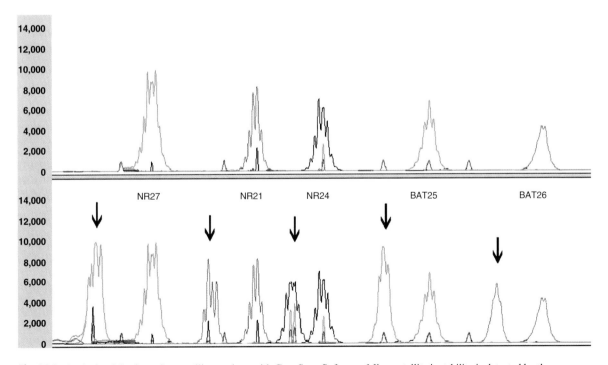

Fig. 28.3 Analysis of the five microsatellite markers with GeneScan Software. Microsatellite instability is detected by the presence of different-sized alleles and is visible as additional peaks in the vicinity of the peak representing normal-sized alleles

- Less than five peaks are visible on electrophero-gram after separation:

  - Multiplex PCR failed to amplify all five fragments. Solution: optimize PCR conditions
  - Fragments reach the detector outside the detection time. Solution: adjust the detection time.

References

1. Dutrillaux B (1995) Pathways of chromosome alteration in human epithelial cancers. Adv Cancer Res 67:59–82
2. Lengauer C, Kinzler KW, Vogelstein B (1997) Genetic instability in colorectal cancers. Nature 386(6625):623–627
3. Sweezy MA, Fishel R (1994) Multiple pathways leading to genomic instability and tumorigenesis. Ann NY Acad Sci 726:165–177
4. Fearon ER, Vogelstein B (1990) A genetic model for colorectal tumorigenesis. Cell 61(5):759–767
5. Vogelstein B, Fearon ER, Hamilton SR, Kern SE, Preisinger AC, Leppert M et al (1988) Genetic alterations during colorectal-tumor development. N Engl J Med 319(9):525–532
6. Aaltonen LA, Salovaara R, Kristo P, Canzian F, Hemminki A, Peltomaki P et al (1998) Incidence of hereditary nonpolyposis colorectal cancer and the feasibility of molecular screening for the disease. N Engl J Med 338(21): 1481–1487
7. Greenson JK, Bonner JD, Ben-Yzhak O, Cohen HI, Miselevich I, Resnick MB et al (2003) Phenotype of microsatellite unstable colorectal carcinomas: Well-differentiated and focally mucinous tumors and the absence of dirty necrosis correlate with microsatellite instability. Am J Surg Pathol 27(5):563–570
8. Gryfe R, Kim H, Hsieh ET, Aronson MD, Holowaty EJ, Bull SB et al (2000) Tumor microsatellite instability and clinical outcome in young patients with colorectal cancer. N Engl J Med 342(2):69–77
9. Gryfe R, Swallow C, Bapat B, Redston M, Gallinger S, Couture J (1997) Molecular biology of colorectal cancer. Curr Probl Cancer 21(5):233–300
10. Thibodeau SN, French AJ, Roche PC, Cunningham JM, Tester DJ, Lindor NM et al (1996) Altered expression of hMSH2 and hMLH1 in tumors with microsatellite instability and genetic alterations in mismatch repair genes. Cancer Res 56(21):4836–4840
11. Lothe RA, Peltomaki P, Meling GI, Aaltonen LA, Nystrom-Lahti M, Pylkkanen L et al (1993) Genomic instability in colorectal cancer: relationship to clinicopathological variables and family history. Cancer Res 53(24):5849–5852
12. Thibodeau SN, Bren G, Schaid D (1993) Microsatellite instability in cancer of the proximal colon. Science 260(5109):816–819
13. Anthoney DA, McIlwrath AJ, Gallagher WM, Edlin AR, Brown R (1996) Microsatellite instability, apoptosis, and loss of p53 function in drug-resistant tumor cells. Cancer Res 56(6):1374–1381
14. Elsaleh H, Joseph D, Grieu F, Zeps N, Spry N, Iacopetta B (2000) Association of tumour site and sex with survival benefit from adjuvant chemotherapy in colorectal cancer. Lancet 355(9217):1745–1750
15. Fink D, Aebi S, Howell SB (1998) The role of DNA mismatch repair in drug resistance. Clin Cancer Res 4(1):1–6
16. Hendriks YM, de Jong AE, Morreau H, Tops CM, Vasen HF, Wijnen JT et al (2006) Diagnostic approach and management of Lynch syndrome (hereditary nonpolyposis colorectal carcinoma): a guide for clinicians. CA Cancer J Clin 56(4):213–225
17. Boland CR, Thibodeau SN, Hamilton SR, Sidransky D, Eshleman JR, Burt RW et al (1998) A National Cancer Institute workshop on microsatellite instability for cancer detection and familial predisposition: development of international criteria for the determination of microsatellite instability in colorectal cancer. Cancer Res 58(22):5248–5257
18. Bocker T, Diermann J, Friedl W, Gebert J, Holinski-Feder E, Karner-Hanusch J et al (1997) Microsatellite instability analysis: a multicenter study for reliability and quality control. Cancer Res 57(21):4739–4743
19. Dietmaier W, Wallinger S, Bocker T, Kullmann F, Fishel R, Ruschoff J (1997) Diagnostic microsatellite instability: definition and correlation with mismatch repair protein expression. Cancer Res 57(21):4749–4756
20. Umar A, Boland CR, Terdiman JP, Syngal S, de la Chapelle A, Ruschoff J et al (2004) Revised Bethesda Guidelines for hereditary nonpolyposis colorectal cancer (Lynch syndrome) and microsatellite instability. J Natl Cancer Inst 96(4):261–268
21. Suraweera N, Duval A, Reperant M, Vaury C, Furlan D, Leroy K et al (2002) Evaluation of tumor microsatellite instability using five quasimonomorphic mononucleotide repeats and pentaplex PCR. Gastroenterology 123(6): 1804–1811
22. Perucho M, (1999) Correspondence re: C.R. Boland et al, A National Cancer Institute workshop on microsatellite instability for cancer detection and familial predisposition: development of international criteria for the determination of microsatellite instability in colorectal cancer. Cancer Res 58:5248–5257, 1998. Cancer Res 59(1):249–256
23. Bacher JW, Flanagan LA, Smalley RL, Nassif NA, Burgart LJ, Halberg RB et al (2004) Development of a fluorescent multiplex assay for detection of MSI-High tumors. Dis Markers 20(4–5):237–250
24. Zhou XP, Hoang JM, Cottu P, Thomas G, Hamelin R (1997) Allelic profiles of mononucleotide repeat microsatellites in control individuals and in colorectal tumors with and without replication errors. Oncogene 15(14):1713–1718
25. Berginc G, Glavac D (2009) Rapid and accurate approach for screening of microsatellite unstable tumours using quasimonomorphic mononucleotide repeats and denaturing high performance liquid chromatography (DHPLC). Dis Markers 26(1):19–26
26. Buhard O, Suraweera N, Lectard A, Duval A, Hamelin R (2004) Quasimonomorphic mononucleotide repeats for high-level microsatellite instability analysis. Dis Markers 20(4–5):251–257
27. Goel A, Nagasaka T, Hamelin R, Boland CR (2010) An optimized pentaplex PCR for detecting DNA mismatch repair-deficient colorectal cancers. PLoS ONE 5(2):e9393

28. Nardon E, Glavač D, Benhattar J, Groenen P, Höfler G, Höfler H et al (2010) A multicenter study to validate the reproducibility of MSI testing with a panel of five quasimonomorphic mononucleotide repeats. Diagn Mol Pathol 19(4):236–242

29. Shibata DK, Arnheim N, Martin WJ (1988) Detection of human papilloma virus in paraffin-embedded tissue using the polymerase chain reaction. J Exp Med 167(1):225–230

30. Muller A, Giuffre G, Edmonston TB, Mathiak M, Roggendorf B, Heinmoller E et al (2004) Challenges and pitfalls in HNPCC screening by microsatellite analysis and immunohistochemistry. J Mol Diagn 6(4):308–315

31. Saiki RK, Gelfand DH, Stoffel S, Scharf SJ, Higuchi R, Horn GT et al (1988) Primer-directed enzymatic amplification of DNA with a thermostable DNA polymerase. Science 239(4839):487–491

32. Sieben NL, ter Haar NT, Cornelisse CJ, Fleuren GJ, Cleton-Jansen AM (2000) PCR artifacts in LOH and MSI analysis of microdissected tumor cells. Hum Pathol 31(11): 1414–1419

33. Bovo D, Rugge M, Shiao YH (1999) Origin of spurious multiple bands in the amplification of microsatellite sequences. Mol Pathol 52(1):50–51

General Protocol for Loss of Heterozygosity Detection

29

Damjan Glavač and Ermanno Nardon

Contents

D. Glavač (✉)
Department of Molecular Genetics, Institute of Pathology,
University of Ljubljana, Ljubljana, Slovenia

E. Nardon
Department of Medical, Surgical and Health Sciences,
University of Trieste, Trieste, Italy

29.1 Introduction and Purpose

Cancer cells often show losses, gains, or rearrangements of chromosomes, and most often the loss of genetic material involves loci harbouring tumour suppressor genes (TSGs) [1]. Many techniques have been developed over time to assess gains and losses of genetic material. Among these, Fluorescence In Situ Hybridization (FISH), Comparative Genomic Hybridization (CGH), and Multiplex Ligation-dependent Probe Amplification (MLPA) can now be consistently used also in formalin-fixed and paraffin-embedded (FFPE) material.

Because of their wide distribution and polymorphic length in a given population, microsatellite sequences (MS) are suitable as markers of deletion at specific chromosomal loci. In fact, the loss of one marker allele but the retention of the other one, the so-called loss of heterozygosity (LOH), is rather frequent because in both sporadic and inherited cancers, the other hit is usually a point mutation [2]. LOH has been extensively adopted for identification of TSGs by deletion mapping and is currently used in the clinical setting as an indicator for disease stratification, prognosis, and response to therapy [3–7]. In designing an LOH assay, prior knowledge of the Smallest Region of Overlapping (SRO) deletions is required and at least two MS markers within this region should be chosen. MS should be polymorphic enough and with elevated heterozygosity (index ≥0.7)[1] in order

---

[1]Heterozygosity refers to the fraction of individuals in a population that is heterozygous for a particular locus. The expected heterozygosity (He) for a given locus is calculated as

$$He = 1 - \sum_{i=1}^{m} (fi)^2 \qquad (29.1)$$

where m is the number of alleles at the target locus, and fi is the allele frequency of the ith allele at the target locus.

G. Stanta (ed.), *Guidelines for Molecular Analysis in Archive Tissues*,
DOI: 10.1007/978-3-642-17890-0_29, © Springer-Verlag Berlin Heidelberg 2011

to minimize the number of uninformative cases where the constitutional DNA is homozygous for the marker. Microsatellite markers can be chosen consistently with the requirements above by referring to suitable web resources.[2] In order to get more reliable results, tri- and tetranucleotide repeats are preferred to dinucleotide repeats as the latter are peculiarly prone to replication slippage during PCR amplification. Since the major problem with FFPE tissues is the extent of degradation of the extracted nucleic acids, it is important to design primers that produce a PCR product shorter than 200–250 bp [8]. Moreover, it is preferable to avoid the choice of markers in which the larger and the smaller allele are too different in size. This strategy is used to avoid the preferential amplification of the shorter allele, which could be interpreted as a deletion of the larger one [9, 10].

This protocol offers a method for LOH assessment in DNA obtained from FFPE samples by providing a standard workflow for PCR amplification and detection of PCR products. Since DNA from FFPE could be quite variable in quality, a control PCR for the assessment of amplifiability and integrity of sample DNA is also suggested. Herein two technical approaches to LOH analysis are proposed:

- A basic method involving singleplex amplification of two microsatellite markers mapping at the 9p21 locus (CDKN2A) and subsequent poliacrylamide gel electrophoresis (PAGE) run and silver stain detection of PCR products;
- A singleplex PCR amplification of the same two markers coupled with capillary electrophoresis analysis of PCR products.

29.2 DNA Extraction from FFPE for LOH Analysis

Total DNA can be extracted following the protocol described in microsatellite instability analysis.

---

[2]Genomic position of the marker can be retrieved from the NCBI GenBank database at http://www.ncbi.nlm.nih.gov/nucleotide/ (Accessed 31 March 2010). Heterozygosity and allele frequencies can be retrieved from the CEPH database at http://www.cephb.fr/en/cephdb/browser.php (Accessed 31 March 2010).

29.2.1 Precautions

For LOH analysis, DNA from both the tumoral and normal tissue of the same patient is required. If both normal and tumour tissues are present in the same paraffin block, microdissection is necessary. Since LOH can be missed because of contamination from normal tissue, a minimal tumour fraction of at least 70% is suggested [11]. To this purpose, manual microdissection by exclusion [12] or laser capture microdissection should be the methods of choice (see Chaps. 4 and 6).

29.3 LOH Analysis: Basic Method

Microsatellite markers amplifications are performed as singleplex PCR. PCR products are then separated on an acrylamide gel, visualized by means of silver staining, and quantified by the use of a gel imaging system. Herein a method for the detection of LOH at the 9p21 locus (CDKN2A) [13] is described. Genetic alterations involving the 9p21 region are common in human cancers, and LOH assessment of this locus may be relevant for prognosis in lung adenocarcinoma and head and neck cancer patients [14].

29.3.1 PCR Amplification

29.3.1.1 Reagents

- *10X PCR buffer with MgCl$_2$*: We report the standard composition of PCR buffer: 15 mM MgCl$_2$, 500 mM KCl, 100 mM Tris pH 8.3 at 25°C
- *dNTPs stock solution pH 8, 100 mM each* (e.g., Amersham): Dilute the dNTP stock solution to prepare 10 mM solution of each dNTP in sterile water
- *Primers*: Lyophilized primers should be dissolved in a small volume of distilled water or 10 mM Tris pH 8 to make a concentrated stock solution. Prepare small aliquots of working solutions containing 30 pmol/µl to avoid repeated thawing and freezing. Store all primer solutions at –20°C (see Table 29.1)
- *AmpliTaq Gold* (e.g., Applied Biosystems, N8080247), 5 U/µl

Table 29.1 LOH primers[a]

Marker name	Genomic position	Sequence 5′–3′	T°m	Range(bp)	Het.[b]
IFNA	9p22	F: TGCGCGTTAAGTTAATTGGTT	61.7	138–150	0.72
		R: GTAAGGTGGAAACCCCCACT	62.3		
D9S171	9p21	F: AGCTAAGTGAACCTCATCTCTGTCT	61.2	159–177	0.80
		R: ACCCTAGCACTGATGGTATAGTCT	59.5		

[a]Purchased from Sigma, lyophilized. Resuspended at 300 pmol/μl in 10 mM Tris pH 8.

[b]Heterozygosity, see footnote 1

F forward primer, *R* reverse primer

- *Negative control* for LOH PCR amplification: see footnote 4 in Chap. 28
- *Positive control* for LOH PCR amplifications: DNA from normal human lymphocytes, 50 ng/μl[3]

29.3.1.2 Equipment

- Adjustable pipettes; range: 2–20 μl, 20–200 μl, 100–1,000 μl
- *Nuclease-free aerosol-resistant pipette tips*
- *0.2 and 1.5 ml nuclease-free microtubes*
- *Sterile laminar flow hood*
- *Tabletop refrigerated centrifuge* suitable for centrifugation of 0.2 ml tubes
- *Thermal cycler*
- *Electrophoresis unit for agarose gel*
- *UV transillumination unit*

29.3.1.3 Method

Operate in a sterile laminar flow hood. Prepare a different master mix for each microsatellite marker (singleplex). PCR is performed in a final volume of 50 μl, containing:

- 5 μl 10X PCR Buffer.................... 1X final
- 1 μl dNTP 10 mM each................ 0.2 mM final
- 0.5 μl forward primer, 30 pmol/μl.................................. 0.3 pmol/μl final
- 0.5 μl reverse primer, 30 pmol/μl .... 0.3 pmol/μl final
- 0.25 μl AmpliTaq Gold, 5 U/μl.... 0.025 U/μl final

- 1 μl of diluted sample DNA
  or
- 1 μl of undiluted negative control or
- 1 μl of positive control for amplification.
- H$_2$O to volume
  - Overlay reaction mixture with 20 μl of mineral oil.
  - Thermal cycling: 94°C 10′+5 × (94°C 60″, 55°C 60″, 72°C 60″)+40 × (94°C 30″, 55°C 30″ 72°C 30″)+72°C 5′.
  - Gel visualization: mix 10 μl of PCR product with 2 μl of 6× loading buffer[4]; load on a 2% agarose gel prepared with TBE 1×, containing 0.5μg/ml ethidium bromide.[5] Include a 100 bp marker ladder (e.g., Amersham 27400701). Run at 80 V constant until bromophenol blue reaches 1/2 of the gel. Inspect under a UV source. A single band should be visible in the sample and in the positive control lanes.

29.3.2 Silver Stain Detection and Interpretation of the Results

29.3.2.1 Silver Stain Detection

The same reagents and equipment described in microsatellite instability analysis can be used (see Chap. 28). The white light transilluminator should be coupled with a gel imaging system (e.g., Bio-Rad Versadoc). Band staining is quantified by means of an optical

---

[3]The use of a high-quality DNA, e.g., a DNA extracted from a cell line, is recommended.

[4]6× loading buffer: 0.25% bromophenol blue, 0.25% xylene cyanol, 30% glycerol in H$_2$O.

[5]Ethidium bromide is a potentially carcinogenic compound. Always wear gloves. Used EtBr solutions must be collected in containers for chemical waste and discharged according to the local hazardous chemical disposal procedures.

density (O.D.) reader. Image acquisition and analysis should be performed according to the manifacturer's instructions.

29.3.2.2 Interpretation

We score an allele as lost if its band signal is reduced by at least 50% with respect to the other allele (Fig. 29.1) [15]. Microsatellite alleles might show very complex band patterns after PAGE separation and silver staining. Polymerase slippage during elongation generates, in addition to the main allele, products referred to as shadow-bands [16, 17]. Sometimes this can make the identification of the true allele cumbersome.

29.3.3 Considerations and Pitfalls

The complete loss of an allele is rarely found in LOH studies, because of contaminating normal tissue and/or genetic heterogeneity within the tumour. Furthermore, preferential amplification of shorter alleles may occur. For these reasons, LOH should be considered on a quantitative basis as a comparison between allelic

Fig. 29.1 Visualization of a representative case with LOH in a silver-stained polyacrylamide gel. An allele is scored as lost if its band signal is reduced by at least 50% with respect to the other allele (*N* normal tissue, *TU* tumour sample)

ratios in neoplastic tissue and in normal control (allelic imbalance, A.I.) [15].

$$A.I. = \frac{\text{O.D. tumor allele 1/O.D. tumor allele 2}}{\text{O.D. normal allele 1/O.D. normal allele 2}} \quad (29.2)$$

When this ratio gives a value higher than one, A.I. is set in an inverted form. A.I. ranges from 0 to 1, indicating a condition from total loss to retained heterozygosity, respectively [11].

Quantification of nucleic acids by means of silver stain should be cautiously evaluated, particularly regarding stain saturation which could result in overestimation of weaker bands. For this reason, [α-^{32}P] dCTP labeling of nucleic acids or primer coupling with fluorochromes in a genescan setting should be preferred for the quantitative analysis of PCR products. Moreover, thanks to the greater sensitivity of the two latter methods with respect to silver staining, PCR amplification can be performed using a minor number of cycles. This allows an easier interpretation of band pattern, as intensity of the shadow or stutter bands decreases by reducing the numbers of PCR cycles [16].

29.3.3.1 Troubleshooting

One or both microsatellite markers fail to amplify: repeat amplification using a different quantity of DNA. Perform the amplification of a control gene (e.g., β-globin, see the protocol for Microsatellite Instability) to check the amplifiability of the sample DNA and to establish the optimal quantity for amplification. Repeat LOH analysis running the sample in duplicate/triplicate. If only one of two/three replicas amplifies, stochastic amplification is suspected and results should be cautiously evaluated. If no replica amplifies, a homozygous deletion of the entire locus can be suspected.

29.4 LOH Analysis: Capillary Electrophoresis Method

Generally, LOH analysis should be performed by methods allowing quantification of both size and extent of amplification products. Fluorescence-based PCR analyzed by capillary electrophoresis optimally fulfills these criteria. The same reagents and equipment described in microsatellite instability analysis with capillary electrophoresis can be used.

Table 29.2 LOH primers

Marker name	Genomic position	Sequence	T°m	Range(bp)	Het.[a]
IFNA	9p22	F: HEX-TGCGCGTTAAGTTAATTGGTT	61.7	138–150	0.72
		R: GTAAGGTGGAAACCCCCACT	62.3		
D9S171	9p21	F: FAM-AGCTAAGTGAACCTCATCTCTGTCT	61.2	159–177	0.80
		R: ACCCTAGCACTGATGGTATAGTCT	59.5		

[a]Heterozygosity, see footnote 1

F forward primer, *R* reverse primer

29.4.1 PCR Reaction

29.4.1.1 Reagents

Note: Reagents from specific companies are reported here, but reagents of equal quality purchased from other companies may be used

- *10X PCR HotMaster™ Buffer with 25 mM MgCl₂* (5 Prime GmbH)
- *Commercial 25 mM stock solution of each dNTP* (Applied Biosystems): dilute the dNTP stock solution to prepare 2.5 mM solution of each dNTP in sterile water.
- *Primers*: Lyophilized primers should be dissolved in a small volume of distilled water or 10 mM Tris pH 8 to make a concentrated stock solution. Prepare small aliquots of working solutions containing 5 pmol/μl to avoid repeated thawing and freezing. Store all primer solutions at –20°C (see Table 29.2).
- *HotMaster™ DNA Taq polymerase* (5 Prime GmbH), 5 U/μl
- *Negative control* for LOH PCR amplification: see footnote 4 in Chap. 28.
- *Positive control* for LOH PCR amplifications: DNA from normal human lymphocytes, 50 ng/μl.[6]

29.4.1.2 Equipment

- *Adjustable pipettes*: range: 2–20 μl, 20–200 μl, 100–1,000 μl
- *Nuclease-free aerosol-resistant pipette tips*
- *0.2 and 1.5 ml nuclease-free microtubes*
- *Sterile laminar flow hood*
- *Tabletop refrigerated centrifuge* suitable for centrifugation of 0.2 ml tubes
- *Thermal cycler*

- *Electrophoresis unit for agarose gel*
- *UV transillumination unit*

29.4.1.3 Method

Operate in a sterile laminar flow hood. Prepare a singleplex master mix for each microsatellite marker.

- PCR is performed in a final volume of 20 μl, containing:
  - 4 μl 10X PCR HotMaster™ Buffer..2X final
  - 2.4 μl dNTP 2.5 mM each,......3 mM final
  - 0.4 μl Forward primer, 5 pmol/μl, ...0.1 pmol/μl final
  - 0.4 μl Reverse primer, 5 pmol/μl ....0.1 pmol/μl final
  - 0.6 μl HotMasterTM DNA Taq polymerase 5 U/μl ....0.15 U/μl final
  - 3 μl of 10 ng/μl DNA sample
  - or
  - 1 μl of undiluted negative control-or
  - 1 μl of positive control for amplification
  - H₂O to volume
- Thermal cycling: 94°C 2'+35 × (94°C 20", 55°C 10", 65°C 30")+65°C 7'.
- Gel visualization: Mix 10 μl of PCR product with 2 μl of 6× loading buffer[7]; load on a 2% agarose gel prepared with TBE 1×, containing 0.5μg/ml ethidium bromide. Include a 100 bp marker ladder (e.g., Amersham 27400701). Run at 80 V constant until bromophenol blue reaches 1/2 of the gel. Inspect under a UV source. One band should be visible in the sample and in the positive control lanes.

---

[6]The use of a high-quality DNA, e.g., a DNA extracted from a cell line or blood (see Chap. 10), is recommended.

[7]See Footnote 4.

29.4.2 Detection by Capillary Electrophoresis and Interpretation of Results

29.4.2.1 Detection by capillary electrophoresis

Reagents, equipment, default run conditions and sample preparation are the same as in the case of microsatellite instability analysis (see Chap. 28).

29.4.2.2 Interpretation

Raw data are analyzed by GeneScan™ software and peak heights are obtained (Fig. 29.2). Loss of the alleles can be precisely determined by calculating the ratio of the peak heights of normal and tumour alleles according to the following formula:

$$LOH = \frac{\text{Peak height of normal allele 2} / \text{Peak height of normal allele 1}}{\text{Peak height of tumour allele 2} / \text{Peak height of tumour allele 1}} \quad (29.3)$$

When this ratio gives a value higher than one, LOH is defined as the reciprocal of the formula above.

LOH is strongly indicated by ratios less than 0.5.

Polymerase slippage during elongation generates products referred to as stutter-peaks. Additional fragments are one to four repeat units shorter than the allele, and when the size of the two alleles differs by one repeat unit, the stutter from the longer allele will contribute significantly to the main peak of the short allele. Such contribution from the stutter peak can be corrected. The height of a stutter peak compared to its main peak is calculated and then the main peak's height, as it would have been without contribution from the neighboring allele's stutter, can be estimated [18].

29.4.2.3 Troubleshooting

- The detected signal is too low to evaluate the results. Solution: increase the injection time or decrease the dilution of the PCR products.
- The detected signals go off-scale. Solution: decrease the injection time or increase the dilution of the PCR products.
- No peaks are visible on electropherogram after separation:
  - PCR failed to amplify microsatellite marker. Solution: optimize PCR conditions
  - Fragments reach the detector outside the detection time. Solution: adjust the detection time.

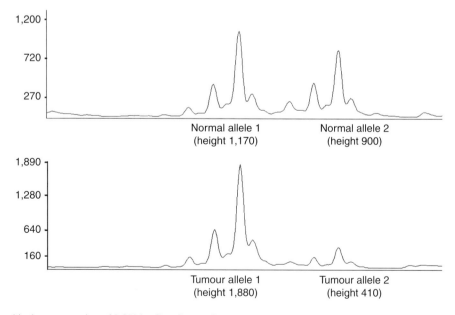

Fig. 29.2 Graphical representation of LOH by GeneScan software. LOH is calculated as the ratio between the allele ratios in tumour and normal DNA. The peak heights are measured in relative fluorescence units. In this example,

$LOH = (900 / 1,170) / (410 / 1,880) = 3.52$, but since the ratio is greater than 1, the LOH value is set to be the inverse. LOH=0.28 is showing a reduction of one allele's intensity, from normal to tumour DNA, by 72% relative to the other allele

References

1. Lasko D, Cavenee W, Nordenskjold M (1991) Loss of constitutional heterozygosity in human cancer. Annu Rev Genet 25:281–314

2. Strachan T, Read A (1999) Human molecular genetics, 2nd edn. Wiley, New York

3. De Schutter H, Spaepen M, Mc Bride WH, Nuyts S (2007) The clinical relevance of microsatellite alterations in head and neck squamous cell carcinoma: a critical review. Eur J Hum Genet 15(7):734–741

4. Fornari D, Steven K, Hansen AB, Jepsen JV, Poulsen AL, Vibits H, Horn T (2006) Transitional cell bladder tumor: predicting recurrence and progression by analysis of microsatellite loss of heterozygosity in urine sediment and tumor tissue. Cancer Genet Cytogenet 167(1):15–19

5. Fromont G, Valeri A, Cher M, Pontes JE, Vallancien G, Validire P, Latil A, Cussenot O (2005) Allelic loss at 16q23.2 is associated with good prognosis in high grade prostate cancer. Prostate 65(4):341–346

6. Jen J, Kim H, Piantadosi S, Liu ZF, Levitt RC, Sistonen P, Kinzler KW, Vogelstein B, Hamilton SR (1994) Allelic loss of chromosome 18q and prognosis in colorectal cancer. N Engl J Med 331(4):213–221

7. Wemmert S, Ketter R, Rahnenfuhrer J, Beerenwinkel N, Strowitzki M, Feiden W, Hartmann C, Lengauer T, Stockhammer F, Zang KD, Meese E, Steudel WI, von Deimling A, Urbschat S (2005) Patients with high-grade gliomas harboring deletions of chromosomes 9p and 10q benefit from temozolomide treatment. Neoplasia 7(10): 883–893

8. Lehmann U, Kreipe H (2001) Real-time pcr analysis of DNA and rna extracted from formalin-fixed and paraffin-embedded biopsies. Methods (San Diego, Calif) 25(4): 409–418

9. Giercksky HE, Thorstensen L, Qvist H, Nesland JM, Lothe RA (1997) Comparison of genetic changes in frozen biopsies and microdissected archival material from the same colorectal liver metastases. Diagn Mol Pathol 6(6):318–325

10. Sieben NL, ter Haar NT, Cornelisse CJ, Fleuren GJ, Cleton-Jansen AM (2000) Pcr artifacts in loh and msi analysis of microdissected tumor cells. Hum Pathol 31(11): 1414–1419

11. Skotheim RI, Diep CB, Kraggerud SM, Jakobsen KS, Lothe RA (2001) Evaluation of loss of heterozygosity/allelic imbalance scoring in tumor DNA. Cancer Genet Cytogenet 127(1):64–70

12. Baisse B, Bian YS, Benhattar J (2000) Microdissection by exclusion and DNA extraction for multiple pcr analyses from archival tissue sections. Biotechniques 28(5):856–858, 860, 862

13. An HX, Niederacher D, Picard F, van Roeyen C, Bender HG, Beckmann MW (1996) Frequent allele loss on 9p21-22 defines a smallest common region in the vicinity of the cdkn2 gene in sporadic breast cancer. Genes Chromosom Cancer 17(1):14–20

14. Tanaka R, Wang D, Morishita Y, Inadome Y, Minami Y, Iijima T, Fukai S, Goya T, Noguchi M (2005) Loss of function of p16 gene and prognosis of pulmonary adenocarcinoma. Cancer 103(3):608–615

15. Zauber NP, Sabbath-Solitare M, Marotta SP, McMahon L, Bishop DT (1999) Comparison of allelic ratios from paired blood and paraffin-embedded normal tissue for use in a polymerase chain reaction to assess loss of heterozygosity. Mol Diagn 4(1):29–35

16. Bovo D, Rugge M, Shiao YH (1999) Origin of spurious multiple bands in the amplification of microsatellite sequences. Mol Pathol 52(1):50–51

17. Smeets HJ, Brunner HG, Ropers HH, Wieringa B (1989) Use of variable simple sequence motifs as genetic markers: application to study of myotonic dystrophy. Hum Genet 83(3):245–251

18. Guo Z, Wu F, Asplund A, Hu X, Mazurenko N, Kisseljov F, Ponten J, Wilander E (2001) Analysis of intratumoral heterogeneity of chromosome 3p deletions and genetic evidence of polyclonal origin of cervical squamous carcinoma. Mod Pathol 14(2):54–61

Qualitative Methylation Status Assessment

30

Damjan Glavač and Ermanno Nardon

Contents

D. Glavač (✉)
Department of Molecular Genetics, Institute of Pathology,
University of Ljubljana, Ljubljana, Slovenia

E. Nardon
Department of Medical, Surgical and Health Sciences,
University of Trieste, Trieste, Italy

30.1 Introduction and Purpose

In the human genome, approximately 5% of cytosines are modified to 5-methylcytosine. This phenomenon occurs only at a CpG dinucleotide; it is a normal part of the development and plays a crucial role in imprinting and X chromosome inactivation [1, 2]. Aberrant promoter methylation is frequently observed in carcinogenesis and is linked to transcriptional inactivation of tumour suppressor genes [3–5]. Methylation detection has been exploited as a marker of tumour cell presence in body fluids or in the bloodstream, for disease stratification and prognosis, and as a predictive tool of response to therapy [6]. Many techniques have been developed to detect the presence of methylation at a given sequence level [6, 7]. Among them, bisulfite conversion-based methods are the most widely used, because they allow the rapid identification of methylated cytosines in any sequence context [6, 8]. After bisulfite modification (BM), unmethylated cytosine residues are deaminated to uracil, whereas methylated cytosine residues are retained as cytosine at CpG sites. Bisulfite modification suffers from two main drawbacks: extensive DNA degradation and incomplete conversion of unmethylated cytosines [9]. Harsher modification conditions (long incubation, high temperatures, high bisulfite molarity) assure complete conversion of cytosine to uracil, but DNA can be degraded to a degree that makes PCR amplification impossible [9]. This is a main concern especially with DNA from formalin-fixed and paraffin-embedded (FFPE) tissues, which is quite variable in quality and requires less aggressive treatments [10]. Methylation-specific PCR (MSP) is the most widely used method to detect the presence of determined methylation patterns with very high sensitivity (detection limit>1:100) [6, 11]. It exploits the specificity of primers for either

G. Stanta (ed.), *Guidelines for Molecular Analysis in Archive Tissues*,
DOI: 10.1007/978-3-642-17890-0_30, © Springer-Verlag Berlin Heidelberg 2011

methylated or unmethylated alleles before chemical conversion. However, since methylation-specific primers always contain cytosines of a CpG site and do not contain cytosines of non-CpG sites at the 3′ position, they may anneal to unconverted or partially unconverted sequences in the bisulfite-treated DNA. This can result in a high rate of false positives especially under mild BM conditions, which is often the case in FFPE tissues [12]. This problem may be overcome by means of a nested PCR design, using primers that do not contain any CpG but only cytosines of non-CpG sites at the 3′ position in the first round of amplification, so that only fully converted sequences can be amplified. The second step uses the conventional MSP, "diagnostic" primer sets.

Recently, a new approach for methylation analysis has been introduced. Methylation-specific multiplex ligation-dependent probe amplification (MS-MLPA) technique is a rapid and easy method that requires only 20 ng of DNA, eliminates the need for bisulfite conversion, and is able to detect changes in up to 40 selected sequences at a time. The discrimination of aberrantly methylated CpG is performed by methylation-specific restriction enzyme HhaI, which cleaves unmethylated CpG island. The method is based on probes that recognize target CpG sequences. Each probe is made of two parts, one short and one long, both containing a target recognition sequence and a universal primer sequence. In addition, the long part contains a stuffer sequence, which is of different length among used probes. After hybridization and simultaneous ligation/digestion reaction a multiplex PCR with universal fluorescently labeled primers is performed. If CpG island is methylated, the restriction enzyme does not cleave the sequence; therefore ligation of both parts of the probe occurs. In the amplification step, a PCR product is formed and can be detected. If CpG island is not methylated, cleavage of target sequence occurs, all subsequent reactions are blocked, and no PCR products are obtained after amplification. This method detects short sequences (~65 bp) and requires low amount of DNA, which is particularly suitable for FFPE (Fig. 30.1) [13].

Herein is proposed a nested, BM-specific MSP design for the qualitative detection of MGMT gene promoter methylation and multilocus MS-MLPA

approach as a rapid, accurate, and simple alternative to conventional methods.

The general frame of BM-specific MSP can be applicable for the methylation assessment of virtually any sequence of interest in the genome. The only restraint could be given by the difficulty of designing consistent primer pairs for BM sequences, especially for those very rich in CpG. The method was previously validated on a case study of lung cancer patients [14] and is currently in use in a laboratory of the IMPACTS group in a diagnostic setting. Along with this method, a protocol for the DNA extraction and for the BM is suggested.

MS-MLPA can be used for detection of any CpG site in the genome that contains HhaI restriction site. Several probe mixes are commercially available; however, one can design probes that correspond to one's region of interest.

30.2 DNA Extraction from FFPE for Methylation Analysis

This protocol can be used for DNA preparation both for BM-specific MSP and MS-MLPA.

The following procedure is an adaptation of a method previously published by researchers from one institution of the IMPACTS group [15]. As protein removal is essential for BM effectiveness [9, 10, 16], the protocol involves an extensive Proteinase K digestion followed by a phenol-chloroform extraction. The time required for the whole procedure is 3 days.

30.2.1 Reagents

Note: Reagents from specific companies are reported here, but reagents of equal quality purchased from other companies may be used.

- *Xylene* (Fluka or Sigma-Aldrich)
- *100% and 70% Methanol* (Fluka or Sigma-Aldrich)
- *Lysis stock solution 10×*: 500 mM KCl, 150 mM Tris-HCl, pH 8, 15 mM $MgCl_2$, Tween 20 (Sigma) 1.5%

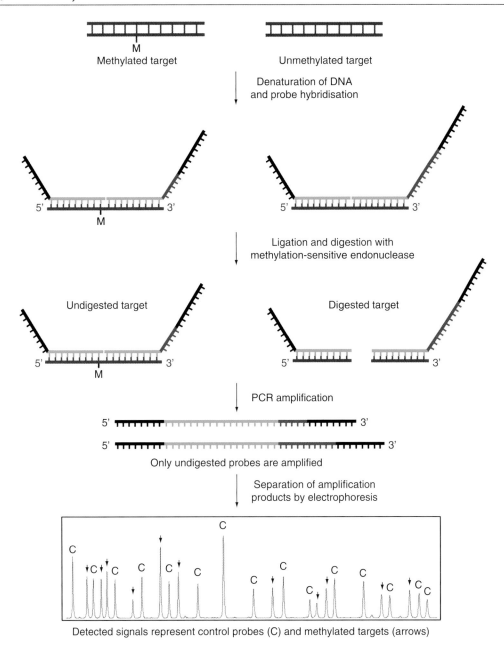

Fig. 30.1 Scheme of methylation-specific multiplex ligation-dependent probe amplification (*MS-MLPA*) technique. MS-MLPA detects the presence of specific methylated sites in a single stranded target nucleic acid and simultaneously the quantification of the target nucleic acid sequence can be performed. The method employs a plurality of probe sets of at least two probes, each including a target-specific region (*in green*) and a noncomplementary region containing a primer binding site (*in black*). The target-specific region of at least one of the probes also includes the sequence of one of the strands of a double stranded recognition site of a methylation-sensitive restriction enzyme. The probes belonging to the same set are ligated together when hybridized to the target nucleic acid sequence. The hybrid is then subjected to digestion by the methylation-sensitive restriction enzyme, resulting in nonmethylated recognition sites being cleaved. The probes of the uncleaved (methylated) hybrid are subsequently amplified by a suitable primer set and separated by electrophoresis according to the length of stuffer sequences (*in blue*)

- *Proteinase K (Sigma P2303)* stock solution, 20 mg/ml. Dissolve Proteinase K in 50% sterile glycerol[1] in DNA-grade water. Store at –20°C
- *Phenol*, equilibrated with 10 mM Tris HCl, pH 8 (Sigma P4557)
- *Chloroform* (Sigma)
- *DNA precipitating solution*: 100% EtOH, 3 M Sodium Acetate
- *Glycogen* (Sigma), diluted at 1 mg/ml in H_2O and used as precipitation carrier
- *70% Ethanol*
- *10× TE buffer*: 100 mM Tris HCl, pH 8, 10 mM EDTA

30.2.2 Equipment

- *Adjustable pipettes*, range: 2–20 μl, 20–200 μl, 100–1,000 μl
- *Nuclease-free aerosol-resistant pipette tips*
- *1.5 ml nuclease-free autoclaved tubes*
- *Microtome, with new blades*
- *Chemical hood*
- *Tabletop refrigerated centrifuge* suitable for centrifugation of 1.5 ml tubes
- *Thermomixer*
- *Spectrophotometer*

30.2.3 Method

- Using a clean, sharp microtome blade,[2] cut five to ten sections[3] of about 5 μm thickness. Discard the first section and displace the other ones in a clean 1.5 ml tube.[4]

- Add 1 ml of xylene,[5] vortex for 10″, and then incubate the tube at RT for approximately 5 min.[6]
- Spin the tube for 10 min at maximum speed (14,000 rpm) in a microcentrifuge and then carefully remove and discard the supernatant.[7] Repeat steps 2–3 with a fresh aliquot of xylene.
- Wash the pellet by adding 1 ml of 100% methanol.[8] Flick the tubes to dislodge the pellet and then vortex the tubes for 10 s. Incubate at RT for approximately 5 min.
- Spin the tube for 10 min at maximum speed (14,000 rpm) in a microcentrifuge and then carefully remove and discard the supernatant.
- Repeat steps 4–5 using first 100% methanol and then 70% methanol
- After removing 70% methanol, allow the tissue pellet to air dry at 37°C for 15–30 min.
- Complete the digestion solution by mixing 100 μl of 10× Lysis solution, 100 μl of 10× TE, 50 μl of 20 mg/ml Proteinase K (final conc. 1 mg/ml), and 775 μl of sterile water/1 ml of final solution. Keep in ice until use.
- Add 200–400 μl of complete digestion solution, depending on the pellet size. Digestion buffer must cover the tissue pellet completely. Pipette the solution until the pellet is resuspended.
- Incubate for at least 24 h in a thermomixer at 55°C shaking moderately.[9]
- Spin down the tubes to collect the condensate from the lids before opening. Add one volume of a 1:1 mixture of phenol-chloroform per volume of digestion solution. Shake vigorously. Leave in ice for 10′.
- Spin the tube for 10 min at maximum speed (14,000 rpm) at 4°C in a tabletop refrigerated microcentrifuge.

---

[1]The solubilization of proteinase K in 50% sterile glycerol avoids the freezing effect at –20°C, maintaining an optimal enzymatic activity.

[2]Clean the microtome with xylene. Cool the paraffin blocks at –20°C or on dry ice in aluminum foil to obtain thin sections.

[3]The number of sections may vary, according to the tissue type and quantity.

[4]Use some sections from a paraffin block without any tissue for the negative controls.

[5]When working with xylene, avoid breathing fumes; it is better to perform the deparaffinization step under a chemical hood.

[6]Wear gloves when isolating and handling DNA to minimize the activity of endogenous nucleases.

[7]Discard the supernatant using a 1micropipette or a glass Pasteur pipette. Disposal of chemicals should be done in keeping with your laboratory safety rules.

[8]Methanol is toxic by inhalation. Work under a chemical hood.

[9]A 48-h digestion might be beneficial to bisulfite modification when contamination from melanin is expected. In such a case, a further Proteinase K addition is advisable after 24 h.

- Transfer the upper, aqueous phase to a new 1.5 ml tube. Avoid touching the interphase disc that contains most of the protein residues.
- Add 5 μl of 1 mg/ml Glycogen solution and three volumes of DNA precipitating solution per volume of recovered aqueous phase.
- Allow DNA precipitation overnight at −20°C.
- Spin the tube for 20 min at maximum speed (14,000 rpm) at 4°C. A small pellet should form at the bottom of the tube.
- Discard the supernatant. Do not touch the pellet.
- Add 100 μl of 70% ice-cold ethanol, without resuspending the pellet.
- Spin the tube for 5 min at maximum speed (14,000 rpm) at 4°C.
- Discard the supernatant. Leave the pellet to air dry at 37°C for almost 15–30 min.
- Resuspend the pellet in 22 μl of 1× TE buffer.
- Determine the DNA concentration spectrophotometrically at 260 and 280 nm. The concentration of dsDNA expressed in μg/μl is obtained as follows: $[DNA] = A_{260} \times dilution\,factor \times 50 \times 10^{-3}$. Absorbances ratio between readings at 260 and 280 nm should be ≥1.6, which is indicative of an acceptable level of contamination from proteins (see Chap. 16).
- Store the DNA at −20°C until use.

30.2.4 Troubleshooting

- If no pellet or "smear" is visible on the tube wall after precipitation, add another 5 μl of glycogen solution and repeat overnight precipitation at −20°C, then centrifuge.

30.3 Bisulfite Modification of DNA from FFPE

This protocol provides a homemade method for DNA modification suitable for subsequent BM-specific MSP PCR analyses. Chemical conditions of modification are relatively mild, in order to reduce the loss and further degradation of DNA. Since this method is suited to FFPE, it could not perform equally well with DNA from culture cells or from body fluids. The time required for the whole procedure is 2 days. Commercial kits for BM of DNA from FFPE are also available,

such as the Zymo EZ DNA methylation kit (Zymo Research), the EpiTect Bisulfite Kit (Qiagen) and the MethylCode™ Bisulfite Conversion Kit (Invitrogen).

30.3.1 Reagents

Note: Reagents from specific companies are reported here, but reagents of equal quality purchased from other companies may be used.

- *3 M NaOH* solution
- *Sodium metabisulfite*[10] (Sigma S1516). Moisture and light sensitive; store in the dark and under vacuum
- *Hydroquinone* (Sigma H9003). Moisture and light sensitive; store in the dark and under vacuum
- *Mineral oil* (Sigma M5904)
- *Binding solution*[11]: 6 M Sodium Iodide, 1 mM β-Mercaptoethanol in H_2O.
- *Herring sperm DNA* (Sigma D7290)
- *QIAexII resin* (Qiagen 20902)
- *Washing buffer:* 50 mM NaCl; 10 mM Tris-HCl, pH 7.5; 2.5 mM EDTA; 50% Ethanol
- *100% and 70% Ethanol*
- *5 M Ammonium acetate solution, pH 7.5*
- *1 mM Tris-HCl buffer, pH 8*

30.3.2 Equipment

- *Heating block*
- *pH meter*
- *Adjustable pipettes*, range: 2–20 μl, 20–200 μl, 100–1,000 μl
- *Nuclease-free aerosol-resistant pipette tips*
- *Sterile glass Pasteur pipettes*
- *15 or 50 ml sterile polypropylene tubes*
- *1.5 ml nuclease-free autoclaved tubes*
- *Chemical hood*
- *Tabletop refrigerated centrifuge suitable for centrifugation of 1.5 ml tubes*
- *Thermomixer (Eppendorf)*

---

[10]We prefer metabisulfite to sodium bisulfite. The precise molarity of the latter cannot be estimated as commercially available sodium bisulfite is a mixture of sodium bisulfite and metabisulfite.

[11]Store in the dark. This solution is stable for weeks at room temperature. Discard if it turns yellow.

30.3.3 Method

- In a sterile polypropylene tube, dissolve 3.8 g of sodium metabisulfite in 8 ml. of sterile H_2O. Bring this solution to pH 5 by adding ~600 µl of freshly made 3 M NaOH. Dissolve 0.22 g of hydroquinone in 10 ml of sterile H_2O.[12] Add 50 µl of the dissolved hydroquinone to the metabisulfite solution and bring to a final volume of 10 ml with H_2O. This results in a final 2 M metabisulfite, 1 mM hydroquinone solution.

- Dilute DNA from FFPE to 0.4 µg/µl in H_2O. In a sterile 1.5 ml tube, add 5 µl of diluted DNA[13] (2 µg) to 40 µl of H_2O. Add 5 µl of 3 M NaOH (final 0.3 M). Incubate at 42°C for 30 min in a heating block.

- Spin tubes briefly in a tabletop microcentrifuge to collect droplets from lids/walls. Add 650 µl of the metabisulfite-hydroquinone solution. At this point, prepare one tube containing all of the above-mentioned solutions but without sample DNA, as negative control. Overlay the mixture with 150 µl of mineral oil and incubate at 55°C for 4 h and 30′ in a heating block. Keep in the dark.

- By means of a glass Pasteur pipette, collect mixture from beneath the mineral oil and transfer it to a new 1.5 ml tube. Add 650 µl of binding solution, 2 µg of herring sperm DNA[14], and 5 µl of QIAexII resin. Mix thoroughly. Incubate for 10 min at RT, inverting tubes occasionally.

- Spin the tubes for 2 min at maximum speed (14,000 rpm) in a tabletop microcentrifuge. Discard the mixture, avoiding touching the resin pellet.

- Add 1 ml of the Washing buffer. Shake to resuspend the resin.

- Spin the tubes for 2 min at maximum speed. Discard the supernatant.

- Resuspend the resin pellet in 1 ml of 70% ethanol.

- Spin the tubes for 2 min at maximum speed. Repeat ethanol washing.

- Leave the pellet to air dry for 5–10 min at 60°C.

- Resuspend the pellet in 40 µl of prewarmed Tris-HCl buffer, pH 8, 1 mM. Incubate for 5 min at 60°C in a thermo mixer, under shaking.

- Spin the tubes for 2 min at maximum speed. Transfer 35 µl of the supernatant to a new 1.5 ml tube. Carefully avoid aspirating the resin. Always keep the tubes with eluted DNA in ice.

- Repeat elution steps 11–12, resuspending the resin pellet in 35 µl of Tris-HCl buffer. Collect the eluted DNA in the same tube.

- Add to the eluted DNA (~70 µl), 7.8 µl of 3 M NaOH (0.3 M fin.). Incubate at 37°C in a heating block for 15 min.

- Centrifuge briefly. Neutralize the solution by adding an equal volume (77.8 µl) of 5 M ammonium acetate, pH 7.5.

- Add three volumes of absolute ethanol per volume of neutralized solution (~470 µl). Allow the mixture to precipitate overnight at −20°C.

- Spin the tube for 20 min at maximum speed (14,000 rpm) at 4°C. A small pellet should be visible at the bottom of the tube.

- Discard the supernatant. Without touching the pellet, add 100 µl of 70% ice-cold ethanol.

- Spin the tube for 5 min at maximum speed (14,000 rpm) at 4°C.

- Discard the supernatant. Leave the pellet to air dry at 37°C for almost 15 min.

- Resuspend the pellet in 40 µl of 1 mM Tris-HCl buffer, pH 8. Store the DNA at −20°C until use. Aliquotation is recommended.

30.3.4 Troubleshooting

- If no pellet or "smear" on the tube wall is visible after DNA precipitation, add 10 µl of glycogen solution and repeat precipitation overnight at −20°C.

30.4 BM-Specific, Nested MSP for Methylation Detection of the MGMT Gene Promoter

We have developed a nested, two-stage PCR approach, mainly based on previously published works [14, 17], adapted to poor-quality, highly degraded DNA from FFPE. The first round of amplification targets a relatively short sequence (177 nucleotides) spanning the promoter region and the first exon of MGMT gene

---

[12]Sodium Metabisulfite and hydroquinone powders are harmful by inhalation. Operate under a chemical hood. To dissolve metabisulfite, shake under a warm water-spout.

[13]Do not exceed this quantity, as the bisulfite amount is the limiting factor if 1.5 ml tubes are used. If you use ≤1 µg of DNA, add Herring DNA (yeast RNA or tRNA from Sigma, cat. R8508) to 2 µg final nucleic acid as a carrier during BM.

[14]Herring DNA is added at this point to enhance the binding of small quantities of DNA to the resin.

Table 30.1 Primer sequences

Primer label	Position at NT_008818.15	Sequence 5′–3′
MGMT_nest_up	2499407–2499425	F: GGATATGTTGGGATAGTT
MGMT_nest_dw	2499565–2499584	F: AAATAAAAACCCCTACAAA
MGMT_M_up	2499448–2499471	F: TTTCGACGTTCGTAGGTTTTCGC
MGMT_M_dw	2499507–2499529	F: GCACTCTTCCGAAAACGAAACG
MGMT_U_up	2499442–2499471	F: TTTGTGTTTTGATGTTTGTAGGTTTTTGT
MGMT_U_dw	2499507–2499535	F: AACTCCACACTCTTCCAAAAACAAAACA

Purchased from Sigma, lyophilized. Resuspended at 300 pmol/µl in 10 mM Tris pH 8

nest nested, *M* methylated, *U* unmethylated

[18]. The primers recognize the bisulfite-modified template but do not discriminate between methylated and unmethylated alleles. The stage-1 PCR products are then subjected to stage-2 PCR in which primers specific to methylated or unmethylated template are used [17]. Along with this method, a simple quality control of BM, with a bisulfite conversion-specific restriction analysis (BCORA) is suggested [12].

30.4.1 Reagents

- *PCR buffer 10× with MgCl₂*: We report the standard composition of PCR buffer: 15 mM MgCl₂, 500 mM KCl, 100 mM Tris, pH 8.3 at 25°C
- *10× PCR buffer II without MgCl₂* (Applied Biosystems)
- *25 mM MgCl₂, solution* (Applied Biosystems)
- *Commercial 100 mM stock solutions of each dNTP (pH8) (Amersham):* dilute the dNTP stock solutions to prepare a 10 mM solution of each dNTP in sterile water
- *Primers*: Lyophilized primers should be dissolved in a small volume of distilled water or 10 mM Tris, pH 8 to make a concentrated stock solution. Prepare small aliquots of working solutions containing 30 pmol/µl to avoid repeated thawing and freezing. Store all primer solutions at −20°C. See Table 30.1
- *Taq DNA Polymerase,* (Amersham) 5 U/µl
- *AmpliTaq Gold* (Applied Biosystems, N8080247), 5 U/µl

30.4.2 Equipment

- *Adjustable pipettes*, range: 2–20 µl, 20–200 µl, 100–1,000 µl
- *Nuclease-free aerosol-resistant pipette tips*

- *0.2 ml nuclease-free microtubes*
- *Sterile hood*
- *Tabletop refrigerated centrifuge* suitable for centrifugation of 0.2 ml tubes
- *Thermal cycler*
- *Electrophoresis unit for agarose gel*
- *Electrophoresis unit for acrylamide gel*
- *UV transillumination unit*

30.4.3 Controls

- Positive control for methylation: DNA from SW48 colon cancer cell line [16], bisulfite modified
- Negative control for methylation: DNA from normal human lymphocytes, bisulfite modified [16].
- No template control (NTC) for PCR reactions: see step 3 in Sect. 30.3.3.
- Quality control for PCR: unmodified DNA from normal human lymphocytes, 50 ng/µl.[15]

30.4.4 Method

30.4.4.1 First Round of Amplification (BM-Specific PCR)

Operate in a sterile hood.
- "Outer" PCR is performed in a final volume of 50 µl, containing:
  - 5 µl 10X PCR Buffer.....................1X final
  - 1 µl dNTPs, 10 mM each.......... 0.2 mM final
  - 0.5 µl primer MGMT_nest_up, 30 pmol/µl ....0.3 pmol/µl final

---

[15]This control can be performed only once, when the method is settled.

- 0.5 µl primer MGMT_nest_dw,
30 pmol/µl ....0.3 pmol/µl final
- 0.25 µl Taq pol, 5 U/µl............0.025 U/µl final
- 5–10 µl of modified DNA[16]
- or
- 5–10 µl of NTC-or
- 5 µl of the quality control for PCR-or
- The appropriate volume of positive and negative controls for methylation.
- H_2O to volume.
- Overlay reaction mixture with 20 µl of mineral oil.
- Thermal cycling:
94°C 4′ + 5 × (94°C 60″, 52°C 60″, 72°C 60″) + 45 × (94°C 30″, 52°C 30″, 72°C 30″) + 72°C 5′
- Gel visualization: mix 10 µl of PCR product with 2 µl of 6× loading buffer[17]; load on a 8% acrylamide gel approx. 10 cm. W × 10 cm. L.; include a 100 bp marker ladder (e.g., Amersham 27400701). Run at 200 V constant until bromophenol blue reaches the bottom of the gel. Stain the gel in 0.5 µg/ml ethidium bromide. Inspect under a UV source. A single band should be visible.

30.4.4.2 Second Round of Amplification (MSP)

- Prepare two distinct master mixes: one containing the primers for the methylated target, the other for the unmethylated one:
- 5 µl 10X PCR Buffer II .... 1X final
- 3 ml 25 mM $MgCl_2$ solution, 1.5 mM final
- 1 µl dNTP, 10 mM each 0.2 mM final
- 0.5 µl primer MGMT_M/U_up, 30 pmol/µl .... 0.3 pmol/µl final
- 0.5 µl primer MGMT_M/U_dw, 30 pmol/µl .... 0.3 pmol/µl final
- 0.25 µl AmpliTaq Gold, 5 U/µl .... 0.025 U/µl final

Table 30.2 Expected results of the second round of amplification

	MSP-M	MSP-U
NTC	No band	No band
Quality control	No band	No band
Methylated control	Single band, 81 bp	No band
Unmethylated control	No band	Single band, 93 bp

- 5 µl of a 1:10–1:1,000 dilution of first-round PCR product[18]
- or
- 5 µl of NTC first-round PCR product-or
- 5 µl of the quality control first-round PCR product
- 5 µl of a 1:10–1:1,000 dilution of positive and negative controls of first-round PCR product
- H_2O to volume
- Overlay reaction mixture with 20 µl of mineral oil.
- Thermal cycling:
95°C 10′ + 35 × (94°C 20″, 62°C 15″, 72°C 20″).
- Gel visualization: mix 10 µl of PCR product with 2 µl of 6× loading buffer[19]; load on a 2.5% agarose gel prepared with TBE 1× containing 0.5 µg/ml ethidium bromide. Include a 100 bp marker ladder. Run at 80 V constant until Bromophenol Blue reaches 1/2 of gel length. Inspect under an UV source. Expected results are reported in table 30.2.

30.4.5 Troubleshooting

- The sample apparently fails to yield a detectable band after the first round of amplification. Use 5 µl of PCR product in the second round anyway. If both MSPs test negative, this may be due to:
  - (most common) high degradation of sample before BM. Solution: repeat DNA extraction or run both BM and PCR in duplicate/triplicate (in the latter case, if only one of two/three replicas tests positive, stochastic amplification of extremely low copies of the target is suspected and results should be cautiously evaluated).
  - Poor sample DNA purification (260/280 ratio <1.6). Solution: repeat extraction, extending Proteinase K digestion to 48 h.

---

[16]As residual resin may be present in DNA suspension, spin the tubes for 2′ at 14,000 RPM before adding the DNA. Do not exceed suggested volumes, as BM DNA may contain PCR inhibitors.

[17]6× loading buffer: 0.25% bromophenol blue, 0.25% xylene cyanol, 30% glycerol in H_2O.

[18]The choice of the dilution relies on band signal intensity at visual inspection. If no band is visible, use 5 ml of undiluted PCR product. Prepare dilutions and PCR master mixes in different rooms. Avoid moving equipment and disposables from one room to the other.

---

[19]See Footnote 17

– Presence of PCR inhibitors in DNA after BM. Solution: repeat BM, or use <5 µl DNA in the PCR.
• NTC or Quality control test positive: check for contamination (usually amplicon carryover).

30.5 Quality Control After MSP

BCORA (bisulfite conversion-specific restriction analysis) allows a qualitative monitoring of conversion status based on gain or loss of specific restriction sites after BM [12]. Briefly, stage-2 PCR products are digested with HhaI that has a recognition sequence (GCG^C) which is present in unconverted DNA but is lost if cytosines are fully converted during BM. Thereafter, the conversion status by BM can be qualitatively monitored by visualization on gel of the digestion products.

30.5.1 Reagents

• *Hha I enzyme (20,000 U/µl)* (New England Biolabs #NEB R0139S)
• *10X NE Buffer 4* (New England Biolabs #NEB R0139S)
• *100X BSA solution, 10 mg/ml* (New England Biolabs #NEB R0139S): dilute to 1 mg/ml with sterile water
• *DNA precipitating solution*: 100% EtOH, 3 M Sodium Acetate
• *Glycogen* (Sigma), diluted at 1 mg/ml in H_2O and used as precipitation carrier
• *70% ethanol*

30.5.2 Equipment

• *Adjustable pipettes,* range : 2–20 µl, 20–200 µl, 100–1,000 µl
• *Nuclease-free aerosol-resistant pipette tips*
• *0.2 ml nuclease-free microtubes*
• *Tabletop refrigerated centrifuge,* suitable for centrifugation of 0.2 ml tubes.
• *Thermal cycler*
• *Electrophoresis unit for acrylamide gel*
• *UV transillumination unit*

30.5.3 Method

• Precipitate 20 µl of MSP product[20] by adding 60 µl of DNA precipitating solution and 5 µl of 1 mg/ml Glycogen (1 h at −20°C should be sufficient).
• Spin in a tabletop refrigerated microcentrifuge for 20′ at 14,000 rpm.
• Wash the pellet once in ice-cold 70% ethanol. Air dry the pellet.
• Resuspend the pellet in 14 µl H_2O.

Perform digestion in 10 µl final volume, containing:
– Hha I 20,000 U/ml 1 µl
– 10X NE Buffer 4,   1 µl
– BSA 1 mg/ml   1 µl
– Resuspended PCR product 7 µl
• Prepare a mix containing all of the above except the enzyme. Also include the amplification product of a sequence containing one or more Hha I sites as digestion control.
• Incubate at 37°C for 2 h in a thermo block or in a thermal cycler.
• Run digestion products in a 10% acrylamide gel. Stain with ethidium bromide.

If only fully converted DNA has been amplified, a single band of 81 bp should be visible. If unconverted DNA has been amplified, two bands (55 and 26 bp for "methylated" MSP) should be visible.

30.6 Methylation Detection by MS-MLPA

The MS-MLPA procedure avails the probe mix prepared by MRC-Holland (Amsterdam, The Netherlands), and contains 40 probe sequences that correspond to 25 genes frequently affected by methylation in tumours, and 15 control sequences that do not contain the HhaI site and that are not influenced by methylation-sensitive HhaI restriction digestion (refer to the method outlined in Figs. 30.1 and 30.2). MLPA is performed according to the manufacturer's protocol. Each sample is analyzed in two parallel reactions, one ligation of hybridized probes and one ligation + digestion reaction. A PCR amplification step of ligated probes follows, and capillary electrophoresis analysis allows the comparison of the abundance

---

[20]Optional: purify PCR product from the agarose gel or directly with a PCR cleanup kit. In the latter case, use 40 µl of PCR product.

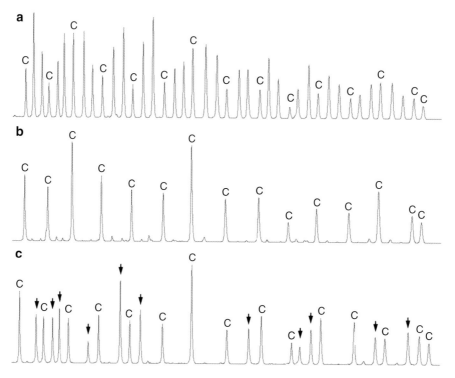

Fig. 30.2 Detection of aberrant methylation pattern in MSI-H colorectal tumours by MS-MLPA using the ME001B probe mixture (MRC-Holland). GeneScan™ electropherogram. The ME001B MS-MLPA probe mix contains 25 sequences that correspond to a set of tumour suppressor genes that are frequently silenced by methylation in different tumours, but are unmethylated in blood-derived DNA of healthy individuals. In addition, 15 different control probes are present that are not influenced by the HhaI digestion. Besides the detection of aberrant methylation, all 40 probes yield information on copy number changes in the analyzed sample. (**a**) Undigested sample. (**b**) HhaI-digested sample from normal tissue. (**c**) HhaI-digested sample from tumour tissue. *c*, control probe without HhaI digestion site. ↓, undigested probes due to the methylation of CpG

and number of PCR products from both reactions. Quality control of each reaction is performed through four control probes that detect DNA quantity and whose amplification is ligation-independent and three ligation-dependent control probes that detect DNA denaturation, which are present in every MLPA probe mix. This method has been recently adopted by one IMPACTS institution in a clinical research setting [19].

30.6.1 Reagents

- *SALSA MS-MLPA kit ME001B* (MRC-Holland), including:
  - *SALSA probe mix*
  - *MLPA buffer*
  - *Ligase-65 buffer A*
  - *Ligase-65 buffer B*
  - *Ligase-65 enzyme*
  - *SALSA PCR primers (FAM labeled)*
  - *SALSA Enzyme Dilution Buffer*
  - *SALSA Polymerase*
- *HhaI enzyme* (Promega R6441, 10 units/μL)
- *Hi-Di™ formamide* (e.g., Applied Biosystems)
- *Performance Optimized Polymer 4™* (POP4; e.g., Applied Biosystems)
- *EDTA Buffer* (10X; Applied Biosystems)
- *GenScan™-500 Rox™ Size Standard* (Applied Biosystems)

30.6.2 Equipment

- *Adjustable pipettes,* range: 2–20 μl, 20–200 μl, 100–1,000 μl
- *Nuclease-free aerosol-resistant pipette tips*
- *0.2 ml nuclease-free microtubes*

- *Tabletop refrigerated centrifuge*, suitable for centrifugation of 0.2 ml tubes.
- *Thermal cycler*
- *ABI PRISM 310 Genetic Analyzer* (Applied Biosystems)
- *Capillaries* (5–47 cm × 50 µm; Applied Biosystems, code 402839)
- *GeneScan™ Software* (Applied Biosystems, code 401734)
- *Coffalyser software* (MRC Holland, The Netherlands)

30.6.3 Method

30.6.3.1 DNA-Denaturation and Hybridization of the SALSA-Probes

- Dilute the DNA-sample (20–200 ng DNA, preferred range 50–200 ng DNA) with TE to 5 µL.
- Heat for 10 min at 98°C; cool to 25°C before opening the thermal cycler.
- Add a mixture of 1.5 µL SALSA Probe-mix + 1.5 µL MLPA buffer to each tube.
- Mix with care. Incubate for 1 min at 95°C, followed by 16 h at 60°C.

30.6.3.2 Ligation + Digestion Reaction

- At RT, add a mix of 3 µL Ligase-buffer A + 10 µL water to eachsample and mix.
- Transfer 10 µL to a second tube.
- Incubate both tubes for at least 1 min at 49°C in the thermal cycler.
- While at 49°C, add 10 µL Ligase-65 mix to the first tube and mix.
- While at 49°C, add 10 µL Ligase-Digestion mix to the second tube and mix (methylation test).
- Incubate both tubes for 30 min at 49°C, then heat for 5 min at 98°C.
- Prepare the "Ligase-65 mix" (made less than 1 h. before use and stored on ice) by mixing 1.5 µL Ligase-65 buffer B + 8.25 µL Water. Add 0.25 µL Ligase-65 enzyme and mix again.
- Prepare the "Ligase-Digestion mix" (made less than 1 h. before use and stored on ice) by mixing 1.5 µL Ligase-65 buffer B + 7.75 µL Water. Add 0.25 µL Ligase-65 enzyme + 0.5 µL HhaI enzyme and mix again.

30.6.3.3 PCR Reaction

- Mix in new tubes: 5 µL of each MLPA ligation or ligation-digestion reaction, with 2 µL SALSA PCR buffer + 13 µL Water.
- While on ice, add 5 µL of Polymerase mix, place in a preheated thermocycler (e.g., 72°C), and immediately start the PCR reaction.
- Prepare the "Polymerase mix" (made less than 1 h. before use and stored on ice): for each PCR reaction mix 1 µL SALSA PCR-primers + 1 µL SALSA Enzyme Dilution buffer + 2.75 µL Water. Add 0.25 µL SALSA Polymerase. Mix well.
- The following cycling programme covers the complete MLPA reaction:
  1. Hybridization reaction: 98°C 10 min, 25°C hold, 95°C 1 min, 60°C hold.
  2. Ligation reaction: 49°C hold, 49°C 30 min, 98°C 5 min, 4°C hold.
  3. PCR reaction: 72°C hold, 35 cycles: 95°C 30 s, 60°C 30 s, 72°C 60 s, 72°C 20 min, 4°C hold

30.6.3.4 Separation of Amplification Products by ABI Prism 310 Genetic Analyzer

- Label: SALSA 6-FAM PCR primer-dNTP mix.
- PCR reaction mix: 0.75 µL of the PCR reaction + 0.75 µL water + 0.5 µL internal standard + 13.5 µL HiDi formamide. Incubate mix for 2 min at 94°C, and cool on ice.
- Settings: Internal standard: 500ROX, filter set D; Capillary length: 47 cm; Polymers: POP-4 (preferable) or POP-6 polymer. Run time: 30 min; Run voltage: 15 kV; Run temperature: 60°C; Capillary fill volume: 184 steps; Prerun voltage: 15 kV; Prerun time: 180 s; Injection voltage: 3.0 kV; Injection time: 10–30 s; Data delay time: 1 s.

30.6.3.5 Interpretation of the Results

After separation of amplification products, raw data are analyzed by GeneScan™ software and the peak areas of each probe are obtained. Quantification of the methylation status of studied CpG sites is performed by Coffalyser software (Fig 30.2). The peak area of each MS-MLPA probe is normalized by dividing it by the combined areas of control probes. Each normalized peak area from digested sample is compared with that obtained in the undigested sample. Aberrant methylation is scored when calculated methylation percentage is >10%.

30.6.4 Troubleshooting

- Unexplainable loss of some signals and increase of others; probably due to incomplete hybridization or denaturation. Solution: check heated lid and thermocycler; perform denaturation step exactly by the protocol.
- All peak signals are too low:
  - If large primer-dimer peaks are visible, the cause is probably incomplete PCR reaction. Solution: repeat PCR reaction; start PCR as soon as possible after adding the polymerase mix.
  - Alternatively, increase capillary injection time or injection voltage; if this doesn't help, repeat PCR with up to five additional cycles or repeat capillary electrophoresis and increase the amount of MLPA product
- All peak signals too high; repeat capillary electrophoresis and decrease amount of used MLPA product or/and decrease capillary injection time/voltage
- Unexplainable extra peaks or loss of some peak signals:
  - DNA denaturation failed. Solution: repeat denaturation exactly by the protocol
  - Omitted denaturation of formamide-marker-PCR sample mix.
- Large differences in relative peak areas between different samples that do not make sense:
  - Impurities in DNA sample. Solution: prepare good-quality DNA
  - The use of extremely large amounts of DNA. Solution: reduce starting DNA amount.

References

1. Ehrlich M, Gama-Sosa MA, Huang LH, Midgett RM, Kuo KC, McCune RA, Gehrke C (1982) Amount and distribution of 5-methylcytosine in human DNA from different types of tissues of cells. Nucleic Acids Res 10(8):2709–2721
2. Li E (2002) Chromatin modification and epigenetic reprogramming in mammalian development. Nat Rev Genet 3(9):662–673
3. Jones PA (1996) DNA methylation errors and cancer. Cancer Res 56(11):2463–2467
4. Jones PA, Baylin SB (2002) The fundamental role of epigenetic events in cancer. Nat Rev Genet 3(6):415–428
5. Robertson KD, Wolffe AP (2000) DNA methylation in health and disease. Nat Rev Genet 1(1):11–19
6. Laird PW (2003) The power and the promise of DNA methylation markers. Nat Rev Cancer 3(4):253–266
7. Fraga MF, Esteller M (2002) DNA methylation: a profile of methods and applications. Biotechniques 33(3):632–634, 636–649
8. Frommer M, McDonald LE, Millar DS, Collis CM, Watt F, Grigg GW, Molloy PL, Paul CL (1992) A genomic sequencing protocol that yields a positive display of 5-methylcytosine residues in individual DNA strands. Proc Natl Acad Sci USA 89(5):1827–1831
9. Grunau C, Clark SJ, Rosenthal A (2001) Bisulfite genomic sequencing: systematic investigation of critical experimental parameters. Nucleic Acids Res 29(13):E65
10. Millar DS, Warnecke PM, Melki JR, Clark SJ (2002) Methylation sequencing from limiting DNA: embryonic, fixed, and microdissected cells. Methods 27(2):108–113
11. Herman JG, Graff JR, Myohanen S, Nelkin BD, Baylin SB (1996) Methylation-specific PCR: a novel PCR assay for methylation status of CpG islands. Proc Natl Acad Sci USA 93(18):9821–9826
12. Sasaki M, Anast J, Bassett W, Kawakami T, Sakuragi N, Dahiya R (2003) Bisulfite conversion-specific and methylation-specific PCR: a sensitive technique for accurate evaluation of CpG methylation. Biochem Biophys Res Commun 309(2):305–309
13. Nygren AO, Ameziane N, Duarte HM, Vijzelaar RN, Waisfisz Q, Hess CJ, Schouten JP, Errami A (2005) Methylation-specific MLPA (MS-MLPA): simultaneous detection of CpG methylation and copy number changes of up to 40 sequences. Nucleic Acids Res 33(14):e128
14. Palmisano WA, Divine KK, Saccomanno G, Gilliland FD, Baylin SB, Herman JG, Belinsky SA (2000) Predicting lung cancer by detecting aberrant promoter methylation in sputum. Cancer Res 60(21):5954–5958
15. Bian YS, Yan P, Osterheld MC, Fontolliet C, Benhattar J (2001) Promoter methylation analysis on microdissected paraffin-embedded tissues using bisulfite treatment and PCR-SSCP. Biotechniques 30(1):66–72
16. Warnecke PM, Stirzaker C, Song J, Grunau C, Melki JR, Clark SJ (2002) Identification and resolution of artifacts in bisulfite sequencing. Methods 27(2):101–107
17. Esteller M, Hamilton SR, Burger PC, Baylin SB, Herman JG (1999) Inactivation of the DNA repair gene o6-methylguanine-DNA methyltransferase by promoter hypermethylation is a common event in primary human neoplasia. Cancer Res 59(4):793–797
18. Harris LC, Potter PM, Tano K, Shiota S, Mitra S, Brent TP (1991) Characterization of the promoter region of the human o6-methylguanine-DNA methyltransferase gene. Nucleic Acids Res 19(22):6163–6167
19. Berginc G, Bracko M, Glavac D (2010) MS-MLPA reveals progressive age-dependent promoter methylation of tumor suppressor genes and possible role of IGSF4 gene in colorectal carcinogenesis of microsatellite instable tumors. Cancer Invest 28(1):94–102

Quantitative Methylation Status Assessment in DNA from FFPE Tissues with Bisulfite Modification and Real-Time Quantitative MSP

31

Ermanno Nardon

Contents

E. Nardon
Department of Medical, Surgical and Health Sciences,
University of Trieste, Cattinara Hospital,
Strada di Fiume 447, Trieste, Italy

31.1 Introduction and Purpose

Because of its simplicity and requirement of only commonly used equipment, methylation-specific PCR (MSP) is the most widely used assay for detection of methylation. Prior to amplification, the DNA is treated with sodium bisulfite to convert all unmethylated cytosines to uracils. The DNA is then amplified using primers that match one particular methylation status of the DNA, such as that in which DNA is methylated at all CpGs sites. This technique is very sensitive and it was originally estimated to be able to detect one methylated target in the presence of a 1,000-fold excess of unmethylated DNA [1]. Nevertheless, MSP suffers from some drawbacks, such as (1) its proneness to false positives in the presence of unconverted or partially unconverted sequences in the bisulfite-treated DNA [2]; (2) its possible dependence to subjective judgment, since the generated PCR product is visualized on a gel; (3) its low throughput.

Real-Time Quantitative MSP is based on the continuous optical monitoring of a fluorogenic PCR and represents a logical implementation of the MSP strategy, in which a dual-labeled, fluorogenic probe is also included (Methylight technology) [3, 4]. One fluorescent dye serves as a reporter, and its emission spectrum is quenched by a second fluorescent dye. During the extension phase, the 5′ to 3′ exonuclease activity of Taq polymerase cleaves the reporter from the probe, thus releasing it from the quencher, resulting in an increase in fluorescence emission [5, 6]. Three main strategies have been developed for the Methylight assays. In the first one, sequence discrimination occurs at PCR amplification only, with primers specific for either the methylated or the unmethylated sequence and a common probe detecting a portion of the

G. Stanta (ed.), *Guidelines for Molecular Analysis in Archive Tissues*,
DOI: 10.1007/978-3-642-17890-0_31, © Springer-Verlag Berlin Heidelberg 2011

amplicon devoid of CpG sites ('Fluorescent MSP'). In the second one, sequence discrimination occurs at probe hybridization level only, with common primers that do not overlap any CpG dinucleotide [7]. In the third approach, primer and probe sets are both designed to detect either fully methylated or fully unmethylated patterns in the targets [3, 4]. This latter is the most widely used strategy and it is the one that assures the highest specificity of methylation detection. In all these assays, data normalization is obtained with an amplification to control for the quantity of input DNA, using primers and probes for a sequence which does not contain any CpG in its unconverted form.

We can take advantage of Methylight technology whenever a quantitative, high-throughput evaluation system is required, e.g. to validate candidate markers arisen from global genome screenings; when a cut-off methylation value needs to be established for a given diagnostic marker, or to stratify a disease according to both the presence and the level of methylation at specific sites [8–11].

Herein is described a Methylight approach to assay for the presence of the CpG island methylator phenotype (CIMP) in human colorectal cancer samples. CIMP affects a significant proportion (15–30%) of sporadic CRC and is characterized by simultaneous aberrant methylation at multiple CpG islands, including several known genes, such as p16, hMLH1, and THBS1 [11, 12]. CIMP-positive CRCs have a distinct clinical, pathologic, and molecular profile, such as association with proximal tumour location, female sex, mucinous and poor differentiated hystology, serrated morphology, microsatellite instability (MSI), and high BRAF and low TP53 mutation rates [10, 13–16]. CIMP+ phenotype association with prognosis is still controversial, but CIMP would be an independent predictor of response to 5-fluorouracil-based treatments [17, 18].

CIMP features can be substantially influenced by CpG islands selected for its detection. In fact, different CIMP panels used in various studies have caused some confusion.

A 6 MINT (Methylated IN Tumor) marker panel was originally developed by Toyota et al. in 1999. CRC tumor population showed a bimodal distribution if stratified according to the number of methylated markers. Tumors testing positive for ≥3 MINTs (CIMP+) displayed a prevalence of KRAS mutations, while tumors with <3 MINTs (CIMP−) showed a prevalence of TP53 mutations [15]. However, subsequent studies failed to observe such a distinct bimodal distribution or to confirm associations with the molecular alterations [14, 19].

These discrepancies may be explained by an overestimation of DNA methylation due to the use of highly sensitive nonquantitative methods such as MSP, by the use of different cutoff values, by the inclusion of other sequences in the panel, or by inadequacy of original panel itself. Recently, an independent genome-wide screening of 195 CpG islands has identified a different set of sequences (CACNA1G, IGF2, NEUROG1, RUNX3, and SOCS1) which identifies CIMP status more reliably than the original set [10]. The new panel, developed on a qRT-PCR platform, detects a heavily methylated subset of CRCs that encompasses almost all BRAF mutants and sporadic MSI-H cancers. MSI and BRAF mutations thus would represent a distinctive feature of this type of tumours [16]. This panel has already been successfully validated in independent large sets of colorectal tumours [20, 21].

The procedure herein described is an adaptation of the Weisenberger's protocol (DOI: 10.1038/nprot.2006.152) for quantitatively assessing the methylation status of CACNA1G, IGF2, NEUROG1, RUNX3, and SOCS1 markers [16] using a TaqMan-based approach. Methylation level at each locus is expressed as a percentage relative to a fully-methylated, M.SssI-treated reference DNA sample (Percentage of Methylated Reference, PMR). A sample is defined as CIMP+ if showing a PMR >10% in at least three of the five markers, while a sample testing positive at two markers or less is considered CIMP negative (CIMP−). The protocol has been divided into four sections:

- DNA extraction from FFPE for methylation analysis
- M.SssI modification
- Bisulfite modification
- Methylight reaction setup and methylation assessment

31.2 DNA Extraction from FFPE tissues for Quantitative Methylation Analysis

Refer to the method described in Sect. 30.2.

31.3 M.SssI Modification

The CpG Methyltransferase M.SssI methylates all cytosine residues within the double-stranded dinucleotide recognition sequence using *S*-adenosyl methionine

(SAM) as a methyl donor. This reaction is performed to create a fully methylated reference sample. Peripheral blood leukocyte DNA (PBL-DNA) is commonly used as a substrate but any genomic DNA can serve the purpose, provided it is of high level purity and bisulfite modification-permissive.

31.3.1 Reagents

- *Human Genomic DNA, 1μg/μl*
- *Nuclease-free water*
- *10X NE Buffer 2* (e.g., New England Biolabs M0226S)
- *32 mM SAM stock* (e.g., New England Biolabs M0226S)
- *M.SssI methyltransferase (4 U/μl)* (e.g., New England Biolabs M0226S)

31.3.2 Equipment

- *Adjustable pipettes*, range: 2–20 μl, 20–200 μl, 100–1,000 μl
- *Nuclease-free aerosol-resistant pipette tips*
- *0.2, 0.5, or 1.5 ml nuclease-free autoclaved tubes*
- *Tabletop refrigerated centrifuge* suitable for centrifugation of 1.5 ml tubes
- *Heating block or Thermal cycler*

31.3.3 Method

- Dilute SAM to 1.6 mM by mixing 2 μl of the 32 mM stock and 38 μl of nuclease-free water. Store unused SAM at −20°C in small aliquots.
- In a 0.5 ml sterile tube add in order[1]:
  - Nuclease-free water .... 118 μl
  - 10X NEBuffer 2 .... 16 μl (1X final conc.)
  - Diluted SAM .... 16 μl (0.16 mM final conc.)
  - Genomic DNA .... 8 μl (0.05 μg/μl final conc.)
  - M.SssI methyltransferase (4 U/μl) ...... 2 μl (0.05 units/μl final conc.)

- Mix, pipette up and down several times.
- Incubate overnight at 37°C in a heating block or in a thermal cycler.
- Add 0.6 μl of M.SssI methylase and 5 μl of 1.6 mM SAM. Mix well.
- Incubate overnight at 37°C in a heating block or in a thermal cycler.
- Stop the reaction by heating at 65°C for 20 min in a heating block or in a thermal cycler.

Modified DNA can now be used for bisulfite conversion. Aliquot unused DNA and store at −20°C.

Since methylation of every CpG site could be incomplete after the described treatment, the whole procedure can be repeated. To verify the completeness of reaction, an aliquot of DNA should be bisulfite-modified and tested with a methylation-specific Methylight reaction (see further on). DNA should be phenol-chloroform extracted before each round of treatment.

31.4 Bisulfite Modification

M.SssI-treated reference and sample DNA are both submitted to Bisulfite Modification (BM) by means of the Zymo EZ DNA methylation kit (Zymo Research, Orange, CA, D5001 or D5002) according to the manufacturer's instructions.[2] For optimal conversion results, the amount of input DNA should be from 200 to 500 ng. DNA amounts from FFPE lower than 200 ng may yield unreliable results in the subsequent Methylight assay.

31.5 Methylight Reaction Setup and Methylation Assessment

The methylation level of each bisulfite-converted sample is detected in real time by a TaqMan chemistry using primers and probes specific for fully methylated sequences at each of the five CIMP-specific loci [16]. Primers and probes are also specific for bisulfite-converted DNA. The amount of fully converted DNA is assessed with a control reaction targeting the repetitive

---

[1]This reaction can be scaled to accommodate different DNA amounts.

[2]Other bisulfite modification kits can be consistently used, e.g. the Qiagen EpiTect™ Bisulfite Kit and the Invitrogen MethylCode™ Bisulfite Conversion Kit.

element ALU-C4 [16],[3] using primers and probes which are specific for bisulfite-converted DNA and do not have any CpG site in their target sequence. Scalar amounts of the M.SssI–treated, bisulfite-modified reference DNA are real-time amplified to set up six standard curves, one for each gene.

For any sample, methylation amount at a given locus is calculated from the standard curve and input DNA amount is obtained from the ALU-C4 reference curve. Sample methylation percentage with respect to the fully methylated reference (PMR) is then computed as:

$$PMR = \frac{\text{Methylation amount}}{\text{Input DNA amount}} \times 100 \quad (31.1)$$

31.5.1 Reagents

- *TaqMan® 1,000 Reactions Gold with Buffer A Pack* (Applied Biosystems, 4304441): includes 10X Buffer, 25 mM MgC_2 and Taq Gold Polymerase
- *Commercial 100 mM stock solution of each dNTP (pH 8)* (e.g., Amersham 27-2035-03). Dilute the dNTP stock solutions to prepare a 10 mM solution of each dNTP in sterile water
- *10× TaqMan Stabilizer solution:* 0.5% gelatin (w/v), 0.1% Tween-20 (v/v)[4]
- *Primers and probes:* lyophilized primers should be dissolved in a small volume of distilled water or 10 mM Tris, pH 8 to make a 300 μM stock solution. Lyophilized probe should be prepared at a final concentration of 100 μM. Dilute the primers and the probe to a working solution of 6 μM (primers) and 2 μM (probe). Make small aliquots of both stock and working solutions and store them at −20°C. In Table 31.1 the specific primer and probe sequences of CACNA1G, IGF2, NEUROG1, RUNX3, and SOCS1 markers are reported.

---

[3]This multi-copy ALU sequence, which is dispersed throughout the genome, is used for normalization reaction, as it is less prone to fluctuations caused by aneuploidy and copy number changes affecting single-copy gene normalization reactions (e.g. MYOD1, ACTB, and COL2A1). In addition, it also allows for sensitive detection of small amounts of DNA.

[4]Prepare in advance a 20% Tween-20 (Sigma, P-9416) solution in water. Weigh out 0.2 g gelatin (Sigma, G-9391) and add it to a 50 ml conical screw capped tube. Add 20 ml of water. Heat to dissolve, then add 0.2 ml of 20% Tween-20 and bring the final volume to 40 ml with nuclease-free water. Store at −20°C.

31.5.2 Equipment

- *Adjustable pipettes*, range: 2–20 μl, 20–200 μl, 100–1,000 μl
- *Nuclease-free aerosol-resistant pipette tips*
- *0.2 and 1.5 ml nuclease-free microtubes*
- *Sterile laminar flow hood*
- *Tabletop refrigerated centrifuge* suitable for centrifugation of 0.2 ml and 1.5 ml tubes
- *Real-Time thermal cycler for standard 96-well format reactions*
- *PCR microplates/strips*
- *PCR strip caps*

31.5.3 Method

31.5.3.1 Samples and Controls

Sample and M.SssI-treated DNAs previously submitted to bisulfite modification by means of the Zymo EZ DNA methylation kit are usually eluted in 10 μl. We suggest diluting ten times this amount to a 100 μl volume, to allow for pipetting errors in subsequent operations. Since bisulfite modification is highly detrimental to nucleic acid integrity, DNA recovery may vary from sample to sample. Thus, the optimal eluate amount to be used in the assay is hardly predictable. In order to obtain reliable quantifications, the amount of bisulfite-treated sample DNA should yield a Ct value within the linearity range generated by the M.SssI-treated reference DNA. Therefore, we suggest performing a preliminary ALU-C4 control reaction using 2 and 10 μl of the above dilution (testing more than one unknown sample is recommended).[5] Sample and M.SssI-treated DNA amounts herein shown are arbitrary and do not reflect an optimized assay.

To check the specificity of the MethyLight Assay to detect bisulfite-converted DNA only, a preliminary experiment should be conducted using unconverted human genomic DNA, which should not give results to the amplification. All methylation quantification experiments should include an NTC (No Template Control), containing all the components of the reaction except for the template.

---

[5]The ALU-C4 reaction is highly sensitive and should generate low C_t values. An unknown DNA amount yielding an ALU-C4 C_t value of less than 20 is usually desirable.

Table 31.1 Primer and probe sequences of the five markers used for the assessment of CIMP

Locus name	GeneBank Number	Sequence (5′–3′)	Amplicon size
CACNA1G (Calcium channel, voltage-dependent, alpha 1 G subunit), 17q22	AC021491	F: TTTTTTCGTTTCGCGTTTAGGT	66 bp
		R: CTCGAAACGACTTCGCCG	
		F:6FAM-AAATAACGCCGAATCCGACAACCGA-BHQ-1	
IGF2 (Insulin-like growth factor 2 (somatomedin A)), 11p15.5	AC132217	F: GAGCGGTTTCGGTGTCGTTA	87 bp
		R: CCAACTCGATTTAAACCGACG	
		F: 6FAM-CCCTCTACCGTCGCGAACCCGA-BHQ-1	
NEUROG1 (Neurogenin 1), 5q23-q31	AC005738	F: CGTGTAGCGTTCGGGTATTTGTA	87 bp
		R: CGATAATTACGAACACACTCCGAAT	
		F: 6FAM-CGATAACGACCTCCCGCGAACATAAA-BHQ-1	
RUNX3 (Runt-related transcription factor 3), 1p36	AL023096	F: CGTTCGATGGTGGACGTGT	116 bp
		R: GACGAACAACGTCTTATTACAACGC	
		F: 6FAM-CGCACGAACTCGCCTACGTAATCCG-BHQ-1	
SOCS1 (Suppressor of cytokine signaling 1), 16p13.13	AC009121	F: GCGTCGAGTTCGTGGGTATTT	83 bp
		R: CCGAAACCATCTTCACGCTAA	
		F: 6FAM-ACAATTCCGCTAACGACTATCGCGCA-BHQ-1	
ALU-C4 (Alu-based normalization control reaction)[a]	Consensus sequence [a]	F: GGTTAGGTATAGTGGTTTATATTTGTAATTTTAGTA	98 bp
		R: ATTAACTAAACTAATCTTAAACTCCTAACCTCA	
		F: 6FAM-CCTACCTTAACCTCCC-MGBNFQ[b]	

[a]ALU is a short nucleotide element interspersed across the whole genome; therefore, no gene locus or number is provided.
[b]*MGBNFQ* Minor Groove Binder, nonfluorescent quencher. *F* forward primer, *R* reverse primer

31.5.3.2 Standard Curve Setup

A 5-point standard curve, consisting of 1:10 serial dilutions of an optimized quantity of the M.SssI-treated reference DNA is required for each of the six considered genes (Table 31.2). Each dilution should be run in duplicate. Dilutions are made in 0.2 ml nuclease-free microtubes. Keep tubes in ice.

31.5.3.3 qReal-Time PCR Setup

Work under a sterile laminar flow hood. We suggest analyzing each unknown sample in duplicate. The herein indicated amount of 1:10-diluted, bisulfite-modified sample does not reflect an optimized assay.

- For each of the six genes and for each duplicate PCR reaction mix the following components in a 0.2 ml nuclease-free microtube:

Table 31.2 Standard curve setup

Curve point	Dilution	Sample	Used for PCR
1	0	M.SssI-treated DNA (1:10 diluted after BM)	10 μl
2	1:10	2.5 μl of 0 dilution + 22.5 μl of sterile H_2O	10 μl
3	1:100	2.5 μl of 1:10 dilution + 22.5 μl of sterile H_2O	10 μl
4	1:1,000	2.5 μl of 1:100 dilution + 22.5 μl of sterile H_2O	10 μl
5	1:10,000	2.5 μl of 1:1,000 dilution + 22.5 μl of sterile H_2O	10 μl

To allow for pipetting errors, it is suggested to prepare each dilution in slight excess, e.g. if 2 μl of M.SssI-treated DNA are theoretically needed for each duplicate, use 2.5 μl instead. The above indicated amounts allow for pipetting errors. For each duplicated standard curve at least 25 μl of M.SssI-treated DNA, 1:10 diluted after BM (bisulfite modification), are needed

- 10X Buffer A............................6.0 µl (1X final)
- 25 mM MgCl₂...................8.4 µl (3.5 mM final)
- 10X stabilizer solution ....6.0 µl (1X final)
- 10 mM dNTPs.................1.2 µl (200 µM final)
- 6 µM forward primer.........3.0 µl (0.3 µM final)
- 6 µM reverse primer..........3.0 µl (0.3 µM final)
- 2 µM probe.......................3.0 µl (0.1 µM final)
- Taq Gold Polymerase ....0.2 µl
- Nuclease-free water...................................9.2 µl
- DNA sample ....20.0 µl

- Mix by pipetting, dispense the mixture on the 96-well plate in duplicates of 30 µl each, avoiding bubble formation. Cover microwells with the strips.
- Refer to instrument software instructions for setting up the real-time run.
    - Use the following Thermal Cycler programme:
        - 1×: 95°C for 10′
        - 40–45×: 95°C for 15″/60°C for 1′

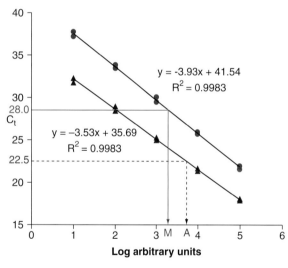

Fig. 31.1 *Example of PMR calculation of an unknown sample - Standard curve method.* Each curve represents the linear regression fitted across C_ts generated by the dilutions of the M. SssI-treated DNA: ▲, Alu C4 reaction; ●, C reaction. Note that since 1,000 copies of the Alu sequence exist per genome, at each control DNA dilution Alu C4 reaction C_ts are considerably lower than the corresponding C_t of the CIMP marker. An unknown sample shows a C_t of 22.5 for the Alu reaction and a C_t of 28.0 for the CIMP marker. From these values sample's log arbitrary amounts A and M are computed from equations of standard curves as $A = (35.69–22.5)/3.53$ and $M = (41.54–28.0)/3.93$ respectively. PMR is then calculated as: antilog $(M–A) \times 100 = 51.13$

31.5.3.4 Data Analysis and Interpretation

Refer to Sect. 25.6 for the relative quantification using a standard curve. For each control reaction, plot the duplicate C_t value generated from the reference DNA against the corresponding log amount, expressed in arbitrary units (Fig. 31.1). Fit the least squares regression line across tabulated points. Convert the C_t value generated by each unknown sample into arbitrary units, using the standard curve regression equation. To obtain the methylation percentage at a given CIMP marker (PMR), divide the calculated arbitrary units by the arbitrary units calculated from the ALU-C4 standard curve and multiply this quotient by 100.[6]

A sample is scored as CIMP+ if showing a PMR >10% at three or more of the five CIMP markers, while a sample testing positive for ≤2 markers is considered CIMP negative (CIMP−).

31.5.4 Troubleshooting

Troubleshooting suggested in Chap. 25 applies also to the qRT-PCR Methylight assay (Table 31.3).

---

[6]This model of quantification does not assume that target and control gene have equal amplification efficiencies and it is fully consistent with the efficiency-corrected model of relative quantification. In fact, it can be easily demonstrated that:

$$\text{PMR} = \frac{(E_\text{target})^{\Delta C_t, (\text{reference}-\text{sample})}}{(E_\text{control})^{\Delta C_t, (\text{reference}-\text{sample})}} \times 100,$$

where E_target and E_control are the amplification efficiencies of the CIMP marker and of the ALU-C4 reaction, respectively, as calculated from the standard curves $E = 10^{-(1/\text{slope})}$; $\Delta C_t (\text{reference-sample})$ is the difference between the C_t value of the fully methylated reference and the C_t value of the unknown sample.

Table 31.3 Troubleshooting

Problem	Possible reason	Solution
The ALU-C4 control reaction yields no amplification signal from unknown samples and/or from reference DNA.	An insufficient amount of bisulfite-modified target DNA is available for amplification. This can be due to:	
	(a) Insufficient DNA submitted to BM: most of the DNA has been degraded during BM.	(a and b): Submit to BM a DNA amount consistent with the instructions provided by the Zymo EZ DNA methylation kit (see Sect. 31.4).
	(b) Too much DNA submitted to BM: poor DNA conversion.	
	(c) Unexhaustive BM due to poor DNA quality.	(c) Repeat DNA extraction. An extensive proteinase K treatment is mandatory to obtain a DNA sample suitable for BM.
CIMP marker(s) reaction yields no amplification signal from the MSss-I-treated reference DNA (the ALU-C4 control reaction is successful).	MSss-I methylation reaction failure. This could be due to:	
	(a) SAM has expired	(a) Use a new batch of SAM. SAM is stable for at least 6 months if stored at −20°C.
	(b) Too much DNA volume in the reaction, which can cause inhibition by changing the pH or salt concentration of the reaction.	(b) Concentrate and desalt the DNA using microcolumns (e.g., Millipore Microcon YM-50)
The C_t value generated by one or more unknown samples is outside the C_t values range of the standard curve.	Nonoptimized assay.	Perform optimization as suggested in Sect. 31.5 of this chapter, using samples whose C_t value is outside the C_t value range of the standard curve.
PMR value of one or more unknown samples is >100%.	Unexhaustive MSss-I methylation treatment of the reference DNA.	MSss-I treatment should be repeated and the DNA Methylight-tested after each round. Refer to Sect. 31.3 of this chapter.

References

1. Herman JG, Graff JR, Myohanen S, Nelkin BD, Baylin SB (1996) Methylation-specific PCR: a novel PCR assay for methylation status of CpG islands. Proc Natl Acad Sci USA 93(18):9821–9826
2. Sasaki M, Anast J, Bassett W, Kawakami T, Sakuragi N, Dahiya R (2003) Bisulfite conversion-specific and methylation-specific PCR: a sensitive technique for accurate evaluation of CpG methylation. Biochem Biophys Res Commun 309(2): 305–309
3. Eads CA, Danenberg KD, Kawakami K, Saltz LB, Blake C, Shibata D, Danenberg PV, Laird PW (2000) Methylight: a high-throughput assay to measure DNA methylation. Nucleic Acids Res 28(8):E32
4. Lo YM, Wong IH, Zhang J, Tein MS, Ng MH, Hjelm NM (1999) Quantitative analysis of aberrant p16 methylation using real-time quantitative methylation-specific polymerase chain reaction. Cancer Res 59(16):3899–3903
5. Heid CA, Stevens J, Livak KJ, Williams PM (1996) Real time quantitative PCR. Genome Res 6(10):986–994
6. Holland PM, Abramson RD, Watson R, Gelfand DH (1991) Detection of specific polymerase chain reaction product by utilizing the 5′–3′ exonuclease activity of thermus aquaticus DNA polymerase. Proc Natl Acad Sci USA 88(16): 7276–7280
7. Zeschnigk M, Bohringer S, Price EA, Onadim Z, Masshofer L, Lohmann DR (2004) A novel real-time PCR assay for quantitative analysis of methylated alleles (qama): analysis of the retinoblastoma locus. Nucleic Acids Res 32(16):e125
8. Jeronimo C, Usadel H, Henrique R, Oliveira J, Lopes C, Nelson WG, Sidransky D (2001) Quantitation of GSTP1 methylation in non-neoplastic prostatic tissue and organ-confined prostate adenocarcinoma. J Natl Cancer Inst 93(22): 1747–1752
9. Laird PW (2003) The power and the promise of DNA methylation markers. Nat Rev Cancer 3(4):253–266
10. Ogino S, Cantor M, Kawasaki T, Brahmandam M, Kirkner GJ, Weisenberger DJ, Campan M, Laird PW, Loda M, Fuchs CS (2006) CpG island methylator phenotype (CIMP) of colorectal cancer is best characterised by quantitative DNA methylation analysis and prospective cohort studies. Gut 55(7):1000–1006
11. Toyota M, Ahuja N, Ohe-Toyota M, Herman JG, Baylin SB, Issa JP (1999) CpG island methylator phenotype in colorectal cancer. Proc Natl Acad Sci USA 96(15):8681–8686
12. Issa JP (2004) CpG island methylator phenotype in cancer. Nat Rev Cancer 4(12):988–993
13. Jass JR (2007) Classification of colorectal cancer based on correlation of clinical, morphological and molecular features. Histopathology 50(1):113–130
14. Samowitz WS, Albertsen H, Herrick J, Levin TR, Sweeney C, Murtaugh MA, Wolff RK, Slattery ML (2005) Evaluation of a large, population-based sample supports a CpG island methylator phenotype in colon cancer. Gastroenterology 129(3):837–845
15. Toyota M, Ohe-Toyota M, Ahuja N, Issa JP (2000) Distinct genetic profiles in colorectal tumors with or without the CpG island methylator phenotype. Proc Natl Acad Sci USA 97(2):710–715
16. Weisenberger DJ, Siegmund KD, Campan M, Young J, Long TI, Faasse MA, Kang GH, Widschwendter M, Weener D, Buchanan D, Koh H, Simms L, Barker M, Leggett B, Levine J, Kim M, French AJ, Thibodeau SN, Jass J, Haile R, Laird PW (2006) CpG island methylator phenotype underlies sporadic microsatellite instability and is tightly associated with BRAF mutation in colorectal cancer. Nat Genet 38(7): 787–793
17. Iacopetta B, Kawakami K, Watanabe T (2008) Predicting clinical outcome of 5-fluorouracil-based chemotherapy for colon cancer patients: is the CpG island methylator phenotype the 5-fluorouracil-responsive subgroup? Int J Clin Oncol. Japan Society of Clinical Oncology 13(6):498–503
18. Van Rijnsoever M, Elsaleh H, Joseph D, McCaul K, Iacopetta B (2003) CpG island methylator phenotype is an independent predictor of survival benefit from 5-fluorouracil in stage III colorectal cancer. Clin Cancer Res 9(8):2898–2903
19. Yamashita K, Dai T, Dai Y, Yamamoto F, Perucho M (2003) Genetics supersedes epigenetics in colon cancer phenotype. Cancer Cell 4(2):121–131
20. Nosho K, Irahara N, Shima K, Kure S, Kirkner GJ, Schernhammer ES, Hazra A, Hunter DJ, Quackenbush J, Spiegelman D, Giovannucci EL, Fuchs CS, Ogino S (2008) Comprehensive biostatistical analysis of CpG island methylator phenotype in colorectal cancer using a large population-based sample. PLoS ONE 3(11):e3698
21. Ogino S, Kawasaki T, Kirkner GJ, Kraft P, Loda M, Fuchs CS (2007) Evaluation of markers for CpG island methylator phenotype (CIMP) in colorectal cancer by a large population-based sample. J Mol Diagn 9(3):305–314

Part IX

Analysis of Copy Number Variations

Comparative Genomic Hybridisation (CGH)

32

Hannelore Kothmaier, Elvira Stacher, Iris Halbwedl, and Helmut H. Popper

Contents

32.1 Introduction and Purpose

Comparative genomic hybridisation (CGH) is a technique used to identify chromosomal aberrations within solid tumours. It provides a global overview of chromosomal gains and losses throughout the whole genome. Kallioniemi et al. were the first to report CGH as a new technique in 1992 [1], shortly followed by Manoir et al. [2].

Tumour DNA (or test DNA) and normal DNA (or reference DNA) are differentially labelled (e.g. test in green and reference in red) with fluorochromes (direct labelling) or haptens (indirect labelling), which are later visualised by fluorescence. Both labelled DNAs are mixed in a 1:1 ratio in the presence of repetitive DNA (Cot-1 DNA) and are hybridised to normal human metaphase chromosomes. The red- and green-labelled DNA fragments compete for the same targets on the metaphase chromosomes. The red to green fluorescence ratio measured along the chromosomal axis represents loss or gain of genetic material in the tumour at that specific locus. These ratios show which chromosomal regions in the test genome are over or under-represented relative to the reference genome. The presence and localisation of chromosomal imbalances can be detected and quantified by digital image analysis.

The following protocol provides a step-by-step guideline to perform CGH analysis (Fig. 32.1 CGH schematic overview [3]) based on genomic DNA extracted from formalin-fixed and paraffin-embedded (FFPE) tissue specimens [4–10].

This CGH protocol is structured into six parts:

- Extraction of genomic DNA
- Metaphase chromosome preparation
- DNA labelling (indirect) by Nick Translation

H. Kothmaier, E. Stacher, I. Halbwedl, and H.H. Popper (✉)
Research Unit for Molecular Lung and Pleura Pathology,
Institute of Pathology, Medical University of Graz,
Graz, Austria

G. Stanta (ed.), *Guidelines for Molecular Analysis in Archive Tissues*,
DOI: 10.1007/978-3-642-17890-0_32, © Springer-Verlag Berlin Heidelberg 2011

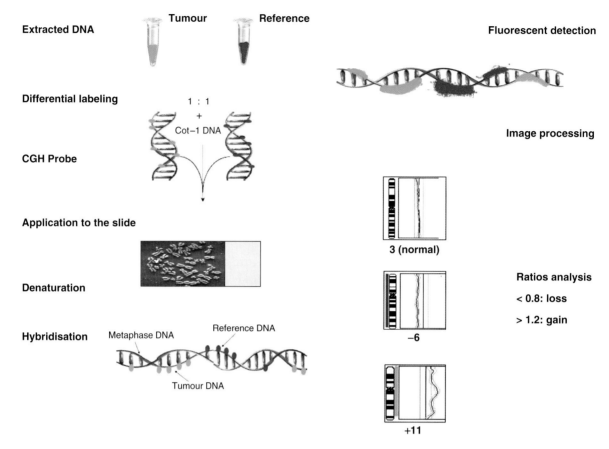

Extracted DNA   **Tumour**   **Reference**   **Fluorescent detection**

Differential labeling   1 : 1
+
Cot–1 DNA

CGH Probe

Image processing

Application to the slide

3 (normal)

Denaturation

Ratios analysis

< 0.8: loss

> 1.2: gain

–6

Hybridisation   Metaphase DNA   Reference DNA

Tumour DNA

+11

Fig. 32.1 Schematic overview of the comparative genomic hybridization (*CGH*) technique. Tumour and reference DNA are labelled with a green and a red fluorochrome, respectively, and hybridised to normal metaphase spreads. After digital imaging the fluorescent signals are captured and the green to red signals ratios are quantified digitally for each chromosomal locus along the chromosomal axis. In the ideogram (*bottom-right*), the *red* lines on the *left* and the *green* lines on the *right* relative to the chromosome indicate regions that are lost or gained, respectively. Threshold for gains and losses were set at 1.2 and 0.8, respectively

- Hybridisation
- DNA detection
- Image Processing

32.2 Extraction of Genomic DNA

Reference DNA is extracted from peripheral blood lymphocytes of a donor showing normal karyogram (see Chap. 10), and test (tumour) DNA is extracted from FFPE tissue specimens by the salting out procedure as described in chapters dedicated to DNA extraction from FFPE tissues.

Before beginning the labelling procedure, the quality of the DNA must be assessed. We routinely take

$0.5–1$ µg sample of the DNA and run it on a 1.5% ethidium bromide-stained agarose gel.

Criteria for sample selection:

- DNA of high molecular weight
- DNA should be quantified using an accurate spectrophotometer
- DNA should be dissolved in water, not TE buffer.

Contamination (or dilution) of the tumour DNA with normal DNA (e.g. from stromal or inflammatory cells) is undesirable but an inevitable problem when analysing neoplastic lesions. If the tumour sample (or test DNA) contains too much normal DNA, CGH analysis may not detect the presence of chromosomal aberrations. In detail, when more than 80% of the cells are neoplastic, whole sections can be processed. Conversely, when the tumour content is less than 80%, the tumour tissue should be manually microdissected by a needle (see Chaps. 4 and 6). In addition, advanced laser microdissection equipment can be used to reduce the number of non-neoplastic cells or separate different populations in a heterogeneous tumour.

32.3 Metaphase Chromosome Preparation

Metaphase chromosomes are prepared using phytohaemagglutinin-stimulated peripheral blood cells (lymphocytes) from a karyotypically normal donor (women or men). Cells are arrested in mitosis by colcemid followed by hypotonic treatment and fixation procedure.

High-quality metaphase preparations for CGH should ideally have little cytoplasm (too much cytoplasm causes high background levels) and minimal overlapping of the chromosomes.

32.3.1 Reagents

Note: Reagents from specific companies are reported here, but reagents of equal quality purchased from other companies may be used.

- RPMI 1640 Medium with 2 mM L-Glutamine, powder (Sigma-Aldrich R6504)
- Foetal Bovine Serum (FBS) heat inactivated (Sigma-Aldrich F4135)
- Sodium bicarbonate ($NaHCO_3$) 2 g/l (Sigma-Aldrich S6297)
- Penicillin/Streptomycin, stock solution 10,000 U/ml/10 mg/ml (Sigma-Aldrich P3539)
- Leucoagglutinin, PHA-L, stock solution 5 µg/ml (Sigma-Aldrich L2769)
- RPMI 1640 Full Medium (1,000 ml): Dissolve RPMI 1,640 powder under stirring in ddH_2O, add 2 g $NaHCO_3$ and adjust the pH to 7.2–7.3 and complement with 100 ml FBS (10%), 12 ml Penicillin/Streptomycin, 700 µl PHA-L. Filter sterilise the full medium with a 0.22 µm filter system. Store full medium at 4°C up to 4 weeks. For longer storage keep at −20°C.

- Colcemid 100 µg/ml stock solution in ddH_2O at 4°C (Demecolcine Sigma-Aldrich D2769)
- Hypotonic solution: 0.075 M potassium chloride (KCl, Merck 1.04936) in ddH_2O
- Fixative: Combine[1] Methanol/glacial acetic acid (Merck) at a 3:1 (vol/vol) ratio, ice-cold. Freshly prepare before use

32.3.2 Equipment

- Sterile 50 ml Falcon tubes (BD Falcon 2098)
- Sterile 15 ml Falcon tubes (BD Falcon 2096)
- Filter system 0.22 µm (e.g. Nalgene 125-0020)
- Sterile pipette tips
- Sterile pipettes, range: 5, 10, 25 ml
- Cell culture flask T-75 (BD Falcon 353112)
- Centrifuge suitable for centrifugation of 50 ml and 15 ml tubes at 150× g
- CO_2 cell culture incubator
- Shaker
- Water bath
- Ethanol-cleaned glass microscopy slides (e.g. Menzel)
- Microscope

32.3.3 Method

Prepare mitotic cells from short-term blood cultures (amounts per 2 ml blood)

- Add 2 ml peripheral blood (anticoagulation by lithium-heparin) to 50 ml RPMI 1,640 full medium in a T-75 culture flask. For incubation stand the culture flask upright for 72 h at 37°C in a cell culture incubator. Gently mix flasks two times per day.
- Add 60 µl colcemid and mix well. Incubate for 15–20 min at 37°C.
- Make 2 aliquots and transfer the cell culture into 50 ml tubes and centrifuge at 150× g for 10 min. Remove medium completely until 5 ml remain.

---

[1]Methanol is a poison, and highly flammable. Use only with adequate ventilation (chemical fume hood). Wear gloves when handling this agent. Contact your waste management department for proper disposal protocol.

Hypotonic treatment[2]

- Resuspend the cells in the remaining medium and gently add 45 ml of the hypotonic solution (pre-warmed at 37°C).
- Incubate the tubes for 12–15 min at 37°C in the water bath. During incubation mix the cell suspension by inverting the tube several times. The cell-hypotonic mixture should appear slightly cloudy, not clear.
- Centrifuge the cells and remove the supernatant as in step 3.

Fixation[3]

- Resuspend the cells in the remaining medium and fix the cells by adding up to 50 ml of the ice-cold fixative while agitating gently on ice. The first 5 ml should be added dropwise.
- Incubate on ice for at least 30 min. For better fixation we recommend placing at −20°C overnight.
- Centrifuge the cells at 150× g for 10 min. Remove the supernatant completely until 5 ml remain.
- After removal and resuspension of the pellet, transfer the cells into a 15 ml tube and wash the cells by adding 10 ml fixative and centrifuge at 150× g for 10 min. Repeat the washing procedure until the cell pellet is white (at least two times).
- Remove the supernatant and resuspend the cells in a small amount (3–5 ml) of fixative.

Storage of the lymphocytes

- If it is not possible to make the slides on the same day, place the lymphocytes for long-term storage at −20°C.

Metaphase spreads onto glass slides[4]

- Take one slide and moisten it by breathing from very close and drop 10–20 µl from the cell suspension onto the ethanol-cleaned slide. It is recommended to drop the cell suspension from a distance of ~40–50 cm onto the slide.
- Slides are air dried at room temperature. Check for chromosome spreading and cytoplasm debris with a light microscope. If necessary, adjust the volume of fixative so that the density of nuclei/metaphases is appropriate.
- Store the slides in a box in a dry location at room temperature for up to 4 days.[5]
- Before starting hybridisation: using a light microscope, select an area for hybridisation that will fit under a 18×18 mm^2 cover glass. Ensure that there is a sufficient number of adequate quality metaphase spreads.

32.4 DNA Labelling by Nick Translation (Indirect Labelling)[6]

The Nick Translation (NT) is used to label and digest the DNA. A small amount of DNA (0.5–1 µg) is sufficient. The NT method is based on the ability of DNase I to introduce randomly distributed nicks into DNA. *E. coli* DNA polymerase I synthesises DNA complementary to the intact strand in a $5' \rightarrow 3'$ direction using the 3'-OH termini of the nick as a primer. The $5' \rightarrow 3'$ exonucleolytic activity of DNA Polymerase I simultaneously removes nucleotides in the direction of synthesis. The polymerase activity sequentially replaces the removed nucleotides with hapten-labelled deoxyribonucleoside triphosphates (biotin or digoxigenin-labelled dNTPs).

32.4.1 Reagents

All the reagents should be kept on ice and mixed.

---

[2]If the cell pellet is just coating the bottom of the tube, add no more than 2 ml hypotonic solution; if there appears to be a volume of 0.5–1 ml of cells at the bottom of the tube, add approximately 50 ml hypotonic solution.

[3]It is crucial to add the fixative very slowly, otherwise the cell pellet will remain full of clumps, the metaphase spreads will become trapped in these clumps and spreading will be compromised.

[4]The success of this step is dependent on many factors such as cell suspension quality and laboratory conditions. Optimal conditions are a room temperature of 24°C and a relative humidity of 60%.

[5]For longer storage the metaphase slides may be kept at −80°C in the presence of silica beads.

[6]We prefer indirect labelling of our samples because it results in brighter fluorescence intensities. It has been suggested that the direct labelling reduces background, although background problems are mostly correlated with the quality of metaphase chromosomes.

- *0.1 M β-Mercaptoethanol solution*: Mix 100 μl β-Mercaptoethanol[7] 14.3 M (Sigma-Aldrich M6250) with 14.4 ml ddH$_2$O. Put an aliquot on ice and store the remaining solution at 4°C for no longer than 2 weeks
- *DNase I stock solution 1 mg/ml*: Dissolve 10 mg DNase I (Roche Applied Science 104159), in 5 g Glycerol, 20 mM Tris HCl pH 7.5, and 1 mM MgCl$_2$, and bring to 10 ml with ddH$_2$O. Store it in aliquots at −20°C
- *DNase I solution*: Dilute DNase I stock solution 1:1,000 in ddH$_2$O and put it immediately on ice
- *Deoxynucleoside Triphosphate mix (dNTP mix)*: dNTP mix, each dNTP 100 mM (Roche Applied Science 11277049001). For the dNTP mix add 5 μl dATP, 5 μl dCTP, 5 μl dGTP, and 1 μl dTTP and complete by adding 984 μl in ddH$_2$O. Aliquot and store at −20°C
- *NT reaction buffer (10×)*: 0.5 M Tris pH 8.0, 50 mM MgCl$_2$, 0.5 mg/ml BSA (Albumin, Bovine Sigma-Aldrich A-7906). Store in aliquots at −20°C
- *DNA Polymerase I 250 U* (Roche Applied Science 642711)
- *Biotin-16-dUTP 1 mM solution* (Roche Applied Science 1093088)
- *Digoxigenin-11-dUTP 1 mM solution* (Roche Applied Science 1093070)
- *Nick Translation Mix (Roche Applied Science 11745824910)*: This mix is used only for labelling the reference DNA
- *EDTA 0.5 M pH 8*

32.4.2 Equipment

- *Disinfected[8] adjustable pipettes*, range: 2–20 μl, 20–200 μl, 100–1,000 μl
- *Nuclease-free aerosol-resistant pipette tips*
- *1.5 and 0.5 ml tubes* (autoclaved)
- *PCR reaction tubes 0.2 ml*

- *Thermocycler* (e.g. Gene Amp PCR System 9700, Applied Biosystems)
- *Centrifuge* suitable for centrifugation at 20,000× g
- *Vortex*

32.4.3 Method

- *Labelling Tumour DNA[9] (or Test DNA)*. All the reagents should be kept on ice and mixed well.
- For one NT reaction 1 μg DNA is used.
- For one NT reaction (20 μl) mix:
  - 2 μl β-Mercaptoethanol solution
  - 2 μl dNTP mix
  - 2 μl NT reaction buffer (10×)
  - 0.1 μl DNase I solution
  - 0.4 μl DNA Polymerase I
  - 1 μl biotin dUTP or digoxigenin dUTP
  - 1 μg DNA in ddH$_2$O
  - Up to 20 μl with ddH$_2$O
- Vortex the sample briefly and spin down (table top centrifuge)
- Incubate the samples immediately for 3–5 h or overnight at 15°C in the PCR machine. The incubation time depends strongly on the length of the extracted DNA.
- Check the extent of fragmentation by gel electrophoresis on an ethidium bromide-stained 1.5% agarose gel (10 μl/sample). Run at 130 V for 40 min. Inspect DNA fragment length[10] with a UV transilluminator. If the DNA fragments have been digested to the right size, stop the reaction as described under point 5.
- Stop the reaction. Chill the reaction on ice and inactivate the enzymes by heating the samples at 75°C for 15 min in the PCR machine. Keep the samples until hybridisation at −20°C.

---

[7]β-Mercaptoethanol is toxic by inhalation, ingestion and through skin contact and is a severe eye irritant. Use only with adequate ventilation, wear gloves and safety glasses. Wasted β-Mercaptoethanol must be placed in a chemical waste container.

[8]Clean the pipette with alcohol or another disinfectant and leave them under the UV lamp for at least 10 min. Alternatively, it is possible to autoclave the pipette depending on the provider instructions for the different pipettes.

[9]In this scheme, the tumour DNA is labelled with biotin and the reference DNA is labelled with digoxigenin. The labelling scheme can be switched: the tumour DNA is labelled with digoxigenin, and the reference DNA with biotin.

[10]The length of DNA fragments should be at least 300 bp. Check the extent of fragmentation by gel electrophoresis. While running the gel, keep the samples on ice. If the fragment sizes are too big, then you can add fresh reagents and continue labelling. If the DNA fragments have been digested to the right size stop the reaction as described under point 5.

- Labelling normal[11] or reference DNA. Only for labelling the reference DNA a commercial mix (Roche Mix; containing DNase I, DNA Polymerase I, and buffer) is used. For one NT reaction 1 µg DNA is used.
- For one NT reaction (20 µl) mix:
  - 2 µl dNTP mix
  - 4 µl Nick translation mix (5×)
  - 1 µl Biotin dUTP or Digoxigenin dUTP
  - 1 µg DNA in ddH$_2$O
  - Up to 20 µl with ddH$_2$O
- Vortex the sample briefly and spin down (table top centrifuge)
- Incubate the samples immediately for 90 min at 15°C in the PCR machine.
- Check the extent of fragmentation by gel electrophoresis on an ethidium bromide-stained 1.5% agarose gel (10 µl/sample). Run at 130 V for 40 min. Inspect DNA fragment length with a UV transilluminator. If the DNA fragments have been digested to the right size stop the reaction as described under point 5.

32.5 Hybridisation

Following the labelling procedure, equal amounts of labelled tumour and labelled reference DNA are precipitated together in the presence of excess human Cot-1 DNA and fish sperm DNA.

This precipitation must be done to remove the dNTPs that have not been incorporated in the labelling reaction. Whereas Cot-1 DNA, which is enriched in repetitive sequences, is added to prevent non-specific binding of labelled DNA to repetitive sequences present in the centromere and heterochromatic regions. Fish-Sperm DNA is used to prevent loss of labelled DNA in the precipitation. Further, the probe is re-suspended in a hybridisation solution and denatured. After denaturation, the probes are allowed to anneal and then added to the denatured slides and allowed to hybridise for 72 h.

32.5.1 Reagents

- Labelled tumour DNA and labelled reference DNA (see Chap. 32.4: Nick Translation)
- Fish Sperm DNA (molecular biology grade), lyophilised (Boehringer Mannheim 1467140)
- Human Cot-1 DNA (Roche Applied Science 1581074)
- 3 M Sodium Acetate pH 5.2 (NaAc, Roth 6773.1)
- 100% Ethanol, 90% Ethanol, 70% Ethanol (Merck 1000.983.2511)
- Deionised Formamide[12] (FA, 100%): Combine 60 ml FA (Merck 1.04008.1000) and 10 g Dowex MR-3 ion exchange resin (Sigma-Aldrich 13686-U) under stirring for 30 min, and filter. Aliquot and store it at -20°C
- 20× SSC (saline sodium citrate): 3 M NaCl and 0.3 M Tri-sodium citrate dihydrate are dissolved in ddH$_2$O under stirring and pH is adjusted to 7 and keep it at 4°C.
- Hybridisation mix: 20% dextran sulphate, 4× SSC pH 7. Dissolve 10 g Dextran sulphate in 10 ml 20× SSC under stirring and heating (set to 50°C), fill up to 50 ml with ddH$_2$O and autoclave the solution. Store it in aliquots at −20°C.
- Denaturation mix: Freshly mix 700 µl FA, 200 µl ddH$_2$O, and 100 µl 20× SSC and store at 4°C until denaturation step

32.5.2 Equipment

- *Disinfected[13] adjustable pipettes*, range: 2–20 µl, 20–200 µl, 100–1,000 µl
- *Nuclease-free aerosol-resistant pipette tips*
- *1.5 and 0.5 ml tubes* (autoclaved)
- *Centrifuge* suitable for centrifugation of 1.5 ml tubes at 20,000× g at 4°C

---

[11]In this scheme, the tumour DNA is labelled with biotin and the reference DNA is labelled with digoxigenin. The labelling scheme can be switched and the tumour DNA is labelled with digoxigenin, and the reference DNA with biotin.

[12]Formamide causes eye and skin irritation and is a known mutagen. Wear gloves when handling this reagent and dispense under a chemical fume hood. Contact your waste management department for proper disposal protocol.

[13]Clean the pipette with alcohol or another disinfectant and leave them under the UV lamp for almost 10 min. Alternatively, it is possible to autoclave the pipette depending on the provider instructions for the different pipettes.

- *Cover glasses*, 60×24 mm^2, 18×18 mm^2 (e.g. Menzel)
- *Rubber cement* (Marabu-Fixogum 2901 17000)
- *Thermomixer* (two) (Eppendorf)
- *Hot plate* for heating up to 80°C
- *Water bath*
- *Staining jars*
- *Metal box* (stainless steel)
- *Diamond pen*

32.5.3 Method

Precipitation of the DNA

- The best recovery of DNA results from leaving the DNA/salt/ethanol mixture at −20°C over night before centrifugation:
- For one precipitation mix:
  20 µl tumour DNA (biotin labelled)
  10 µl reference DNA (digoxigenin labelled)
  40 µl Cot-1 DNA (40-fold excess)
  1 µl Fish Sperm DNA
  7.1 µl 3 M NaAc (1/10 vol)
  180 µl 100% Ethanol (2.5 × vol)
  Place the tubes at −20°C overnight
- Centrifuge the samples at 20,000× g at 4°C for 30 min. Remove the supernatant and wash the pellet with 180 µl 70% ethanol and centrifuge at 20,000× g at 4°C for 15 min. Remove the supernatant and air dry the pellet.
- Resuspend the DNA pellet in 7 µl FA and incubate at 37°C in the thermomixer with shaking at 1,400 rpm for 60–90 min.
- Gently add 7 µl hybridisation solution and incubate at 37°C in the thermomixer with shaking (1,400 rpm) for at least 10 min.

Denaturation of the genomic DNA

- Denature the DNA for 5 min at 85°C.

Pre-annealing

- Immediately after denaturation spin down the condensate and place the probe back into a different thermomixer at 37°C for 1 h. During pre-annealing, start the metaphase chromosome slide preparation!

Metaphase chromosome slide preparation

- Inspect the slides with the metaphase spreads in a light microscope and select the best region for hybridisation that will fit under an 18×18 mm^2 cover glass with a diamond pen.

Slide-denaturation[14]

- Begin the slide-denaturation step by adding 130 µl of the denaturation mix to a 24×60 mm^2 cover glass. Invert the slide (with chromosome spreads) slowly on the drop of the denaturation mix, re-invert the slide, metaphase-spread-side up (avoid air bubbles) and incubate for 10 min at room temperature and then denature the slide onto a hot plate at 78°C for 75–85 s.[15]
- Dehydrate the slides by transferring them into an ascending concentration of ethanol in separate staining jars. First place the slide vertically to remove the cover slip and then immerse the slides in ice-cold[16] 70%, 90%, and 100% ethanol for 2 min each.
- Briefly air dry the slide warming up to room temperature and then place it horizontally to complete the drying process with nitrogen gas.

Probe hybridisation on the slide

- Centrifuge the tube with the pre-annealed DNA (see point of Pre-annealing) to remove vapour and fluid from the lid.
- Carefully apply the entire content of the probe (14 µl) onto the area selected for the hybridisation, and gently place the cover glass (18×18 mm^2) onto the region containing the probe.[17]
- Seal the edges of the cover glass with rubber cement.
- Work quickly and place the slide in a pre-warmed metal box (37°C) and keep the metal box in a water bath at 37°C for 72 h.

---

[14]Denaturation of the slide is a major determinant of the success of the hybridization of CGH probes. It is important that the hot plate (or slide warmer) is set to the right temperature. The age of slides, amount of cytoplasm and humidity at which cells were dropped onto the slides determine the time required for denaturation.

[15]The correct timing is often determined empirically.

[16]Immersing the slides in ice-cold ethanol after slide denaturation immediately arrests the heat denaturation process and prevents the chromosomes from re-annealing. Cool the ethanol to −20°C by storing a tightly sealed staining jar inside a −20°C freezer before use. When ready to use, keep the staining jars in an ice bucket.

[17]Taking care not to scratch the slide during this process and taking care to remove air bubbles.

32.6 Detection

After hybridisation for 72 h, the unbound DNA is washed from the slide using formamide and stringent washes of SSC. The slides are also incubated with blocking solution to suppress any unspecific binding of the antibodies. After detection by the fluorescently labelled antibodies FITC (fluorochrome-conjugated avidin-fluorescein isothiocyanate antibody) and TRITC (sheep antidigoxigenin-tetramethyl rhodamine isothiocyanate), respectively, the slides are washed again to remove any unbound antibodies and counterstained with DAPI. This produces a banding pattern, which enables chromosomes identification and karyotyping.

32.6.1 Reagents

- *Formamide deionised (FA, 100%)*: 60 ml FA (Merck 1.04008.1000) and 10 g ion exchange resin Dowex MR3 (Sigma Aldrich I-9005) are mixed and stirred for 30 min, and filtered. Aliquot and store it at −20°C.
- *20× SSC:* 3 M NaCl and 0.3 M tri-sodium citrate dihydrate are dissolved in ddH$_2$O under stirring and pH is adjusted to 7. Keep the solution at 4°C.
- *DAPI*[18] *(4,6-diamidino-2phenylinodole dihydrochloride) solution 0.4 µg/ml*: For DAPI stock solution add 10 mg DAPI (Roche Applied Science Nr. 236276) to 2 ml ddH$_2$O at a concentration of 5 mg/ml. Store aliquots at −20°C. For DAPI working solution add 2 µl DAPI stock solution to 100 ml 2× SSC (1:10 dilution of 20× in ddH$_2$O). Store aliquots at −20°C. Keep solution in use in light-protected staining jar.
- *Formamide/SSC 1:1*: For final volume of 300 ml. Combine 75 ml 4× SSC, 75 ml ddH$_2$O, and 150 ml FA and adjust the pH to 7.0–7.3 with 1 N HCl. Prepare three staining jars and preheat them to 37°C in the water bath.
- *4× SSC:* Mix 100 ml 20× SSC pH 7 with 400 ml in ddH$_2$O. Prepare four staining jars with 4× SSC and preheat 1 to 37°C and the remaining 3 to 45°C in the water bath

- *4× SSC/0.1% Tween 20*: (Merck 822184 S21990 715)
- *0.1× SSC*: Mix 10 ml 4× SSC with 390 ml ddH$_2$O. Fill three staining jars with 0.1× SSC and preheat them to 60°C in the water bath.
- *Blocking solution*: 3% BSA (Albumin, Bovine Sigma-Aldrich A-7,906) in 4× SSC/ 0.1% Tween. Pre-warm the blocking solution to room temperature.
- *TRITC* (Anti-Digoxigenin-Rhodamine, Fab Fragment Roche 12077509)
- *Streptavidin FITC* (Fluorescein DCS from Vector Laboratories A-2011)
- *Fluorochrome solution:* Before adding the fluorochromes,[19] TRITC and FITC, to the blocking solution centrifuge each for 3 min at 20,000× g. After centrifugation combine 14 µl TRITC and 7 µl Streptavidin FITC with 1 ml blocking solution and pre-warm it at 37°C. (Prepare just before beginning detection).
- *DABCO (1.4-diazabicyclo2.2.2octane)*: Dissolve 23 mg DABCO (Sigma-Aldrich D2522) in 10 ml glycerol/Tris (9 vol glycerol and 1 vol 0.2 M Tris pH 8).

32.6.2 Equipment

- *Disinfected*[20] *adjustable pipettes*, range: 2–20 µl, 20–200 µl, 100–1,000 µl
- *Nuclease-free aerosol-resistant pipette tips*
- *1.5 and 0.5 ml tubes* (autoclaved)
- *Centrifuge* suitable for centrifugation of 1.5 ml tubes at 20,000× g at 4°C
- *Plastic cover slips* (e.g. Appligene Oncor S1370-14)
- *Cover glasses*, 60 × 24 mm^2 (e.g. Menzel)
- *Thermomixer* (e.g. Eppendorf)

---

[18]DAPI is a known mutagen; avoid contact with eyes and skin. Protect DAPI solution from light; it can be stored at 4°C for as long as 2 months. For longer storage, aliquot and keep it at −20°C.

[19]Do not repeatedly freeze and thaw antibodies used in the detection protocols. Store aliquots at 4°C according to the manufacturer's directions and record expiration dates. Keep the antibodies out of light when using.

[20]Clean the pipette with alcohol or another disinfectant and leave them under the UV lamp for almost 10 min. Alternatively, it is possible to autoclave the pipette depending on the provider's instructions for the different pipettes.

- *Water bath*
- *Staining jars*
- *Moist chamber*
- *Incubator* (e.g. Heraeus)

32.6.3 Method

- After the 72-h hybridisation, carefully remove the rubber cement and the cover glass from the slide, taking care not to drag the cover glass across the slide, thereby scratching the metaphase preparations.
- Wash the slides in Formamide/SSC (1:1) at 37°C for 3 min in the water bath. Repeat this step two more times using fresh solution each time.
- Wash the slides in 0.1× SSC for 2 min and 30 s at 60°C.[21] Repeat this step two more times using fresh solution each time.
- Keep the slides in 4× SSC/0.1% Tween at 37°C for 5 min.
- *Blocking step*: Cover each slide with130 µl blocking solution and a plastic cover slip and incubate for 10 min in a moist chamber at 37°C within an incubator.
- Remove cover glass and place the slides immediately in 4× SSC/0.1% Tween (pre-warmed to 37°C) and wash the slides by gently shaking for 5 min.
- *Prepare the fluorochrome solution*. Cover each slide with 130 µl of the fluorochrome solution and a cover slip and incubate for 30 min in a moist chamber at 37°C (protecting from light).
- Gently remove the cover slip. Wash the slides in 4× SSC/0.1% Tween for 3 min at 45°C. Repeat this washing step two more times using a fresh solution each time (protecting from light).
- *DAPI counterstain*.[22] Incubate the slides for 5 min in the DAPI solution at room temperature. Keep the slides out of direct light throughout the DAPI staining procedure.
- Briefly immerse the slides in 2× SSC/50 µl Tween 20.

- Place the slide horizontally and blow it off with nitrogen gas.
- Mount the slides using DABCO. Add 35 µl DABCO per glass slide. Carefully remove any air bubble. The slides can be imaged immediately or alternatively, can be stored at 4°C under dark conditions until microscopy observation.[23]

32.7 Image Processing

Images are acquired with a Zeiss Axioplan microscope (Jena, Germany) equipped with a cooled charge-coupled device camera (Micromax; Princeton Instruments, Trenton, NY) controlled by the IPlab software (Scanalytics, Vienna, USA), and DAPI (blue), FITC (green), and TRITC (red) band pass filters.

Furthermore objective lenses (plan apochromatic, 63× and 100×) and camera should be appropriate and apochromatic without autofluorescence.

For CGH image analysis, the Quips 3.1 software (produced by Vysis Downers Grove, IL, USA and now available by Applied Imaging, United Kingdom) is used.

For the identification of chromosomal gains and losses, it is essential that the chromosomes are correctly classified. For each case, the results of analysis of multiple metaphase spreads are averaged; we prefer to analyse at least ten metaphase spreads. Special attention should be paid to the G-C-rich regions 1p, 9p, 16, 19, and 22, which are known as problem areas because they show aberrations in a negative control experiment and indicate suboptimal CGH (Fig. 32.2).

Similar to other molecular genetic techniques, the proper controls should be included. As a control, male versus female hybridisations should be performed in order to test stringency and sensitivity. In control experiments the confidence interval (99%) should be within the set thresholds (1.2 and 0.8) with the exception of the sex chromosomes in a sex-mismatch control. For an internal control for CGH use control DNA of the other sex. To detect chromosomal imbalances of the X chromosome in the analysed samples, CGH should be performed using tumour and control DNA of the same gender.

---

[21]Take care: these washing steps are very important; the slides should be washed in adequate time temperature, and without any agitation.

[22]It is frequently described combining DAPI in the antifade solution. However, this often results in a transient weak blue fluorescence background. We prefer to stain the chromosomes by DAPI diluted in 2× SSC.

---

[23]Store the slides in the dark at 4°C ideally no longer than a week until microscopy observation.

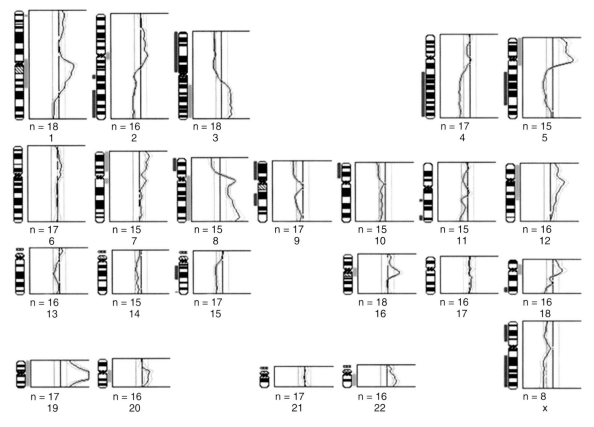

Fig. 32.2 Quantitative analysis of a CGH profile (lung, pleomorphic carcinomas) by relative copy number karyotype. On the average, ten metaphase spreads were evaluated per case. The mean red to green fluorescence ratio of the chromosome of multiple metaphases is plotted on a graph corresponding to the chromosome ideograms. Thresholds for chromosomal gains and losses were set to 1.2 and 0.8, respectively. Losses of genetic material are displayed in *red* to the *left*, while gains are pictured in *green* to the *right* of each chromosome

In our lab the detection thresholds for gains and losses of chromosomal regions are set at 1.2 and 0.8, respectively (Fig. 32.2). Ratios below 0.8 indicate a loss, whereas ratios over 1.2 indicate a gain of chromosomal regions. The signals should be smooth because a granular signal results in large standard deviations after interpretation. Varieties in signal intensity in a banding-like pattern probably result from inadequate denaturation and therefore not useful. Background should be low and homogenous.

32.7.1 Troubleshooting

- *Chromosome metaphase preparations on the slide have excess cytoplasm.* Hypotonic treatment was too short. The time of exposure to the hypotonic solution should be prolonged. Hypotonic treatment causes a swelling of the cells; the optimal time of treatment varies for different blood donors and must be determined empirically.
- *If large amounts of residual cytoplasm are still visible on the slides*, the fixation step should be repeated a few times.
- *Metaphase spreads are trapped in clumps.* Ensure that before starting the fixation step the cell pellet (after hypotonic treatment) is fully resuspended in the remaining solution and then very slowly add the fixative.
- *Small number of metaphase spreads.* The quality of metaphase spreading is dependent upon a number of factors, including humidity, air-flow, and cell concentration.
- *DNA fragments are still too large after NT process I.* Add more DNase I and DNA polymerase I

(approximately 3 μl), and incubate at 15°C for 15–30 min. For optimum hybridisation conditions the DNA fragments should be at least 300 bp (visible as a smear). Check fragment size by gel electrophoresis and inactivate the enzymes as described.

- *DNA fragments are still too large after NT process II*. Another possibility is that there is something inhibiting the enzymatic reaction (residual salts from the DNA extraction; DNA is resuspended in TE buffer and not in water).
- *DNA fragments are too small after NT*. Check the DNA quality of the unlabelled sample; the starting DNA might have been already degraded. Use only DNA of high molecular weight for a CGH experiment. Small fragments can result in non-specific hybridisation, whereas long fragments can result in halos around the chromosomes and an increased background.
- *Optimal-sized DNA fragments* of labelled DNA can be acquired by an optimally sized ratio of DNase I and DNA polymerase I, and constant temperature at 15°C.
- *Weak signal*. If both the tumour and reference DNA samples are weak, then it could be possible that there was not sufficient haptens incorporation into the DNA. Check the DNA fragment size. Further consider the possibility that the DNA polymerase has lost its activity.
- *Signal intensity is weak*:
  - If the metaphase spreads are not sufficiently denatured, then the labelled DNA probes cannot obtain access to these sites for hybridisation. Check the age of the slides, older slides are more resistant to heat denaturation.
  - Pre-annealing time was too extensive. Reduce time.
- *No signal*:
  - Quality of metaphase (slides are over 14 days).
  - Probe and slide denaturation temperature. Check the actual temperature of the equipment used to denature the probe and slide.
  - Detection reagents. The detection reagents used in the process are approaching expiration or were improperly diluted.
  - The wrong filter was used during imaging.
- *Weak DAPI staining*. Decrease denaturation time. Weak DAPI staining is a hint that the DNA on the metaphase spreads was denatured too long. Over-

denatured chromosomes show very bright-staining centromere. Reduce the time for denaturation process.

- *Non-specific background* (haze of red and green around the metaphases).
  - Residual cytoplasm can prevent successful access of the probe to the chromosomes.
  - Blocking step was not sufficient. The incubation time can be increased.
  - Washing conditions. Change stringency of the wash solutions (salt concentration and temperature).
- *High background*
  - Wash solutions are not prepared freshly and stored for a long period of time. Always make fresh solutions and check the pH.
  - BSA was not properly washed off.
  - Slides were not denatured long enough.
  - Insufficient hybridisation time.
- *Poor chromosome banding*. The chromosomes might be over-denatured, and a new batch of chromosomes, less formamide in the hybridization buffer, or lower temperatures for denaturation might be tested.

References

1. Kallioniemi A, Kallioniemi OP, Sudar D, Rutovitz D, Gray JW, Waldman F, Pinkel D (1992) Comparative genomic hybridization for molecular cytogenetic analysis of solid tumors. Science 258(5083):818–821
2. du Manoir S, Speicher MR, Joos S, Schrock E, Popp S, Dohner H, Kovacs G, Robert-Nicoud M, Lichter P, Cremer T (1993) Detection of complete and partial chromosome gains and losses by comparative genomic in situ hybridization. Hum Genet 90(6):590–610
3. Jeuken JW, Sprenger SH, Wesseling P (2002) Comparative genomic hybridization: practical guidelines. Diagn Mol Pathol 11(4):193–203
4. Blaukovitsch M, Halbwedl I, Kothmaier H, Gogg-Kammerer M, Popper HH (2006) Sarcomatoid carcinomas of the lung – are these histogenetically heterogeneous tumors? Virchows Arch 449(4):455–461
5. Halbwedl I, Ullmann R, Kremser ML, Man YG, Isadi-Moud N, Lax S, Denk H, Popper HH, Tavassoli FA, Moinfar F (2005) Chromosomal alterations in low-grade endometrial stromal sarcoma and undifferentiated endometrial sarcoma as detected by comparative genomic hybridization. Gynecol Oncol 97(2):582–587
6. Padilla-Nash HM, Barenboim-Stapleton L, Difilippantonio MJ, Ried T (2006) Spectral karyotyping analysis of human and mouse chromosomes. Nat Protoc 1(6):3129–3142
7. Stacher E, Ullmann R, Halbwedl I, Gogg-Kammerer M, Boccon-Gibod L, Nicholson AG, Sheppard MN, Carvalho L,

Franca MT, Macsweeney F, Morresi-Hauf A, Popper HH (2004) Atypical goblet cell hyperplasia in congenital cystic adenomatoid malformation as a possible preneoplasia for pulmonary adenocarcinoma in childhood: a genetic analysis. Hum Pathol 35(5):565–570

8. Ullmann R, Bongiovanni M, Halbwedl I, Fraire AE, Cagle PT, Mori M, Papotti M, Popper HH (2003) Is high-grade adenomatous hyperplasia an early bronchioloalveolar adenocarcinoma? J Pathol 201(3):371–376

9. Ullmann R, Petzmann S, Klemen H, Fraire AE, Hasleton P, Popper HH (2002) The position of pulmonary carcinoids within the spectrum of neuroendocrine tumors of the lung and other tissues. Genes Chromosom Cancer 34(1): 78–85

10. Ullmann R, Petzmann S, Sharma A, Cagle PT, Popper HH (2001) Chromosomal aberrations in a series of large-cell neuroendocrine carcinomas: unexpected divergence from small-cell carcinoma of the lung. Hum Pathol 32(10): 1059–1063

Multiplex Ligation-dependent Probe Amplification (MLPA)

33

Ana Pilar Berbegall, Eva Villamón, Samuel Navarro, and Rosa Noguera

Contents

33.1 Introduction and Purpose

To date, many different methods have been used for the detection of genetic alterations, including conventional genetic analysis, Southern blot, Fluorescence In Situ Hybridization (FISH), Metaphasic-Comparative Genomic Hybridization (mCGH), and real-time quantitative PCR. More recently, array-CGH (a-CGH) and Single Nucleotide Polymorphism (SNP)-based platforms have been applied in order to detect copy number changes. Such methods improve resolution and sensitivity; these are very robust techniques for the detection of cryptic chromosome rearrangements, but they are labor intensive and require expensive equipment.

The Multiplex Ligation-dependent Probe Amplification (MLPA) method was first described in 2002 [1] and is a PCR-based method that allows determining the copy number status of up to 40 DNA sequences at once. This method relies on sequence-specific probe hybridization genomic DNA followed by multiplex-PCR amplification of the hybridized probe and a semiquantitative analysis of the resulting PCR products. MLPA has several advantages over currently used techniques. The first advantage is the number of loci that can be analyzed in one reaction. Furthermore, it is a sensitive and relatively fast technique; only a small amount of DNA is required (20 ng is sufficient for one reaction in which 40 loci are tested) and results are available within 2 days. Since then different studies have been published, the majority using commercial MLPA assays to detect gene deletions and duplications in different disorders [2–8].

The critical factors when performing MLPA analyses from formalin-fixed paraffin-embedded (FFPE) tissues are DNA integrity and purity; for this reason, a suitable DNA extraction method must be chosen. Formalin treatment should not hinder the MLPA technique since each MLPA probe hybridizes to a very small DNA target

A.P. Berbegall, E. Villamón, S. Navarro, and R. Noguera (✉)
Department of Pathology, Medical School,
University of Valencia, Valencia, Spain

G. Stanta (ed.), *Guidelines for Molecular Analysis in Archive Tissues*,
DOI: 10.1007/978-3-642-17890-0_33, © Springer-Verlag Berlin Heidelberg 2011

sequence (less than 100 nucleotides). However the form-aldehyde reactivity leads to the formation of protein–nucleic acid and protein–protein cross-links [5, 9]. These chemical modifications have a stronger influence on MLPA results than on common nonquantitative PCR ones. In addition, the extremely low pH during the fixation period provokes DNA fragmentation [10]. Related with the purity, the DNA extracted from FFPE tissues generally contains traces of substances such as formalin or xylene that can interfere with DNA, affecting its purity and reducing the polymerase activity in the PCR reaction. Important practical details and considerations are discussed in this chapter to obtain an improved DNA quality extraction with the highest multiplex ligation probe amplification efficiency.

This MLPA protocol provides a methodology for analyzing the genetic copy number changes from FFPE tissues. The critical steps include: DNA extraction from tissue sections, and MLPA procedure. In our protocol, all MLPA reagents have been supplied by the MRC-Holland biotech company. The whole procedure requires 2 days, approximately 1 h of bench work on the first day and 2–3 h on the second one.

33.2 DNA Extraction

33.2.1 Reagents

- *100% and 70% ethanol*
- *3 M Sodium acetate (NaAc) pH 5.2*
- *P-Buffer*: Made by mixing 425 μL sterile water + 25 μL 1 M Tris–HCl pH 8.5 + 10 μL 5 M NaCl + 1 μL 0.5 M EDTA pH 8.0 + 10 μL 1 M DTT + 25 μL of a mixture of 10% Tween-20 + 10% NP40
- *Proteinase K solution, 20 mg/mL*: Dissolve 50 mg proteinase K (Sigma-Aldrich, cat. No P6556) in 2.5 mL of PBS buffer. Prepare 100 μL aliquots and store the enzyme at −20 °C
- *TE buffer*: 10 mM TrisHCl pH 8.2 + 1 mM EDTA

33.2.2 Equipment

- *Automatic pipettes*. Range: 10–20 μL, 100–1000 μL
- *Bath at 90 °C, 56 °C, and 37 °C*
- *Centrifuge* suitable for 1.5 mL tubes

- *Nanodrop spectrophotometer*
- *Vortex*
- *Thermometer*
- *Timer*

33.2.3 Method

- Cut around ten sections of 5 μm using a microtome and collect them in a 1.5 mL microcentrifuge tube. A minimal tumor cell percentage of 50–70% is needed to obtain reliable MLPA results.
- Add 500 μL of P-Buffer to each tube. Vortex the tubes for 15 s and ensure that the buffer covers the tissue slices.
- Incubate the tubes at 90 °C for 15 min to melt the paraffin in a water bath. At this stage, it is possible to observe a paraffin layer at the top of the digestion solution.[1]
- Incubate the samples at 56 °C in a water bath.
- Once the temperature of the samples has decreased add 12.5 μL of proteinase K solution and continue the tissue digestion at 56 °C overnight (at least 16 h) in a water bath.[2]
- Pipette 20 μL of proteinase K to each tube and incubate them at 37 °C for other 24 h. Again, add extra proteinase K (20 μL) to each tube and incubate them at 37 °C for 24 h.
- Centrifuge the tubes at ~13,000 rpm for 10 min.
- Carefully transfer the supernatant to a clean tube avoiding the paraffin residues.
- Repeat the last two steps to ensure that the maximum amount of paraffin residues is discarded.[3]
- Add two volumes of 100% ethanol and 0.1 volume of NaAc 3 M to the digestion solution and let the DNA precipitate overnight at −20 °C.
- Centrifuge the tubes at ~13,000 rpm for 30 min.
- Remove the supernatant.
- Wash the DNA with 500 μL of 70% ethanol.[4]

---

[1] The pH of the Tris solution will decrease during the heat treatment at elevated temperatures, thus it may result in DNA depurination induced by low pH. For this reason it is recommended using Tris pH 8.5 during the heat pretreatment.

[2] Incubation of the samples at 56 °C for 48 h or longer is a suitable option to increase the PCR amplifiable DNA yield.

[3] It is important to remove the paraffin as much as possible by centrifuging after the tissue digestion instead of using any solvent.

[4] The MLPA PCR reaction may be hampered by the presence of impurities. In order to reduce the impurities repeat the ethanol wash when enough DNA is available.

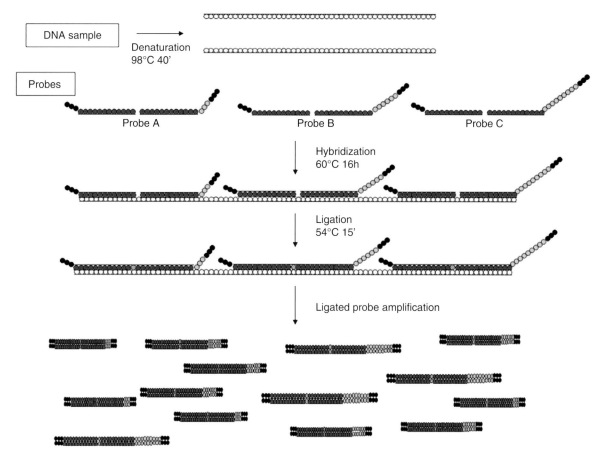

Fig. 33.1 General overview of the MLPA reaction. Hybridization sequences appear in *red*. PCR primer sequences are in *black* and stuffer sequence in *blue*

- Centrifuge the tubes at ~13,000 rpm for 15 min.
- Remove the supernatant.
- Evaporate the ethanol (i.e. 37°C, vacuum dry or thermoblock with tube open cap).
- Resuspend the DNA in water or TE buffer.[5]
- The DNA concentration is measured by using a NanoDrop spectrophotometer. Dilute the DNA to working stocks of ~20 ng/μL using TE.
- Store at 4°C or −20°C for long storage periods.

33.3 MLPA Procedure

There are different MLPA probemix available, applied in human genetics and cancer research. MLPA can be used to determine genetic copy number changes, to

detect known point mutations and SNPs (using the so-called P-products), methylation status of imprinted and promoter regions (MS-MLPA; ME products), and relative quantification of mRNAs (RT-MLPA; R-products).

The following MLPA protocol provides a methodology for using P-products. Details about all MLPA P-products are available at MRC-Holland company website (http://www.mlpa.com). The SALSA MLPA kit includes the MLPA probemix to detect alterations in around 40 genes or chromosomal regions. Each MLPA probe is complementary to a given target sequence. MLPA probes consist of two oligonucleotides that hybridize immediately adjacent to each other, each containing one of the sequences recognized by the PCR primer pair (Fig. 33.1). After hybridization to their adjacent targets, the two hemi probes are ligated by a specific ligase enzyme. The resulting ligation product contains both PCR primer sequences in one fragment and hence will be

---

[5]Resuspending the DNA in TE is more appropriate than water because TE keeps the DNA more stable.

exponentially amplified during the PCR reaction. Identical PCR fluorescently (6-FAM) labeled primers are used for all probes. In contrast, probe oligonucleotides that are not ligated will contain only a single primer sequence. As a consequence, nonligated hemi probes will not be exponentially amplified and will not generate a signal. One of the two oligos presents a stuffer sequence of variable length (19–370 pb, 3 pb increments) specific for each probe. This stuffer sequence is responsible for the different size of PCR products (between 130 and 481 bp). The specific amplification products size allows identifying them by capillary electrophoresis.

33.3.1 Reagents

- *MLPA kit (MRC-Holland)*: the kit contains SALSA MLPA buffer, SALSA Probemix solution, Ligase-65, Ligase-65 buffers A and B, SALSA PCR buffer, SALSA PCR Primer mix (universal PCR primer pair+dNTPs), SALSA Polymerase, and SALSA Enzyme dilution buffer
- *Highly deionized formamide* (Applied Biosystems, cat No. 4311320).
- *Labeled size standard* ROX, TAMRA, D1-500, or LIZ-500 (GeneScan, Applied Biosystems, cat No. 4322682)
- *Suitable gel separation matrix for MLPA* (Applied Biosystems: POP4, suggested, or POP7)

33.3.2 Equipment

- *Thermocycler* with heated lid
- *Vortex*
- *Automatic pipettes*. Range: 0.2–1 μL, 1–10 μL, 10–20 μL, 100–1,000 μL
- *1.5 mL microcentrifuge tubes*
- *0.2 mL tubes or PCR plates*
- *Capillary Sequencer* with fragment analysis software (GeneMapper software package of Applied Biosystems or GeneScan of BIOMED-2-GS)
- *Coffalyser MLPA-DAT software* (MRC-Holland)/ *MLPA-Vizard* (Austrian Research Centres) software for neuroblastoma kits

33.3.3 Method

33.3.3.1 MLPA Reaction

- Add 5 μL of DNA working stock to each tube or water for negative control. Spin the tubes briefly in a centrifuge to collect the sample DNA at the bottom.[6]
- Place the tubes in the thermocycler and start the denaturation for 40 min at 98°C and cool the samples at 25°C.[7]
- Make a mix of 1.5 μL of MLPA buffer and 1.5 μL of the Probemix solution for each reaction. Vortex briefly.
- Once the samples have reached 25°C add 3 μL of the mix to each tube and mix well.
- Incubate the samples at 95°C for 1 min and allow them to hybridize for 16 h at 60°C.[8]
- Prepare the ligase mix: 25 μL dH_2O + 3 μL of buffer A + 3 μL of Buffer B + 1 μL of Ligase 65 for each reaction. Vortex with care.[9]
- At 54°C pipette 32 μL of the ligase mix to each tube and mix well. Incubate for 15 min at 54°C, followed by 5 min at 98°C and hold at 4°C.
- Prepare the PCR buffer mix by mixing: 2 μL of PCR buffer and 3 μL of dH_2O for each reaction. Add 5 μL of the mix to each of the clean tubes.
- Add 15 μL of the ligation product to each clean tube and mix it by pipetting. Spin the tubes briefly to make sure that all the mix remains at the bottom. Place the tubes on a preheated thermocycler (60–72°C).
- Prepare the PCR master mix: 1 μL primer mix + 1 μL enzyme dilution buffer + 2.25 μL dH_2O for each reaction. Prepare it on ice and mix well by pipetting up and down.

---

[6]Sample volumes should not exceed 5 μL.

[7]Some contaminants like ions or salts can cause low denaturation efficiency. Incomplete DNA denaturation results in low signal of probes located close to CpG islands. We recommend extending the first MLPA denaturation step up to 40 min when using DNA from FFPE samples to be sure that the DNA denaturation is complete.

[8]The hybridization time can be anywhere between 12 and 24 h. The hybridization of probes to their targets should be nearly complete after 12 h. A hybridization time between 16 and 18 h is recommended.

[9]When performing MLPA on large sample numbers, multichannel pipettes are recommended.

- Add 0.25 μL of polymerase for each reaction and mix it well.
- Add 5 μL of the PCR mix to each tube and start the PCR as soon as possible.[10]

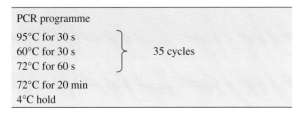

PCR programme	
95°C for 30 s	
60°C for 30 s	35 cycles
72°C for 60 s	
72°C for 20 min	
4°C hold	

33.3.3.2 MLPA Capillary Electrophoresis

Amplification products must be identified and quantified by capillary electrophoresis on a genetic analyzer.

- The amount of the MLPA PCR reaction required for analysis by capillary electrophoresis depends on the instrument and fluorescent label used (settings details are available in www.mlpa.com in the support section under technical MLPA protocols). The following specifications are applicable when using ABI 3730 DNA analyzer (Applied Biosystems):
- Mix 1–3 μL of the PCR product + 0.044 μL internal size standard + 5 μL of formamide on a plate. Incubate the mix for 2 min at 80°C, and hold at 4°C for 5 min. Place the injection plate in your capillary device and run the appropriate run module. The electrophoresis conditions are:
  - Initial injection voltage: 1.6 kV
  - Initial injection time: 20 s
- Peak pattern evaluation: MLPA fragment data generated on ABI 3730 are visualized using a data analysis software as GeneMapper software package (Applied Biosystems) or GeneScan (BIOMED-2-GS). Each MLPA probemix contains nine control fragments (Q-fragments, D-fragments, X- and Y-specific fragments). Q-fragments serve as a DNA concentration check. Since they are not sample DNA-dependent, when too low DNA concentration is used, their amplification products will be visible at 64, 70, 76, and 82. However their peak

size will be barely visible when more than 100 ng of DNA is added. D-fragments differ from the Q-fragments as they are sample DNA and ligation dependent. Thus, their small amplicons of 88, 92, and 96 nt length will be present when denaturation and ligation have been successfully completed. Chromosome X and Y probes, at 105 and 118 nt, can be used for sex determination and to locate possible sample switch. A comparison between the experimental peak pattern and that of reference samples will indicate which sequences show aberrant copy numbers (Fig. 33.2).

33.3.3.3 MLPA Data Analysis

Data analysis is performed with the Coffalyser MLPA-DAT software (MRC-Holland, Amsterdam, the Netherlands) to provide a reliable normalization for MLPA fragment data files (Fig. 33.3). The software generates the normalized peak value or the so-called probe ratio. To analyze neuroblastoma kits, the MLPA-Vizard (Austrian Research Centres) software has been used by the European Neuroblastoma Quality Assessment (ENQUA) Group [11].

33.3.3.4 Result Interpretation

MLPA interpretation enables the detection of structural aberrations (amplifications and segmental chromosomal aberrations like gains and losses) as well as numeric chromosome changes. Probe ratios below 0.7 or above 1.3 are regarded as indicative of a heterozygous deletion or duplication, respectively, according to manufacturer's instructions. To determine whether the produced ratios are reliable, sufficient knowledge about the MLPA technique and the studied application is essential. Our experience in neuroblastoma kits shows that these ratios are 0.75 and 1.25, respectively.

To define the genetic profile of the tumors we follow the definitions described by the INRG Biology Committee [11]. Table 33.1 summarizes the most frequent chromosome and gene status analyzed by MLPA in neuroblastic tumors.

Two cases of neuroblastoma tumor analyzed by MLPA Vizard are shown as examples (Fig. 33.4).

---

[10]Minimize time between creating the polymerase mix and starting the PCR reaction to avoid the primer–dimer formation (large 50 bp peak).

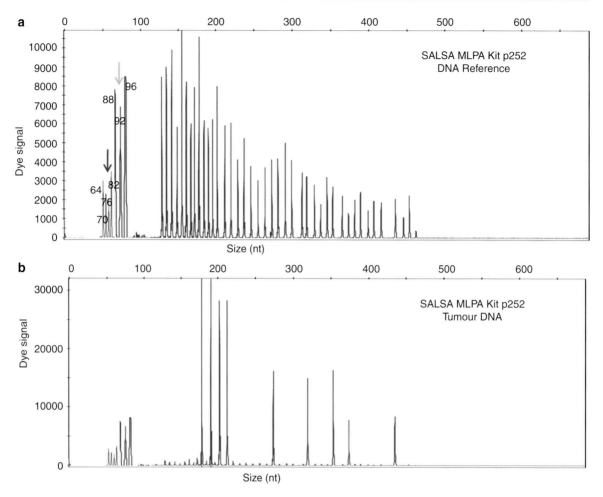

Fig. 33.2 Representation of different peak patterns visualized with Genemapper software. (**a**) MLPA product separation from a reference sample. In each experiment, at least three DNA samples isolated from healthy tissue donors must be included as references. The *red arrow* points to the four DNA quantity control fragments (Q-fragments) and the *green arrow* to the DNA and ligation-dependent fragments (D-fragments). Maximum height dye signals around 10,000. (**b**) MLPA product separation from a neuroblastoma tumour with *NAG* (three probes), *MYCN* (two probes), and *DDX* (four probes) gene amplification. Maximum height dye signals around 30,000; for this reason, the peaks of other analyzed genes are not evident and are near the base line

33.3.4 Troubleshooting

- Problems in MLPA data normalization: in order to get an adequate data normalization it is important to reduce the experimental variations, analyze only samples that have been extracted by the same method and run within the same experiment. Adding the same amount of DNA is also recommended. Using at least three references samples in each run is strongly recommended; references should preferably derive from the same type of tissue as the probe of study and thus minimize variation.

- Primer-dimer formation: in order to avoid them, the PCR should start as soon as possible after addition of the polymerase mix to the samples. In case large numbers of samples are tested, prepare the PCR reaction on ice. After addition of the

a

P251 MLPA probemix lot 0407.txt

Sample name	ROD-AL6-14-07-1-53 AM.fsa^GMsTXT.txt
Analysis Date	14/16/2007 9:10
normalisation method	Coffalysor - LS
Average/Median	Median on Normalisation Factor
Nr Pfobe Signals	43
Nr Control Probe Signals	5
Ligation Probe Found?	YES
Male/Female?	Female
PPMC(pearson); ne probes	0.97.31
MAD all: nr probes	0.04 :5
DNA concentration ok?	OK

Operator	cotta
Kit number	

Reference runs (<7)
C1-26-13-07-10-26 PM.fsa^GMsTXT.txt
C16-13-07-1-41 PM.fsa^GMsTXT.txt
C2-26-13-07-11-00 PM.fsa^GMsTXT.txt
C26-13-07-2-22 PM.fsa^GMsTXT.txt
C4-26-13-07-11-35 PM.fsa^GMsTXT.txt

Gene		Chr pos	Length (bp)	MV36	Ratio	sidov	95% Range	Odds Gain	Odds Deletion
KIF1B probe 4682-L4060	1	01p36.2	219	01.001.0	0.97	0.03	0.91-1.03	1044:1	47899:1
TNFRSF4 probe 3062-L8961	2	01p36	391	01.001.1	1.18	0.06	1.05-1.3	2:1	1692:1
SCNN1D probe 4692-L4070	3	01p36.3	202	01.001.2	0.88	0.05	0.78-0.99	62:1	684:1
GNB1 probe 2890-L7968	4	01p36.33	178	01.001.8	0.93	0.02	0.89-0.98	545:1	14717:1
GABRD probe 4690-L7968	5	01p36	166	01.001.9	0.97	0.03	0.93-1	2952:1	135763:1
TP73 probe 1682-L1262	6	01p36.3	328	01.003.6	0.92	0.03	0.85-0.99	228:1	4953:1
DFFB probe 4656-L4074	7	01p36	319	01.003.8	0.97	0.04	0.89-1.05	502:1	22960:1
NPHP4 probe 4700-L4078	8	01p36.22	283	01.005.9	1.01	0.02	0.97-1.04	14354:1	1290976:1
PARK7 probe 2188-L1686	9	01p36.23	436	01.008.0	1.03	0.05	0.94-1.11	271:1	29216:1
PLOD1 probe 4686-L4064	10	01p36.2	310	01.011.9	1.01	0.03	0.95-1.07	1742:1	152972:1
PTAFR probe 2267-L1425	11	01p35.3	355	01.028.4	1.03	0.05	0.94-1.12	230:1	26501:1
PPAP2B probe 2876-L2343	12	01p32	160	01.056.8	1.05	0.03	1-1.11	317:1	57017:1
NTNG1 probe 8504-L6014	13	01p13.3	463	01.107.7	0.98	0.09	08-1.17	39:1	1691:1
PDE4DIP probe 5712-L5712	14	01q21.1	130	01.143.7	1	0.03	0.95-1.05	3966:1	327525:1
LHX4 probe 7233-L6883	15	01q25.2	274	01.178.5	0.97	0.03	0.92-1.02	1629:1	75500:1
LIN9 probe 4253-L3618	16	01q42.12	292	01.224.5	0.96	0.05	0.9-1.07	465:1	26682:1
VHL probe 1161-L0717	17	03p25.3	418	03.001.0	0.99	0.05	0.89-1.1	310:1	20105:1
TGFBR2 probe 3861-L3610	18	03p23	196	03.030.7	0.94	0.03	0.89-1	550:1	16795:1
CTNNB1 probe 0673-L0117	19	03p21	454	03.041.2	0.95	0.04	0.87-1.03	351:1	12138:1
RASSF1 probe 3991-L3258	20	03p21.3	400	03.050.3	1	0.02	0.96-1.03	17558:1	1342703:1
SEMA3B probe 3210-L2625	21	03p21.3	364	03.050.3	1	0.03	0.94-1.05	3810:1	285910:1
ZMYND10 probe 3207-L2622	22	03p21.3	445	03.050.4	0.92	0.05	0.83-1.02	137:1	2982:1
CASR probe 2683-L2148	23	03q13	481	03.123.4	0.99	0.02	0.96-1.03	18721:1	1343457:11
PIK3CA probe 3826-L5027	24	03q26.3	211	03.180.4	1.66	0.07	0.93-1.19	34:1	5412:1
ST5 probe 6679-L6257	25	11p15	142	11.008.8	1	0.03	0.95-1.05	5433:1	424033:1
CD44 probe 2245-L1731	26	11p13	265	11.035.2	0.97	0.06	0.85-1.1	133:1	5947:1
PTPRJ probe 5918-L5363	27	11p11.2	190	11.048.1	0.98	0.03	0.91-1.04	1172:1	61358:1
GSTP1 probe 6819-L7011	28	11q13	172	11.067.1	1.33	0.03	1.28-1.38	162:1	612:1
CNTN5 probe 8310-L8960	29	11q21	256	-099.1 Exon (	0.66	0.04	0.59-0.74	7:1	1:1
CNTN5 probe 8311-L8180	30	11q21	382	-099.1 Exon (	0.73	0.02	0.69-0.77	13:1	8:1
CNTN5 probe 8312-L8181	31	11q22	229	-099.1 Exon (	0.71	0.02	0.67-0.74	10:1	4:1
CNTN5 probe 8313-L8182	32	11q23	301	-099.1 Exon 2	0.73	0.01	0.71-0.76	13:1	9:1
CASP1 probe 0559-L0128	33	11q22.3	184	11.104.4	0.7	0.04	0.63-0.77	9:1	3:1
ATM probe 2664-L2131	34	11q22	337	11.107.7	0.74	0.02	0.69-0.78	13:1	10:1
IGSF4 probe 1640-L1178	35	11q23.2	238	11.114.7	0.72	0.02	0.68-0.76	12:1	6:1
MLL probe 1637-L1175	36	11q23	427	11.117.9	0.63	0.03	0.57-0.68	6:1	1:5
HMBS probe 1662-L1237	37	11q23.3	136	11.118.5	0.62	0.06	0.5-0.74	5:1	1:5
THY1 probe 4777-L4125	38	11q23.3	409	11.118.8	0.7	0.06	0.65-0.73	10:1	3:1
APC probe 2065-L1586	39	05q22	148	c	1.02	0.23	0.58-1.47	4:1	93:1
FBN1 probe 3919-L3374	40	15q21.1	373	c	1.04	0.02	0.99-1.09	599:1	105660:1
MVP probe 0550-L0372	41	16p13	346	c	1	0.02	0.96-1.04	12592:1	998092:1
SLC12A3 probe 4566-L3955	42	16q13	154	c	0.93	0.06	0.82-1.03	111:1	2432:1
KCNE1 probe 5068-L4468	43	21q22.12	247	c	0.95	0.05	0.84-1.05	157:1	4910:1

Fig. 33.3 Data and graphical output generated by Coffalyser MLPA-DAT software from a neuroblastoma tumor. Results from SALSA MLPA kit P251. (**a**) Data sheet showing in *blue* the normal ratio values and in *red* and *green* the aberrant values (losses and gains, respectively). (**b**) According to manufacturer's instructions. Graphic representation of the ratio values for each gene analyzed

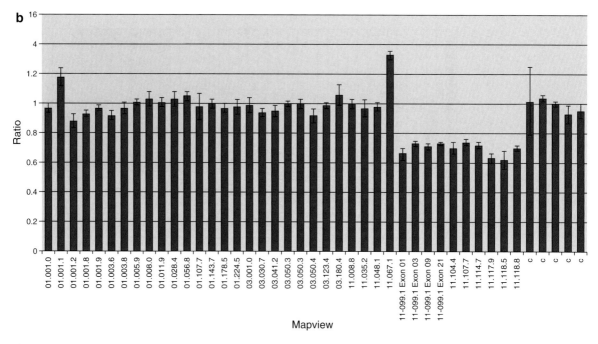

Fig. 33.3 (continued)

Table 33.1 Terms and definitions of chromosome aberrations analyzed by MLPA in neuroblastic tumors

Status	Description
Gain	Unbalanced ratio (low signal excess) between the signals of the chromosomal region of interest and the reference signals of the chromosomal region of interest
Amplification	Unbalanced ratio (high signal excess) between the signals of a gene and all other probes located on the same chromosome
Deletion	Unbalanced ratio between the signals of the chromosomal region of interest and the reference signals

polymerase mix to the last sample, the PCR reactions should be transferred to a preheated thermocycler (60–72°C) and the PCR should be started immediately.

- For large number of samples, the use of multichannel pipettes is suggested in order to reduce the time required to start the PCR.

- Q-fragments: detection of the DNA Quantity control fragments indicates that low or no DNA was used or that ligation failed. Repeating the PCR with higher amount of DNA or repeating the ligation should solve the problem.
- Too low peak signal: increasing the PCR cycles up to 40 and/or increasing the amount of the injected PCR product will result in an improved peak signal. If no improvement is noticed, checking the DNA quality is recommended.
- DNA quality problems: when using DNA samples of uncertain quality it is recommended to add less DNA amount to get better results.
- Low size standard peaks: one of the most frequent reasons is the deterioration of the gel used for capillary electrophoresis, especially after prolonged exposure to temperatures over 25°C. Remove the gel from the instrument after use and store it at 4°C.

Acknowledgments This work was supported by grants: RD06/0020/0102 from RTICC, ISCIII and ERDF and 396/2009. from FAECC.

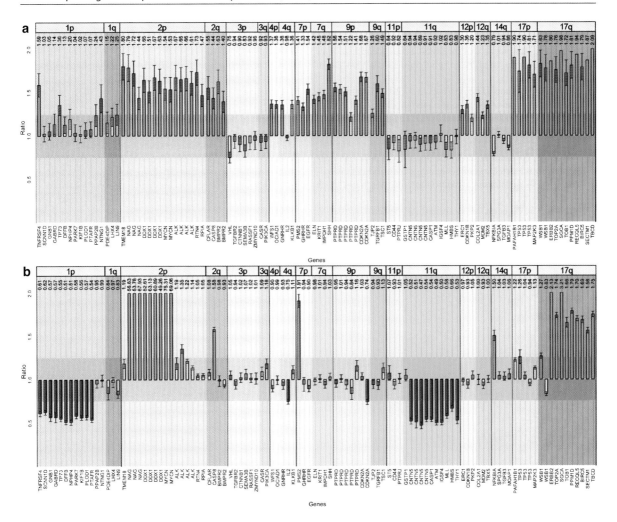

Fig. 33.4 Graphic results obtained using the MLPA-Vizard (Austrian Research Centres) software and neuroblastoma probemixes P251, P252, and P253, in two different cases of neuroblastoma tumor. (**a**) Case with numerical chromosomal changes (+2, +4, +7, +9, +12, +17). (**b**) Case with segmental chromosomal changes (1 p deletion, *NAG, DDX,* and *MYCN* gene amplification, 11 q deletion and 17 q gain). Normal values are shown in *yellow bars*, gains and amplifications in *green bars,* and losses in *red bars*

References

1. Schouten JP, McElgunn CJ, Waaijer R, Zwijnenburg D, Diepvens F, Pals G (2002) Relative quantification of 40 nucleic acid sequences by multiplex ligation-dependent probe amplification. Nucleic Acids Res 30(12):e57
2. Alibakhshi R, Kianishirazi R, Cassiman JJ, Zamani M, Cuppens H (2008) Analysis of the CFTR gene in Iranian cystic fibrosis patients: identification of eight novel mutations. J Cyst Fibros 7(2):102–109.
3. Erlandson A, Samuelsson L, Hagberg B, Kyllerman M, Vujic M, Wahlstrom J (2003) Multiplex ligation-dependent probe amplification (MLPA) detects large deletions in the MECP2 gene of Swedish Rett syndrome patients. Genet Test 7(4):329–332.
4. Hogervorst FB, Nederlof PM, Gille JJ, McElgunn CJ, Grippeling M, Pruntel R, Regnerus R, van Welsem T, van Spaendonk R, Menko FH, Kluijt I, Dommering C, Verhoef S, Schouten JP, van't Veer LJ, Pals G (2003) Large genomic deletions and duplications in the BRCA1 gene identified by a novel quantitative method. Cancer Res 63(7):1449–1453

5. Jeuken J, Cornelissen S, Boots-Sprenger S, Gijsen S, Wesseling P (2006) Multiplex ligation-dependent probe amplification: a diagnostic tool for simultaneous identification of different genetic markers in glial tumors. J Mol Diagn 8(4):433–443.

6. Nakagawa H, Hampel H, de la Chapelle A (2003) Identification and characterization of genomic rearrangements of MSH2 and MLH1 in lynch syndrome (HNPCC) by novel techniques. Hum Mutat 22(3):258.

7. Villamón E, Piqueras M, Mackintosh C, Alonso J, de Alava E, Navarro S, Noguera R (2008) Comparison of different techniques for the detection of genetic risk-identifying chromosomal gains and losses in neuroblastoma. Virchows Arch 453(1):47–55.

8. Villamón E, Piqueras M, Berbegall AP, Tadeo I, Castel V, Navarro S, Noguera R (2011) Comparative study of mlpa-fish to determine DNA copy number alterations in neuroblastic tumors. Histol. Histopathol. 26(3):340–350

9. Puchtler H, Meloan SN (1985) On the chemistry of formaldehyde fixation and its effects on immunohistochemical reactions. Histochemistry 82(3):201–204

10. Lehmann U, Kreipe H (2001) Real-time PCR analysis of DNA and RNA extracted from formalin-fixed and paraffin-embedded biopsies. Methods 25(4):409–418

11. Ambros PF, Ambros IM, Brodeur GM, Haber M, Khan J, Nakagawara A, Schleiermacher G, Speleman F, Spitz R, London WB, Cohn SL, Pearson AD, Maris JM (2009) International consensus for neuroblastoma molecular diagnostics: report from the International Neuroblastoma Risk Group (INRG) Biology Committee. Br J Cancer 100(9): 1471–1482.

Fluorescence In Situ Hybridization (FISH) on Formalin-Fixed Paraffin-Embedded (FFPE) Tissue Sections

34

Marta Piqueras, Manish Mani Subramaniam, Samuel Navarro, Nina Gale, and Rosa Noguera

Contents

34.1 Introduction and Purpose

In Situ Hybridization (ISH) is a powerful technique for localizing specific nucleic acid targets directly in the fixed tissue or cells. Currently, two main methods are used to visualize DNA and RNA targets in situ: Fluorescence In Situ Hybridization (FISH) and Chromogenic In Situ Hybridization (CISH). In this chapter the FISH method is described. In particular, the following FISH protocol provides a methodology for analyzing the cytogenetic alterations in formalin-fixed-paraffin-embedded (FFPE) tissue sections. The critical steps include deparaffinization of tissue sections, optimal pretreatment (target retrieval and protein digestion), and probe hybridization. The technique requires 2 days, as an overnight incubation of the FISH probes is needed for optimal hybridization. Approximately, 2–3 h of bench work is involved each day[1] [1–5].

34.2 Protocol

34.2.1 Reagents

Note: Reagents from specific companies are reported here, but reagents of equal quality purchased from other companies may be used.

Xylene (Merck)

- *Absolute Ethanol* (Merck). Prepare 70% and 85% dilutions from the absolute ethanol. They can be reused up to 2 weeks if stored in well-sealed containers

M. Piqueras, M.M. Subramaniam, S. Navarro, and R. Noguera (✉)
Department of Pathology, Medical School, University
of Valencia, Valencia, Spain

N. Gale
Department of Pathology, Medical School, University
of Ljubljana, Ljubljana, Slovenia

---

[1]FISH (and CISH) can also be performed using automated slide staining systems, such as the BenchMark ULTRA system by Ventana Medical Systems (http://www.ventanamed.com/) or the Hybridizer by Dako (http://www.dakousa.com/).

G. Stanta (ed.), *Guidelines for Molecular Analysis in Archive Tissues*,
DOI: 10.1007/978-3-642-17890-0_34, © Springer-Verlag Berlin Heidelberg 2011

- 1× *Pretreatment buffer.* Add 5 ml of 10× target retrieval solution (DAKO) in 45 ml of distilled water
- *Phosphate buffer saline (PBS).* Dissolve 8 g of NaCl, 1.43 g of $N_2HPO_4 \cdot 2H_2O$, 0.2 g of KCl, and 0.2 g of KH_2PO_4 in 1 l of distilled water. Adjust pH to 7.5. Store at 4°C
- *Proteinase K 10 mg/ml* (stock solution). Dissolve 50 mg proteinase K (Sigma-Aldrich) in 5 ml of PBS buffer. Store at −20°C in aliquots of 0.25 ml
- *Digestion solution.* Add 0.25 ml of stock solution of proteinase K in 50 ml of PBS at 37°C. Prepare 15 min before use and discard after use
- *20× SSC buffer* (Sigma). We use the commercially available solution. The noncommercial version is reported in Chapter 23
- *2× SSC buffer solution.* Mix 30 ml of 20× SSC buffer with 270 ml of distilled water. Store at room temperature up to 6 months
- *10% buffered Formalin* (VWR International)
- *Fixogum* (Marabu GmbH)
- *2× SSC/NP-40 0.3% posthybridization buffer.* Add 0.3 ml solution of NP-40 (AppliChem) to 100 ml 2× SSC buffer solution. Store at room temperature up to 6 months
- *Probe/probe mix*: DNA commercial probes are supplied ready for use or concentrated and must be diluted following the manufacturers instructions. We usually perform FISH using POSEIDON Repeat-Free probes (Kreatech Biotechnology, Amsterdam, the Netherlands), LSI probes (Vysis, Abbot Molecular, Wiesbaden, Germany), and FISH DNA probes (DAKO DENMARK, Glostrup, Denmark).

34.2.2 Equipment

- *Automatic pipettes.* Range: 1–2 µl, 2–10 µl, 10–20 µl, 100–1,000 µl
- *Microtome*
- *Coplin jars*
- *Timer*
- *Bath* at 37°C and 72°C
- *Thermometer*
- *Incubator*
- *Vortex*
- *Centrifuge* suitable for 1.5 ml tubes

- *Hybrite, Thermobrite or hotplate* (with accurate temperature control up to 95°C)
- *Fluorescence microscope* equipped with 63× or 100× objectives and suitable filters (DAPI, FITC, Texas Red or Rhodamine) to view probes signals

34.2.3 Method

34.2.3.1 Sections

Tissue sections (1 section/ probe) of 3–5 µm thickness are cut and transferred to polylysine-treated slides. Sections are then dried overnight in an oven at 60°C.

34.2.3.2 Deparaffinization

Immerse the tissue sections in a coplin jar containing 50 ml of xylene for 10 min. This step is repeated twice with new xylene followed by absolute ethanol wash for 10 min. The latter is also repeated with two washes of fresh ethanol. The sections are hydrated for 5 min each in decreasing concentrations of alcohol (85%, 70%). Incubate them in distilled water for 5 min in a coplin jar.

34.2.3.3 Pretreatment (Heat-Induced Epitope Retrieval, HIER)

Transfer the slides to a coplin jar with the pretreatment buffer and place it into the pressure cooker, closing the pressure cooker lid. As soon as a 1.5 atm pressure is reached, wait for 3 min. Turn off the pressure cooker and depressurize it slowly (see manufacturer's recommendations). Open the lid and cool the sections in running tap water for 20 min. Incubate the slides in distilled water for 5 min.

34.2.3.4 Protein Digestion

A coplin jar with PBS solution is prewarmed in a water bath at 37°C 30 min prior to use. An amount of 0.25 ml of proteinase K stock solution is pipetted into the coplin jar just before use. The sections are digested for a total of 15 min followed by washing them using 50 ml of 2× SSC buffer to arrest the proteolytic action.

34.2.3.5 Prehybridization

The tissue sections are air dried for 10 min. Fix the tissue sections in 10% buffered formalin for 10 min and wash them for 5 min with 50 ml of 2× SSC.

The subsequent steps are carried out in the dark to avoid loss of probe fluorescence.

34.2.3.6 Concomitant Codenaturation and Hybridization[2, 3]

Prepare the fluorescent probe mix for each tissue section according to the manufacturer's instructions. Ten microliter of probe mix is applied on the tissue area, covered with plastic slip, sealed with Fixogum, and air dried for about 5 min. Tissue sections are denatured for 4 min at 90°C and subjected to overnight hybridization at 37°C in a humidified incubator.

34.2.3.7 Posthybridization Wash[4]

This step involves two posthybridization washes; one at room temperature and the other in water bath at 72°C. Coverslips are removed and tissue sections are incubated in posthybridization wash solution in a coplin jar first at room temperature for 5 min and subsequently at 72°C for 2 min in a water bath. Sections are air dried for 10 min.

Counterstaining of the tissue sections is performed using 10 μl of DAPI/Antifade. The sections are coverslipped and sealed with Fixogum. The slides are stored momentarily at 4°C for 30 min before observation at the fluorescence microscope.

---

[2]Probes and counterstain vials must be stored in the dark. We recommend working without direct light to avoid loss of probe fluorescence.

[3]WARNING: DNA probes and hybridization buffers contain formamide, which is teratogen. Do not breathe or allow skin contact. Wear gloves and a lab coat when handling DNA probes.

[4]WARNING: DNA probes and DAPI counterstain contain formamide which is teratogen. Do not breathe or allow skin contact. Wear gloves and a lab coat when handling DNA probes and DAPI/antifade counterstain.

34.2.3.8 Interpretation of Results

The FISH scoring is performed independently by at least two investigators. At least 200 nonoverlapping nuclei/core should be scored to achieve optimal evaluation. We recommend analyzing the sections at multiple focal planes to ensure the visualization of all signals. A simplified scoring system enables to group the cells, based on their pattern of FISH signals, into distinct categories in normal tissues and tumor samples. In addition, it also helps to differentiate true chromosomal aberrations from false positives.

Here are described two scoring schemes that can be used for evaluation of cytogenetic aberrations such as gene disruption (translocation) and gene-specific deletions.

1. *Scoring scheme for gene rearrangement using break-apart FISH probe* [6, 7].
   The "dual-color break-apart gene rearrangement (translocation) probe" is a mixture of two DNA FISH probes. One of them extends distally telomeric from the target gene (i.e., labeled in green), while the other flanks the centromeric region (i.e., labeled in red). Tissues not harboring the gene rearrangement exhibit a pair (red and green) of fused or tightly colocalized signals, while a rearrangement of the gene is indicated by split signals separated by a distance equivalent to three signals. The total numbers of paired and unpaired signals are enumerated for each sample. The FISH scoring scheme defines different groups, represented by the different possible planes of section of cell nuclei and the resulting nuclear fragments for break-apart FISH probes (Fig. 34.1). Group 1 has two paired signals, representing cells without gene rearrangement (Fig. 34.1a). Group 2 has only one paired signal, signifying either normal cells with loss of one copy of paired signals due to nuclear truncation (Fig. 34.1b from normal cells) or monosomy of the chromosome under study. However, group 2 includes tumor cells with true gene disruption displaying one paired signal with loss of the unpaired signals (derivatives) due to sectioning of the nuclei (Fig. 34.1b from tumor cells). Group 3 is composed of ·tumor cells showing true gene disruption (Fig. 34.1c–e). The presence of centromeric signal (i.e., labeled in red), situated proximal to the breakpoint, is considered a reliable indication of the

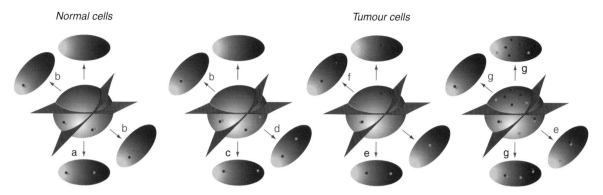

Fig. 34.1 Schematic representation of the scoring scheme displaying different possible signal patterns that could arise due to truncated nuclei in a tissue section

derivate chromosome. Group 4, containing one paired signal and one telomeric signal (i.e., labeled in green) (Fig. 34.1f), also indicates a likely gene rearrangement; some of them could represent true gene disruption but others may not, as the gene break-apart probe can neither detect the other derivative chromosome nor localize the telomeric end if it is translocated to any other chromosome. Group 5 represents gene disruption associated with increased paired and unpaired signals (Fig. 34.1g). The percentage of groups 3, 4, and 5 are summed to obtain the total number of tumor cells exhibiting gene rearrangement. Translocation is diagnosed in cases with ≥60% of cells with genetic aberrations. For cases with less than 60% of cells with disruption, we calculate the difference between the sum of groups 3 and 5 and the sum of groups 2 and 4. We consider a sample positive for gene rearrangement when the difference is more than 15%.

2. *Scoring scheme for evaluation of gene deletions/ amplifications using chromosome enumeration probes (CEP) and locus-specific probes* [5, 6, 8, 9]. The "cocktail DNA probes" are composed of a mixture of a reference or CEP probe (as a chromosomal centromere), indicating the status of chromosome, and a gene specific probe for detection of gains, amplifications, or deletion of gene or genomic loci. In the example shown below (Fig 34.2), the normal cell should have a pair each of red (gene-specific probe) and green (centromere or CEP probes) signals, while a cell with gene deletion should display either one red signal and two green signals if the deletion is heterozygous (HT) or absent red signals and two green signals if the deletion is homozygous

(HM). The FISH scoring scheme applied for tumor cells harboring gene deletions defines the following groups of cells:

Group 1: Cells without any gene alteration possessing the same number of CEP/gene-specific signals (Fig 34.2a and e);

Group 2: Cells with possible gene deletion, having altered numbers of CEP/gene-specific signals (Fig 34.2c, d and f);

Group 3: Cells with cutting artifacts, including nuclear fragments generated from sectioning of the nuclei; these display less CEP signals compared to gene-specific signals in case of gene amplifications or gains; more gene-specific signals compared to CEP signals in case of gene deletion or imbalance (Fig 34.2b).

The net percentage of cells with genetic alteration (group 4) is obtained by subtracting the total percentages of cells with cutting artifacts (group 3) from the cellular population belonging to group 2, thus eliminating all possible false positives. We consider a tumour to be positive for a particular genetic alteration when the percentage of tumour cells in group 4 is more than 15% (mean + 3SD).

Table 34.1 details the FISH criteria for genetic alterations and chromosomal anomalies for some genetic markers [10]. A FISH scoring scheme can be defined as above for each situation.

34.2.3.9 Recommendations

- Single band-pass filters are used for individual color visualization (DAPI, FITC, Texas Red,

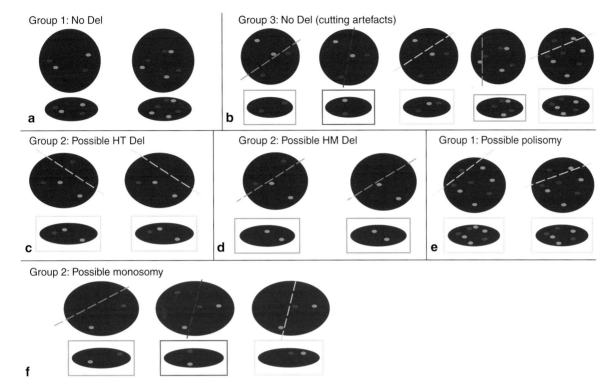

Fig. 34.2 Schematic representation of different patterns of FISH signals arising due to nuclear truncation observed in tumors with and without gene deletions

Table 34.1 Recommendations for the terminology and definitions of the molecular genetic results that have been formulated for neuroblastic tumors by the ENQUA group [10]

Marker	Status	Description
MYCN	No amplification	Same number of gene signals and control probe signals
	Gain	Number of gene signals is between one and four times higher than the number of control probe signals
	Amplification	Number of gene signals is four times higher than the number of control probe signals
1p36	No deletion	Same number of 1p36 region signals and control probe signals
	Imbalance	Less signals (but minimum two) of 1p36 region than control probe signals
	Deletion	One signal from 1p36 region with at least two control probe signals
17q	No gain	Same number of region signals and control probe signals
	Gain	Number of region signals is between one and four times higher than the number of control probe signals

Aqua, Gold,…), whereas dual band-pass filters (FITC/Texas Red) may be used to view multiple colors

- The use of a thermometer is strongly recommended for measuring temperatures of solutions, water baths, and incubators, because these temperatures are critical for optimum product performance.

34.2.4 Troubleshooting

- Temperature and buffer concentration (stringency) of hybridization and washing are important. Lower stringency can result in nonspecific binding of the probe to other sequences, whereas higher stringency can result in a lack of signal.

- Lack of signal could be due to an incomplete denaturation of target DNA. So repeat the process increasing the temperature or time of hybridization.
- Broken small signals could be due to an overdenaturation or overdigestion of target DNA. In this case, repeat the process decreasing the temperature, the time of digestion, or the time of hybridization.
- If excessive background appears after detection, repeat posthybridization wash and counterstain steps.
- If nuclei appear silver with blue filter (DAPI) and no signals are detected, probably the quality of the tissue is low and consequently the hybridization is not possible.

Acknowledgments This work was supported by grants: RD06/0020/0102 from RTICC, ISCIII & ERDF and 396/2009. from FAECC.

References

1. Bunyan DJ, Skinner AC, Ashton EJ, Sillibourne J, Brown T, Collins AL, Cross NC, Harvey JF, Robinson DO (2007) Simultaneous MLPA-based multiplex point mutation and deletion analysis of the dystrophin gene. Mol Biotechnol 35(2):135–140
2. Petersen BL, Sorensen MC, Pedersen S, Rasmussen M (2004) Fluorescence in situ hybridization on formalin-fixed and paraffin-embedded tissue: optimizing the method. Appl Immunohistochem Mol Morphol 12(3):259–265
3. Summersgill B, Clark J, Shipley J (2008) Fluorescence and chromogenic in situ hybridization to detect genetic aberrations in formalin-fixed paraffin embedded material, including tissue microarrays. Nat Protoc 3(2):220–234
4. Weremowicz S, Schofield DE (2007) Preparation of cells from formalin-fixed, paraffin-embedded tissue for use in fluorescence in situ hybridization (fish) experiments. Curr Protoc Hum Genet Chapter 8:Unit 8.8
5. Piqueras M, Navarro S, Cañete A, Castel V, Noguera R (2010) Prognostic value of partial genetic instability in neuroblastoma with ≤50% neuroblastic cell content. Histopathology (in press)
6. Noguera R, Machado I, Piqueras M, Lopez-Guerrero JA, Navarro S, Mayordomo E, Pellin A, Llombart-Bosch A (2008) Tissue microarrays: applications in study of p16 and p53 alterations in ewing's cell lines. Diagn Pathol 3 suppl 1:S27
7. Subramaniam MM, Noguera R, Piqueras M, Navarro S, Carda C, Pellin A, Lopez-Guerrero JA, Llombart-Bosch A (2007) Evaluation of genetic stability of the SYT gene rearrangement by break-apart fish in primary and xenotransplanted synovial sarcomas. Cancer Genet Cytogenet 172(1):23–28
8. Combaret V, Turc-Carel C, Thiesse P, Rebillard AC, Frappaz D, Haus O, Philip T, Favrot MC (1995) Sensitive detection of numerical and structural aberrations of chromosome 1 in neuroblastoma by interphase fluorescence in situ hybridization. Comparison with restriction fragment length polymorphism and conventional cytogenetic analyses. Int J Cancer 61(2):185–191
9. Subramaniam MM, Noguera R, Piqueras M, Navarro S, Lopez-Guerrero JA, Llombart-Bosch A (2006) P16INK4A (CDKN2A) gene deletion is a frequent genetic event in synovial sarcomas. Am J Clin Pathol 126(6):866–874
10. Ambros IM, Benard J, Boavida M, Bown N, Caron H, Combaret V, Couturier J, Darnfors C, Delattre O, Freeman-Edward J, Gambini C, Gross N, Hattinger CM, Luegmayr A, Lunec J, Martinsson T, Mazzocco K, Navarro S, Noguera R, O'Neill S, Potschger U, Rumpler S, Speleman F, Tonini GP, Valent A, Van Roy N, Amann G, De Bernardi B, Kogner P, Ladenstein R, Michon J, Pearson AD, Ambros PF (2003) Quality assessment of genetic markers used for therapy stratification. J Clin Oncol 21(11):2077–2084

Introduction on Genome-wide Expression Profiling from Formalin-Fixed Paraffin-Embedded Tissues Using Microarrays

35

Manfred Dietel, Jan Budczies, Wilko Weichert, and Carsten Denkert

Contents

35.1 Introduction and Purpose

Worldwide the routine workflow for the preparation of diagnostic biopsies taken from different organ sites includes immediate formalin fixation and subsequent paraffin embedding. This is true also for the vast majority of surgical specimens obtained at therapeutic surgical interventions. This is because formalin-fixed paraffin-embedded (FFPE) samples are easy to handle, stable, and particularly suitable for diagnostic histology, immunohistochemistry, (IHC) and in situ hybridization (ISH). Together with clinical parameters such as tumor stage, pattern of metastases, clinical lymph node status, and the histopathological characteristics including histotyping, grading, and biomarker analyses have strong implications on therapy selection. Lately, it has been suggested that these "classic" parameters can and should be complemented by molecular high-throughput methods to further improve the basis of targeted or individualized therapy of cancer.

35.2 Clinical Applications

In recent years, numerous gene expression microarray studies have revealed clinically relevant information about tumor biology and correlated gene expression signatures with tumor behavior, such as response to chemotherapy and prognosis of patients [1–9]. Large prospective studies such as MINDACT and TAILORx are currently being conducted to test whether molecular profiling can contribute to a refined diagnosis, for example, of breast cancer and help to stratify breast cancer patients for a more individually tailored, personalized therapy.

M. Dietel (✉), J. Budczies, W. Weichert, and C. Denkert
Institute of Pathology, Charité University Hospital,
Berlin, Germany

G. Stanta (ed.), *Guidelines for Molecular Analysis in Archive Tissues*,
DOI: 10.1007/978-3-642-17890-0_35, © Springer-Verlag Berlin Heidelberg 2011

The differential diagnosis of hematological malignancies can be very tricky. A landmark study including gene expression profiling with DNA microarrays as an extension of histopathological classification has shown the possibility to differentiate between acute myeloid leukemia (AML) and acute lymphoblastic leukemia (ALL) and to improve the precision of prognosis prediction [10]. More recently, it has been shown that using oligonucleotide microarrays, 12 clinically relevant subtypes of leukemia can be recognized and be distinguished from nonleukemia with an accuracy of 95% [11].

Cis-platinum (cDDP) resistance was shown by microarray analysis to be accompanied with altered expression of genes coding for membrane proteins and a glycoprotein hormone subunit [12], not previously known to play any role in cDDP resistance. This opens the possibility to predict cDDP efficiency.

In an excellent study by Zembutsu et al. [13], 85 human cancer xenografts were tested with regard to characteristic expression profiles in response to 9 anticancer drugs often used in clinical therapy. The expression profile of more than 1,500 genes has been correlated in some way with chemosensitivity. The authors identified sets of genes which could partly be associated with chemosensitivity of particular tumor types (colon, breast, NSCLC, etc.) to the different drugs applied.

In addition, bone marrow samples from 19 patients with ALL were investigated with regard to resistance to an ABL tyrosine kinase inhibitor [14]. On the basis of 95 differentially expressed genes, it appeared to be possible to distinguish responder from nonresponder. Raponi et al. [15] showed the possibility to detect the pathways modulated by inhibitory drugs in AML elucidating the mechanisms of drug action.

A further step toward clinical application was published by our group in 2005 [16]. A gene expression signature was derived by comparing doxorubicin-resistant and -sensitive cell lines and applied to a group of 44 breast cancer cases described by Sorlie et al. [4] showing a correlation with overall survival of the patients.

Although a number of experiments still have to be conducted, it might become possible to predict chemoresistance and to avoid noneffective drugs and unnecessary side effects for the patients. The discrimination between responders and nonresponders prior to therapy will further stimulate the development of an individualized therapeutic strategy with a personalized combination of drugs.

Recently, two independent groups have published gene expression signatures that refine the histological grade classification of breast cancers. By comparing G3 with G1 breast cancers, Sotiriou et al. [17, 18] have identified a signature of 242 probe sets ("genomic grade signature"), while Ivshina et al. [19] have identified signatures of 6 ("SWS genetic grade signature") and 18 ("PAM genetic grade signature") probe sets. These signatures have been shown to stratify G2 breast cancers in two subgroups with different prognosis.

Most of the gene expression studies cited above are based on RNA extracts from frozen tissue samples that were collected during cancer surgery. Exemplified for frozen breast cancer biopsies it was shown that microarray-based gene expression profiling is feasible and helps to uncover important properties of tumor biology [20–22]. However, due to the fact that frozen tissue samples are hardly available in routine diagnostic and therapeutic workflows in the vast majority of hospitals these studies are of limited value for most cancer patients. It would be of great advantage to perform the full spectrum of molecular diagnostics using routinely collected FFPE tissue.

35.3 Basics on the FFPE Technologies

For most of the molecular methods, including expression profiling with microarrays, it is standard to use fresh or frozen tissues as RNA source. These samples usually have a high quality of nucleic acid preservation. By contrast, formalin fixation produces significant chemical modifications of the RNA which depend on fixation conditions such as the quality of formalin, exactness of buffer concentration, etc. size of the samples, and time in the fixation solution [23]. First and of crucial importance, fixation conditions such as formalin concentration, temperature, and incubation time should be kept as uniform as possible. We have recently worked on protocols that would allow the standardized collection and handling of biospecimens in the context of translational research in clinical trials [24].

During the last few years, different manual protocols for RNA extraction from FFPE tissue have been developed with various results. Most of the protocols use a similar approach including deparaffinization, cell disruption, release of RNA by proteinase K as well as RNA extraction using silica columns or beads and chaotropic salts. However, the differences between FFPE samples from different tissue types compromise the standardization of RNA extraction protocol. Thus, the use of the existing methods in a routine laboratory and for high-throughput analyses is difficult and labor-intensive. Therefore, RNA extracts from FFPE tissues can be of suboptimal quality and it appears to be difficult to compare between samples and to ensure standardization. Recent studies have reported that only 25–55% of unselected FFPE cancer samples aged 1–8 years provided RNA of sufficient quality for successful gene expression analysis with microarrays [25, 26]. On the other hand, integration of frozen tissue collection in the diagnostic interdisciplinary workflow is severely hampered by logistic problems, e.g. availability of liquid nitrogen as well as transport issues, slicing

and manufacturing of the frozen tissues, etc. Further, taking small diagnostic biopsies or core needle biopsies is an invasive procedure which delivers limited amounts of biomaterial that preferably should not be subdivided into frozen and paraffin-embedded tissue parts.

Recently, several RNA extraction techniques suitable for profiling genes in FFPE tissue samples [27] and FFPE core biopsies [28] have been described. We are currently evaluating the performance of microarray hybridizations with RNA extraction from FFPE breast cancer biopsies (Fig. 35.1). Difficulties connected with

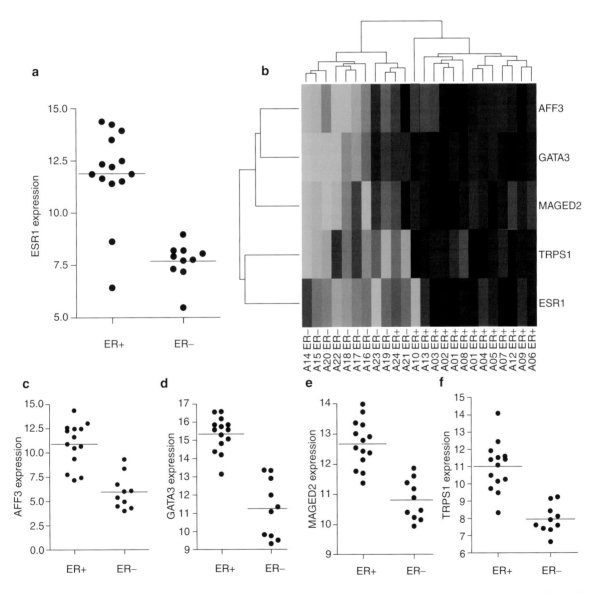

Fig. 35.1 Comparison of the gene expression profiles of 24 ER+ and ER– breast cancer biopsies. ESR1 as well as the four genes AFF3, GATA3, MAGED2, and TRPS1 are differentially expressed at high significance level ($p < 0.00001$). (**a**), (**c–f**) The expression levels of these genes in the 24 biopsies. (**b**) Clustering of the biopsies with respect to the five genes reproduces the result for ER receptor status from immunohistochemistry with high accuracy. Conflicting results arise only for a single ER-positive biopsy (A24) that clusters together with the immunohistologically ER-negative biopsies

the limited amount of biomaterial were overcome with an amplification procedure that has been validated before [29]. RNA was extracted from two 10 μm sections of the FFPE core needle biopsies, amplified with PCR, and further processed for hybridization to Affymetrix GeneChip. A first check of diagnostic quality was performed by comparing ER and the HER2 transcripts with state-of-the-art IHC results. A high level of concordance was observed when comparing ER-positive and ER-negative biopsies (Fig. 35.1a, $p < 0.00001$, Welch's t-test). Second, 346 differentially expressed genes could be identified from ER-positive and ER-negative FFPE biopsies. A high proportion of these genes (63%) could be confirmed by an independent gene expression study where frozen tissue samples were used as RNA source. A heatmap of the five most significant differentially expressed genes is shown in Fig. 35.1b. Further validation studies and a detailed publication are on the way.

Recently, another high-throughput method for gene expression profiling has become feasible in FFPE by the use of the cDNA-mediated Annealing, Selection, extension and Ligation (DASL) assay (Illumina, San Diego, CA), which allows profiling randomly fragmented mRNA extracted from FFPE tissue. Briefly, fragmented FFPE-RNA is converted into cDNA using random primers. For each target site on the cDNA, a pair of query oligos separated by a single nucleotide is annealed to the cDNA, and, the gap between the query oligos is extended and ligated to generate a PCR template. Universal PCR primers are then used for amplification, and linearly amplified PCR products are hybridized to a bead microarray, which is then scanned by a BeadArray Reader (Illumina, San Diego, CA) [30].

Acknowledgments This project was funded by the BMBF, grant 01ES0725 NEO-PREDICT and by the European Commission, FP7 grant 200327 METAcancer.

References

1. Buyse M, Loi S, Van't Veer L, Viale G, Delorenzi M, Glas AM, d'Assignies MS, Bergh J, Lidereau R, Ellis P, Harris A, Bogaerts J, Therasse P, Floore A, Amakrane M, Piette F, Rutgers E, Sotiriou C, Cardoso F, Piccart MJ (2006) Validation and clinical utility of a 70-gene prognostic signature for women with node-negative breast cancer. J Natl Cancer Inst 98(17):1183–1192

2. Desmedt C, Piette F, Loi S, Wang Y, Lallemand F, Haibe-Kains B, Viale G, Delorenzi M, Zhang Y, d'Assignies MS, Bergh J, Lidereau R, Ellis P, Harris AL, Klijn JG, Foekens JA, Cardoso F, Piccart MJ, Buyse M, Sotiriou C (2007) Strong time dependence of the 76-gene prognostic signature for node-negative breast cancer patients in the transbig multicenter independent validation series. Clin Cancer Res 13(11):3207–3214

3. Perou CM, Sorlie T, Eisen MB, van de Rijn M, Jeffrey SS, Rees CA, Pollack JR, Ross DT, Johnsen H, Akslen LA, Fluge O, Pergamenschikov A, Williams C, Zhu SX, Lonning PE, Borresen-Dale AL, Brown PO, Botstein D (2000) Molecular portraits of human breast tumours. Nature 406(6797):747–752

4. Sorlie T, Perou CM, Tibshirani R, Aas T, Geisler S, Johnsen H, Hastie T, Eisen MB, van de Rijn M, Jeffrey SS, Thorsen T, Quist H, Matese JC, Brown PO, Botstein D, Eystein Lonning P, Borresen-Dale AL (2001) Gene expression patterns of breast carcinomas distinguish tumor subclasses with clinical implications. Proc Natl Acad Sci USA 98(19):10869–10874

5. Sorlie T, Tibshirani R, Parker J, Hastie T, Marron JS, Nobel A, Deng S, Johnsen H, Pesich R, Geisler S, Demeter J, Perou CM, Lonning PE, Brown PO, Borresen-Dale AL, Botstein D (2003) Repeated observation of breast tumor subtypes in independent gene expression data sets. Proc Natl Acad Sci USA 100(14):8418–8423

6. Sorlie T, Wang Y, Xiao C, Johnsen H, Naume B, Samaha RR, Borresen-Dale AL (2006) Distinct molecular mechanisms underlying clinically relevant subtypes of breast cancer: gene expression analyses across three different platforms. BMC Genomics 7:127

7. van de Vijver MJ, He YD, Van't Veer LJ, Dai H, Hart AA, Voskuil DW, Schreiber GJ, Peterse JL, Roberts C, Marton MJ, Parrish M, Atsma D, Witteveen A, Glas A, Delahaye L, van der Velde T, Bartelink H, Rodenhuis S, Rutgers ET, Friend SH, Bernards R (2002) A gene-expression signature as a predictor of survival in breast cancer. N Engl J Med 347(25):1999–2009

8. Van't Veer LJ, Dai H, van de Vijver MJ, He YD, Hart AA, Mao M, Peterse HL, van der Kooy K, Marton MJ, Witteveen AT, Schreiber GJ, Kerkhoven RM, Roberts C, Linsley PS, Bernards R, Friend SH (2002) Gene expression profiling predicts clinical outcome of breast cancer. Nature 415(6871):530–536

9. Wang Y, Klijn JG, Zhang Y, Sieuwerts AM, Look MP, Yang F, Talantov D, Timmermans M, Meijer-van Gelder ME, Yu J, Jatkoe T, Berns EM, Atkins D, Foekens JA (2005) Gene-expression profiles to predict distant metastasis of lymph-node-negative primary breast cancer. Lancet 365(9460):671–679

10. Golub TR, Slonim DK, Tamayo P, Huard C, Gaasenbeek M, Mesirov JP, Coller H, Loh ML, Downing JR, Caligiuri MA, Bloomfield CD, Lander ES (1999) Molecular classification of cancer: class discovery and class prediction by gene expression monitoring. Science 286(5439):531–537

11. Haferlach T, Kohlmann A, Schnittger S, Dugas M, Hiddemann W, Kern W, Schoch C (2005) Global approach to the diagnosis of leukemia using gene expression profiling. Blood 106(4):1189–1198

12. Higuchi E, Oridate N, Furuta Y, Suzuki S, Hatakeyama H, Sawa H, Sunayashiki-Kusuzaki K, Yamazaki K, Inuyama Y,

Fukuda S (2003) Differentially expressed genes associated with cis-diamminedichloroplatinum (II) resistance in head and neck cancer using differential display and cDNA microarray. Head Neck 25(3):187–193

13. Zembutsu H, Ohnishi Y, Tsunoda T, Furukawa Y, Katagiri T, Ueyama Y, Tamaoki N, Nomura T, Kitahara O, Yanagawa R, Hirata K, Nakamura Y (2002) Genome-wide cDNA microarray screening to correlate gene expression profiles with sensitivity of 85 human cancer xenografts to anticancer drugs. Cancer Res 62(2):518–527

14. Hofmann WK, de Vos S, Elashoff D, Gschaidmeier H, Hoelzer D, Koeffler HP, Ottmann OG (2002) Relation between resistance of Philadelphia-chromosome-positive acute lymphoblastic leukaemia to the tyrosine kinase inhibitor STI571 and gene-expression profiles: a gene-expression study. Lancet 359(9305):481–486

15. Raponi M, Belly RT, Karp JE, Lancet JE, Atkins D, Wang Y (2004) Microarray analysis reveals genetic pathways modulated by tipifarnib in acute myeloid leukemia. BMC Cancer 4:56

16. Gyorffy B, Serra V, Jurchott K, Abdul-Ghani R, Garber M, Stein U, Petersen I, Lage H, Dietel M, Schafer R (2005) Prediction of doxorubicin sensitivity in breast tumors based on gene expression profiles of drug-resistant cell lines correlates with patient survival. Oncogene 24(51):7542–7551

17. Loi S, Haibe-Kains B, Desmedt C, Lallemand F, Tutt AM, Gillet C, Ellis P, Harris A, Bergh J, Foekens JA, Klijn JG, Larsimont D, Buyse M, Bontempi G, Delorenzi M, Piccart MJ, Sotiriou C (2007) Definition of clinically distinct molecular subtypes in estrogen receptor-positive breast carcinomas through genomic grade. J Clin Oncol 25(10):1239–1246

18. Sotiriou C, Wirapati P, Loi S, Harris A, Fox S, Smeds J, Nordgren H, Farmer P, Praz V, Haibe-Kains B, Desmedt C, Larsimont D, Cardoso F, Peterse H, Nuyten D, Buyse M, Van de Vijver MJ, Bergh J, Piccart M, Delorenzi M (2006) Gene expression profiling in breast cancer: understanding the molecular basis of histologic grade to improve prognosis. J Natl Cancer Inst 98(4):262–272

19. Ivshina AV, George J, Senko O, Mow B, Putti TC, Smeds J, Lindahl T, Pawitan Y, Hall P, Nordgren H, Wong JE, Liu ET, Bergh J, Kuznetsov VA, Miller LD (2006) Genetic reclassification of histologic grade delineates new clinical subtypes of breast cancer. Cancer Res 66(21):10292–10301

20. Rody A, Karn T, Gatje R, Ahr A, Solbach C, Kourtis K, Munnes M, Loibl S, Kissler S, Ruckhaberle E, Holtrich U, von Minckwitz G, Kaufmann M (2007) Gene expression profiling of breast cancer patients treated with docetaxel, doxorubicin, and cyclophosphamide within the GEPARTRIO trial: HER-2, but not topoisomerase II alpha and microtubule-associated protein tau, is highly predictive of tumor response. Breast 16(1):86–93

21. Rody A, Karn T, Gatje R, Kourtis K, Minckwitz G, Loibl S, Munnes M, Ruckhaberle E, Holtrich U, Kaufmann M, Ahr A (2006) Gene expression profiles of breast cancer obtained from core cut biopsies before neoadjuvant

docetaxel, adriamycin, and cyclophosphamide chemotherapy correlate with routine prognostic markers and could be used to identify predictive signatures. Zentralbl Gynäkol 128(2):76–81

22. Rody A, Karn T, Solbach C, Gaetje R, Munnes M, Kissler S, Ruckhaberle E, Minckwitz GV, Loibl S, Holtrich U, Kaufmann M (2007) The erbB2+ cluster of the intrinsic gene set predicts tumor response of breast cancer patients receiving neoadjuvant chemotherapy with docetaxel, doxorubicin and cyclophosphamide within the GEPARTRIO trial. Breast 16(3):235–240

23. Ribeiro-Silva A, Zhang H, Jeffrey SS (2007) RNA extraction from ten year old formalin-fixed paraffin-embedded breast cancer samples: a comparison of column purification and magnetic bead-based technologies. BMC Mol Biol 8:118

24. Leyland-Jones BR, Ambrosone CB, Bartlett J, Ellis MJ, Enos RA, Raji A, Pins MR, Zujewski JA, Hewitt SM, Forbes JF, Abramovitz M, Braga S, Cardoso F, Harbeck N, Denkert C, Jewell SD (2008) Recommendations for collection and handling of specimens from group breast cancer clinical trials. J Clin Oncol 26(34):5638–5644

25. Linton KM, Hey Y, Saunders E, Jeziorska M, Denton J, Wilson CL, Swindell R, Dibben S, Miller CJ, Pepper SD, Radford JA, Freemont AJ (2008) Acquisition of biologically relevant gene expression data by Affymetrix microarray analysis of archival formalin-fixed paraffin-embedded tumours. Br J Cancer 98(8):1403–1414

26. Penland SK, Keku TO, Torrice C, He X, Krishnamurthy J, Hoadley KA, Woosley JT, Thomas NE, Perou CM, Sandler RS, Sharpless NE (2007) RNA expression analysis of formalin-fixed paraffin-embedded tumors. Lab Invest 87(4):383–391

27. Paik S, Shak S, Tang G, Kim C, Baker J, Cronin M, Baehner FL, Walker MG, Watson D, Park T, Hiller W, Fisher ER, Wickerham DL, Bryant J, Wolmark N (2004) A multigene assay to predict recurrence of tamoxifen-treated, node-negative breast cancer. N Engl J Med 351(27):2817–2826

28. Chang JC, Makris A, Gutierrez MC, Hilsenbeck SG, Hackett JR, Jeong J, Liu ML, Baker J, Clark-Langone K, Baehner FL, Sexton K, Mohsin S, Gray T, Alvarez L, Chamness GC, Osborne CK, Shak S (2008) Gene expression patterns in formalin-fixed, paraffin-embedded core biopsies predict docetaxel chemosensitivity in breast cancer patients. Breast Cancer Res Treat 108(2):233–240

29. Klur S, Toy K, Williams MP, Certa U (2004) Evaluation of procedures for amplification of small-size samples for hybridization on microarrays. Genomics 83(3):508–517

30. Hoshida Y, Villanueva A, Kobayashi M, Peix J, Chiang DY, Camargo A, Gupta S, Moore J, Wrobel MJ, Lerner J, Reich M, Chan JA, Glickman JN, Ikeda K, Hashimoto M, Watanabe G, Daidone MG, Roayaie S, Schwartz M, Thung S, Salvesen HB, Gabriel S, Mazzaferro V, Bruix J, Friedman SL, Kumada H, Llovet JM, Golub TR (2008) Gene expression in fixed tissues and outcome in hepatocellular carcinoma. N Engl J Med 359(19):1995–2004

Part

XI

Protein Extraction in Archive Tissues

Introduction

36

Valentina Faoro, Karl-Friedrich Becker and Giorgio Stanta

Contents

V. Faoro
Department of Medical, Surgical and Health Sciences,
University of Trieste, Trieste, Italy

K.-F. Becker (✉)
Institute of Pathology, Technical University of Munich,
Munich, Germany

G. Stanta
Department of Medical Sciences,
University of Trieste, Cattinara Hospital, Strada di Fiume 447,
Trieste, Italy

The identification of differentially expressed proteins may elucidate possible molecular alterations regulating the progression of a disease and define some useful biomarkers for better diagnosis and improved therapies.

In a tissue sample a complex mixture of thousands of proteins that greatly differ in their physiochemical and biological properties are typically present. Clinical tissue samples (i.e. archive tissues) represent a complete source of protein expression profiles associated with diseases such as cancer.

Tissues of bioptic or surgical origin are routinely formalin fixed and paraffin embedded (FFPE) in every hospital in the Department of Pathology. After few sections are cut for the histopathological examination, the tissues are stored for decades in the hospital archives.

Molecular analysis performed using archive tissues represents an important resource, especially because these tissues maintain a strong correlation with the clinical behaviour of the disease, such as resistance to chemotherapeutics, recurrences, etc. However, proteomics studies on FFPE samples are impaired due to the formalin fixation cross-links of proteins, making protein investigations more difficult. Nevertheless, FFPE tissue specimens have gained attention and several studies have focused on this subject. Successful protein extraction was done by a number of authors using different tissue types and moreover, these methods have also been extended to samples fixed with formalin-free solutions [1–27]. Very recently, successful protein extraction even from over-fixed and long-term stored formalin-fixed tissues was reported [28].

The following chapters are dedicated to extraction and analysis of proteins isolated from archival tissues and report the general methods that can be applied to proteins extracted from archive tissues treated with conventional formalin as well as from samples fixed with some new commercial protein-preserving fixatives.

G. Stanta (ed.), *Guidelines for Molecular Analysis in Archive Tissues*,
DOI: 10.1007/978-3-642-17890-0_36, © Springer-Verlag Berlin Heidelberg 2011

36.1 Protein Handling and Management

36.1.1 Sample Preparation

Appropriate sample preparation is essential for good proteomic results [29–31]. Due to the great diversity of protein sample types and origins, the optimal sample preparation procedure will result in the complete solubilization, disaggregation, denaturation, and chemical reduction of the proteins in the sample. Postextraction chemical modification, such as enzymatic or chemical degradation of the proteins, should be avoided by maintaining the samples on ice during the experiments and using, for example, chilled solutions. Interfering molecules such as nucleic acids, polysaccharides, lipids, salts, and other particulate material must be removed by protein precipitation (for example, using trichloroacetic acid (TCA), TCA/acetone or by "salting out" procedure), as described in the related chapter. Moreover, during cell disruption hydrolytic enzymes, especially proteases, can be released and activated. In order to prevent their activity, solubilization of proteins in a strong denaturing environment and the use of protease inhibitors are suggested.

36.1.2 Protein Storage

Proteins need to be put rapidly at low temperature to avoid proteolysis and protein degradation. All the experimental steps should be performed at low temperature; you should keep the sample on ice as long as possible and use chilled solutions.

The extracted proteins can be stored for up to 1 week at 4°C. For longer-term storage, aliquot the extracted proteins and store at −80°C, avoiding repeated freeze–thaw cycles.

36.1.3 Protein Management

Contamination by keratin could be a problem in proteomics [32]. Careful sample handling is important, because everything in the environment is a potential source of keratins, especially the operator's skin cells and hair. In order to minimize keratin contamination

during the experiment, be careful to reduce the exposure of all equipment and supplies to primary sources of keratins, such as skin, hair, clothing, but also dust and particulates. Wearing a clean lab coat and handling all materials with powder-free nitrile or vinyl (no latex) gloves is necessary in order to ensure a keratin-free environment. The reagents should be prepared in a laminar flow hood and it is recommended to limit exposure of reagents to open laboratory environment as much as possible. All the solutions should be filtered and stored in well-cleaned containers. Wash all glass plates thoroughly with detergent and finally with 70% ethanol prior to casting an SDS-PAGE gel and use clean plastic or glass gel containers.

References

1. Becker KF, Metzger V, Hipp S, Hofler H (2006) Clinical proteomics: new trends for protein microarrays. Curr Med Chem 13(15):1831–1837
2. Becker KF, Schott C, Hipp S, Metzger V, Porschewski P, Beck R, Nahrig J, Becker I, Hofler H (2007) Quantitative protein analysis from formalin-fixed tissues: implications for translational clinical research and nanoscale molecular diagnosis. J Pathol 211(3):370–378
3. Belief V, Boissiere F, Bibeau F, Desmetz C, Berthe ML, Rochaix P, Maudelonde T, Mange A, Solassol J (2008) Proteomic analysis of RCL2 paraffin-embedded tissues. J Cell Mol Med 12(5B):2027–2036
4. Bellet V, Boissiere F, Bibeau F, Desmetz C, Berthe M, Rochaix P, Maudelonde T, Mange A, Solassol J (2007) Proteomic analysis of RCL2 paraffin-embedded tissues. J Cell Mol Med 12(5B):2027–2036
5. Chu WS, Furusato B, Wong K, Sesterhenn IA, Mostofi FK, Wei MQ, Zhu Z, Abbondanzo SL, Liang Q (2005) Ultrasound-accelerated formalin fixation of tissue improves morphology, antigen and mRNA preservation. Mod Pathol 18(6):850–863
6. Crockett DK, Lin Z, Vaughn CP, Lim MS, Elenitoba-Johnson KS (2005) Identification of proteins from formalin-fixed paraffin-embedded cells by LC-MS/MS. Lab Invest 85(11): 1405–1415
7. Fowler CB, Cunningham RE, O'Leary TJ, Mason JT (2007) 'Tissue surrogates' as a model for archival formalin-fixed paraffin-embedded tissues. Lab Invest 87(8):836–846
8. Fowler CB, Cunningham RE, Waybright TJ, Blonder J, Veenstra TD, O'Leary TJ, Mason JT (2008) Elevated hydrostatic pressure promotes protein recovery from formalin-fixed, paraffin-embedded tissue surrogates. Lab Invest 88(2):185–195
9. Gräntzdörffer I, Yumlu S, Gioeva Z, von Wasielewski R, Ebert MPA, Röcken C (2010) Comparison of different tissue sampling methods for protein extraction from formalin-fixed and paraffin-embedded tissue specimens. Exp Mol Pathol 88(1):190–196

10. Guo T, Wang W, Rudnick PA, Song T, Li J, Zhuang Z, Weil RJ, DeVoe DL, Lee CS, Balgley BM (2007) Proteome analysis of microdissected formalin-fixed and paraffin-embedded tissue specimens. J Histochem Cytochem 55(7):763–772

11. Hood BL, Conrads TP, Veenstra TD (2006) Unravelling the proteome of formalin-fixed paraffin-embedded tissue. Brief Funct Genomic Proteomic 5(2):169–175

12. Hood BL, Darfler MM, Guiel TG, Furusato B, Lucas DA, Ringeisen BR, Sesterhenn IA, Conrads TP, Veenstra TD, Krizman DB (2005) Proteomic analysis of formalin-fixed prostate cancer tissue. Mol Cell Proteomics 4(11):1741–1753

13. Hwang SI, Thumar J, Lundgren DH, Rezaul K, Mayya V, Wu L, Eng J, Wright ME, Han DK (2007) Direct cancer tissue proteomics: a method to identify candidate cancer biomarkers from formalin-fixed paraffin-embedded archival tissues. Oncogene 26(1):65–76

14. Ikeda K, Monden T, Kanoh T, Tsujie M, Izawa H, Haba A, Ohnishi T, Sekimoto M, Tomita N, Shiozaki H, Monden M (1998) Extraction and analysis of diagnostically useful proteins from formalin-fixed, paraffin-embedded tissue sections. J Histochem Cytochem 46(3):397–403

15. Kaplan B, Martin BM, Livneh A, Pras M, Gallo GR (2004) Biochemical subtyping of amyloid in formalin-fixed tissue samples confirms and supplements immunohistologic data. Am J Clin Pathol 121(6):794–800

16. Kothmaier H, Rohrer D, Quehenberger F, Becker K-F, Popper HH (2010) Comparison of formalin-free tissue fixatives:A proteomic study testing their application for routine pathology and research Archives of Pathology & Laboratory Medicine. (in pres)

17. Kroll J, Becker KF, Kuphal S, Hein R, Hofstadter F, Bosserhoff AK (2008) Isolation of high quality protein samples from punches of formalin fixed and paraffin embedded tissue blocks. Histol Histopathol 23(4):391–395

18. Lemaire R, Desmons A, Tabet JC, Day R, Salzet M, Fournier I (2007) Direct analysis and maldi imaging of formalin-fixed, paraffin-embedded tissue sections. J Proteome Res 6(4):1295–1305

19. Lykidis D, Van Noorden S, Armstrong A, Spencer-Dene B, Li J, Zhuang Z, Stamp GW (2007) Novel zinc-based fixative for high quality DNA, RNA and protein analysis. Nucleic Acids Res 35(12):e85

20. Nirmalan NJ, Harnden P, Selby PJ, Banks RE (2009) Development and validation of a novel protein extraction methodology for quantitation of protein expression in formalin-fixed paraffin-embedded tissues using western blotting. J Pathol 217(4):497–506

21. Ostasiewicz P, Zielinska D, Mann M, Wiśniewski JR (2010) The phosphoproteome is quantitatively preserved in formalin-fixed paraffin-embedded tissue and analyzable by high-resolution mass spectrometry. J Proteome Res 9(7):3688–3700

22. Palmer-Toy DE, Krastins B, Sarracino DA, Nadol JB Jr, Merchant SN (2005) Efficient method for the proteomic analysis of fixed and embedded tissues. J Proteome Res 4(6):2404–2411

23. Röcken C, Wilhelm S (2005) Influence of tissue fixation on the microextraction and identification of amyloid proteins. J Lab Clin Med 146(4):244–250

24. Shi SR, Liu C, Balgley BM, Lee C, Taylor CR (2006) Protein extraction from formalin-fixed, paraffin-embedded tissue sections: quality evaluation by mass spectrometry. J Histochem Cytochem 54(6):739–743

25. Stanta G, Mucelli SP, Petrera F, Bonin S, Bussolati G (2006) A novel fixative improves opportunities of nucleic acids and proteomic analysis in human archive's tissues. Diagn Mol Pathol 15(2):115–123

26. Tian Y, Zhang H (2010) Isolation of proteins by heat-induced extraction from formalin-fixed, paraffin-embedded tissue and preparation of tryptic peptides for mass spectrometric analysis. Curr Protoc Mol Biol 26:11–17, Chapter 10:Unit 10

27. Vincek V, Nassiri M, Nadji M, Morales AR (2003) A tissue fixative that protects macromolecules (DNA, RNA, and protein) and histomorphology in clinical samples. Lab Invest 83(10):1427–1435

28. Wolff C, Schott C, Porschewski P, Reischauer B, Becker KF (2011) Successful protein extraction from over-fixed and long-term stored formalin-fixed tissues. PLoS One 6(1):e16353.

29. Bodzon-Kulakowska A, Bierczynska-Krzysik A, Dylag T, Drabik A, Suder P, Noga M, Jarzebinska J, Silberring J (2007) Methods for samples preparation in proteomic research. J Chromatogr B 849(1–2):1–31

30. Canas B, Pineiro C, Calvo E, Lopez-Ferrer D, Gallardo JM (2007) Trends in sample preparation for classical and second generation proteomics. J Chromatogr A 1153(1–2):235–258

31. Westermeier R, Naven T, Höpker HR (2008) Proteomics in practice. A guide to successful experimental design, 2nd edn. Weinheim, Germany

32. Shevchenko A, Tomas H, Havlis J, Olsen JV, Mann M (2007) In-gel digestion for mass spectrometric characterization of proteins and proteomes. Nat Protoc 1(6):2856–2860

Protein Extraction from Formalin-Fixed Paraffin-Embedded Tissues

37

Karl-Friedrich Becker and Christina Schott

Contents

37.1 Introduction and Purpose

The Qproteome FFPE Tissue Kit (Qproteome FFPE Tissue Kit, QIAGEN, cat No. 37625) is used for extracting full-length proteins from unstained FFPE tissues.[1] Extraction efficiency is comparable to that seen in frozen tissues, and the extracted proteins are suitable for several downstream applications, such as western blot or protein array analysis [1, 2].

FFPE tissue samples cut directly from an FFPE sample block or unstained FFPE sections mounted on a microscope slide (e.g., sections from a series of FFPE tissue sections that could be used for histological or immunohistological analysis but have not been stained, for example with hematoxylin/eosin) are the most suitable starting material for protein extraction.

Two variants of the extraction protocol are available, according to the starting material. The first one is used with unstained FFPE sections mounted onto a microscope slide, while the second one is required for samples cut directly from a block. The former is the method of choice and should be preferred when possible, because the deparaffinization is improved. Histological stains can decrease the yield of the extracted proteins. For better results a guided protein extraction method may be used, which consists of an adjacent hematoxylin/eosin-stained tissue section to control dissection of an unstained specimen for subsequent protein extraction and quantification [3]. This is just one of the commercial kits available for protein extraction. Of course you can find other available commercial kits for this kind of extraction in several companies' websites.

K.-F. Becker (✉) and C. Schott
Institute of Pathology, Technical University of Munich, Munich, Germany

[1] For more details, see the *QProteome FFPE Tissue Handbook*, QIAGEN.

G. Stanta (ed.), *Guidelines for Molecular Analysis in Archive Tissues*,
DOI: 10.1007/978-3-642-17890-0_37, © Springer-Verlag Berlin Heidelberg 2011

37.2 Deparaffinization and Protein Extraction from Slide-Mounted FFPE Sections

37.2.1 Reagents

- *Xylene*
- *100%,[2] 96%, 70% ethanol (v/v, in distilled water)*
- *QProteome FFPE Tissue Kit* (QIAGEN, cat. No. 37623)
- *Reagents required for the preparation of the SDS PAGE* (see Chap. 41)
- *Reagents for quantification* (e.g., Bio-Rad DC Protein Assay Kit, cat. No. 500-0111 or Pierce Micro BCA Protein Assay Kit, cat. No. 23235)[3]

37.2.2 Equipment

- *Microtome*
- *Staining dishes for deparaffinization of FFPE sections*
- *Adjustable pipettes*, range: 2–20 µl, 20–200 µl, 100–1000 µl
- *Microcentrifuge (with rotor for 1.5 ml tubes)*
- *Vortex*
- *Water bath*
- *Thermomixer*
- *Equipment required for the preparation of the SDS PAGE* (see Chap. 41)

37.2.3 Method

37.2.3.1 Sample Preparation

- Cut 5–10 µm-thick sections using a microtome.[4] If necessary, perform microdissection (see Chaps. 4 and 6) and perform deparaffinization directly on the slides.

37.2.3.2 Deparaffinization

- Transfer the sections to a staining dish containing fresh xylene. The slide should be completely covered. Incubate for 10 min at room temperature (15–25°C). Xylene washes should be performed in a fume hood. Repeat this step twice, using fresh xylene each time.
- Transfer the slide to a staining dish containing fresh 100% ethanol[5] for 10 min at room temperature (15–25°C). Repeat this step twice using fresh 100% ethanol.
- Transfer the slide to a staining dish containing fresh 95% ethanol and incubate for 10 min. Repeat this step twice using fresh 96% ethanol.
- Transfer the slide to a staining dish containing fresh 70% ethanol and incubate for 10 min. Repeat this step twice using fresh 70% ethanol.
- Transfer the slide to a staining dish containing fresh double-distilled water and immerse for 30 s. Remove the slide and soak up excess water by tapping the slide carefully on a paper towel. Slides may stay in water for 1 min for up to 3 h.
- Do not touch the section with the paper towel. Ensure that the sections do not dry out.
- Excise areas of interest with a needle (or a blade) and transfer into a 1.5 ml collection tube containing the appropriate volume of Extraction Buffer (EXB). Put the samples on ice.[6]

37.2.3.3 Protein Extraction

- Mix by vortexing. Seal the collection tube with a Collection Tube Sealing Clip (supplied in the kit).
- Incubate on ice for 15 min and mix by vortexing.[7]
- Incubate the tube in a water bath at 100°C for 20 min.
- Using a Thermomixer, incubate the tube at 80°C for 2 h with agitation at 750 rpm.

---

[2]100% isopropanol could also be used instead of ethanol.

[3]Other quantification methods can be used, such as the Bradford one (Bio-Rad Protein Assay Kit II, cat. No. 500-0002), the EZQ Protein Quantitation Kit (Molecular Probes Invitrogen, cat. No. R33200), or Quant-iT Protein Assay Kit (Molecular Probes Invitrogen, cat. No. Q33211).

[4]The paraffin blocks should be kept at −20°C or on dry ice in aluminum foil to obtain thin sections. Clean the microtome with xylene after each cutting.

[5]Alternatively, 100% isopropanol can be used.

[6]The volume of the extraction buffer should be adjusted according to the size of the tissue area. For an area of about 0.5 cm in diameter use 100 µl buffer.

[7]Be sure that collection tubes are properly sealed with a Collection Tube Sealing Clip before performing the next step.

- After incubation, place the tube at 4°C for 1 min and remove the Collection Tube Sealing Clip.
- Centrifuge the tube for 15 min at 14,000× g at 4°C. Transfer the supernatant containing the extracted proteins to a new 1.5 ml collection tube.
- The extracted proteins can be stored for up to 1 week at 4°C. For longer-term storage, aliquot the extracted proteins and store at−20°C. Avoid repeated freeze–thaw cycles.
- Verify the quality of the extraction by loading 1/10 of the lysate into a SDS-PAGE (see Chap. 41).

37.3 Deparaffinization and Protein Extraction from Free FFPE Sections

37.3.1 Reagents

- *Xylene*
- *100%[5], 96%, 70% ethanol* (v/v, in distilled water)
- *QProteome FFPE Tissue Kit* (QIAGEN, cat. no. 37623)
- *Reagents required for the preparation of the SDS PAGE* (see Chap. 41)
- *Reagents for quantification* (e.g., Bio-Rad DC Protein Assay Kit, cat. No. 500-0111 or Pierce Micro BCA Protein Assay Kit, cat. No. 23235)[8]

37.3.2 Equipment

- *Microtome*
- *1.5 ml tubes*
- *Adjustable pipettes*, range: 2–20 μl, 20–200 μl, 100–1,000 μl
- *Microcentrifuge (with rotor for 1.5 ml tubes)*
- *Vortex*
- *Water bath*
- *Thermomixer*
- *Equipment required for the preparation of the SDS PAGE* (see Chap. 41)

---

[8]Other quantification methods can be used, such as the Bradford one (Bio-Rad Protein Assay Kit II, cat. No. 500-0002), the EZQ Protein Quantitation Kit (Molecular Probes Invitrogen, cat. No. R33200), or Quant-iT Protein Assay Kit (Molecular Probes Invitrogen, cat. No. Q33211).

37.3.3 Method

37.3.3.1 Sample Preparation

- Cut 3–5 sections up to 10 μm thick, using a microtome.[9]
- Place the sections in a 1.5 ml collection tube.[10]

37.3.3.2 Deparaffinization

- Pipette 1 ml of xylene into the tube. Close the tube tightly, vortex vigorously for 10 s, and incubate for 10 min at room temperature (15–25°C).
- Centrifuge the tube in a microcentrifuge at 14,000× g for 2 min. The tissue will form a soft pellet on the bottom of the tube.
- Carefully remove the supernatant using a pipette and discard it. Do not decant the supernatant and do not disturb the pellet.[11]
- Repeat wash with xylene twice, using 1 ml fresh xylene each time.
- Pipette 1 ml of 100% ethanol into the tube containing the pellet, close tightly, and mix by vortexing. Incubate for 10 min at room temperature (15–25°C). Centrifuge the tube at full speed for 2 min.
- Carefully remove the supernatant using a pipette and discard. Do not disturb the pellet.
- Repeat wash using 1 ml fresh 100% ethanol.
- Repeat washes using 96% ethanol twice and 70% ethanol twice.

37.3.3.3 Protein Extraction

- Pipette the appropriate volume of EXB (supplied by the Qproteome kit) into the tube containing the pellet and mix by vortexing.[12] Seal the Collection

---

[9]The paraffin blocks should be kept at −20°C or on dry ice in aluminum foil to obtain thin sections. Clean the microtome with xylene after each cutting.

[10]The cut sections should be stored at −20°C before performing the deparaffinization step if not immediately used.

[11]If the pellet detaches from the wall of the tube, repeat centrifugation step to enable removal of any residual ethanol. Do not disturb or remove the pellet.

[12]The volume of the extraction buffer should be adjusted according to the size of the tissue area. For an area of about 0.5 cm in diameter use 100 μl.

Tube with a Collection Tube Sealing Clip (supplied in the kit).

- Incubate on ice for 15 min and mix by vortexing.[13]
- Incubate the tube in a water bath at 100°C for 20 min.
- Using a Thermomixer, incubate the tube at 80°C for 2 h with agitation at 750 rpm.
- After incubation, place the tube at 4°C for 1 min and remove the Collection Tube Sealing Clip.[14]
- Centrifuge the tube for 15 min at 14,000× g at 4°C. Transfer the supernatant containing the extracted proteins to a new 1.5 ml Collection Tube.
- The extracted proteins can be stored for up to 1 week at 4°C. For longer-term storage, aliquot the extracted proteins and store at −20°C. Avoid repeated freeze–thaw cycles.
- Verify the quality of the extraction by loading 1/10 of the lysate into a SDS-PAGE (see Chap. 41).
- Quantify the protein yield[15] by Lowry or BCA method (or others) following the manufacturer's instructions

---

[13]Be sure that Collection Tubes are properly sealed with a Collection Tube Sealing Clip.

[14]Be sure that Collection Tube Sealing Clip has been removed before starting the centrifugation step.

[15]Due to the presence of detergents, the lysate should be diluted with the same volume of distilled water before performing the protein quantification; otherwise you should have some interference.

37.3.4 Troubleshooting

If the protein yield is low there could be four possible causes:

- Poor quality of starting material. Samples that have been fixed for more than 24 h or stored for very long periods may lead to an incomplete extraction of protein.
- Too little starting material. In this case, the number of starting sections should be increased.
- Insufficient deparaffinization. The paraffin may be more difficult to remove from thicker sections. We recommend using 10 µm-thick sections. If samples contain large amounts of paraffin, the xylene treatment should be repeated for additional two times.
- Loss of the pellet or incomplete resuspension. During the deparaffinization steps, do not pour the supernatant, but aspire it with a pipette to avoid disturbing the pellet itself.

References

1. Becker KF, Schott C, Hipp S, Metzger V, Porschewski P, Beck R, Nahrig J, Becker I, Hofler H (2007) Quantitative protein analysis from formalin-fixed tissues: implications for translational clinical research and nanoscale molecular diagnosis. J Pathol 211(3):370–378
2. Wolff C, Schott C, Porschewski P, Reischauer B, Becker KF (2011) Successful protein extraction from over-fixed and long-term stored formalin-fixed tissues. PLoS One 6(1):e16353
3. Becker KF, Schott C, Becker I, Höfler H (2008) Guided protein extraction from formalin-fixed tissues for quantitative multiplex analysis avoids detrimental effects of histological stains. Proteomics Clin Appl 2(5):737–743

Rapid Protein Extraction from Formalin-Fixed Paraffin-Embedded Tissue Samples

38

Valentina Faoro and Giorgio Stanta

Contents

V. Faoro (✉)
Department of Medical, Surgical and Health Sciences,
University of Trieste, Trieste, Italy

G. Stanta
Department of Medical Sciences,
University of Trieste, Cattinara Hospital,
Strada di Fiume 447, Trieste, Italy

38.1 Introduction and Purpose

This protocol describes a rapid and low-cost method based on SDS-containing-Laemmli buffer that allows an efficient extraction of proteins from FFPE [1]. The presence of SDS and the exposure to high temperature allow an efficient reversal of protein cross-link on FFPE. In literature, several works proposed different heat-induced antigen retrieval strategies for a good formaldehyde cross-link reversal, enabling proteomics analyses also in archive tissues [2–5]. The protocol described herein provides a feasible and quick option for protein extraction not only from FFPE but also from formalin-free fixatives.

38.2 Protocol

38.2.1 Reagents[1]

- *Xylene* (Fluka or Sigma-Aldrich)
- *100%, 96%, 70%* ethanol (v/v)
- *Laemmli Buffer* (100 mM Tris-HCl in distilled water, pH 6.8, 2% (w/v) SDS, 20% (v/v) glycerol, 4% (v/v) β-mercaptoethanol)[2]
- *Bromophenol blue stock solution* (1% (w/v) Bromophenol blue, 50 mM Tris-base)
- *10% (v/v) glycerol*

---

[1] If not specified, all the buffers are made in distilled water.

[2] It is better to prepare this solution under a chemical hood, because b-Mercaptoethanol is very toxic.

G. Stanta (ed.), *Guidelines for Molecular Analysis in Archive Tissues*,
DOI: 10.1007/978-3-642-17890-0_38, © Springer-Verlag Berlin Heidelberg 2011

- *EZQ Protein Quantitation Kit* for protein quantification (Molecular Probes Invitrogen, cat. No. R33200)[3]
- *Reagents required for preparation of the SDS PAGE* (see Chap. 41).

38.2.2 Equipment

- *Adjustable pipettes,* range: 2–20 μl, 20–200 μl, 100–1,000 μl
- *1.5 ml tubes*
- *Microcentrifuge* (with rotor for 1.5 ml tubes)
- *Microtome*
- *Vortex*
- *Thermomixer*
- *Equipment required for preparation of the SDS PAGE* (see Chap. 41)

38.2.3 Method

38.2.3.1 Sample Preparation

- Cut about three sections up to 10 μm thickness (~25 mm[2] each) using a microtome[4].
- Place the sections in a 1.5 ml collection tube with the help of a scalpel or a little brush[5].

38.2.3.2 Deparaffinization

- Pipette 1 ml of xylene into the tube[6]. Close the tube tightly, vortex vigorously.
- Incubate for 5 min at room temperature (15–25°C).
- Centrifuge the tube in a microcentrifuge at maximum speed for 5 min at room temperature.

- Carefully remove the supernatant using a pipette and discard it. To avoid disturbing the pellet, do not pour the supernatant.
- Repeat the previous steps using 1 ml fresh xylene.
- Pipette 1 ml of 100% ethanol into the tube containing the pellet, close tightly, vortex vigorously.
- Incubate for 5 min at room temperature (20–25°C).
- Centrifuge the tube in a microcentrifuge at maximum speed for 5 min at room temperature and then remove the supernatant using a pipette and discard it.
- Repeat the steps using 1 ml of 96% and then 70% ethanol.

38.2.3.3 Protein Extraction

- Resuspend the pellet in 150 μl[7] of Laemmli Buffer[8].
- Using a thermomixer, incubate the tube at 99°C for 20 min.
- Spin down at maximum speed for 5 min at 4°C.
- For protein quantification use the EZQ Protein Quantitation Kit following the manufacturer's instructions[9].
- Add to the supernatant 0.02% of Bromophenol blue stock solution and load it (at least 10 μl) into an SDS-PAGE gel (see Chap. 41) to verify the quality of the extraction[10].
- Store the lysate at –20°C.

38.2.4 Troubleshooting

If the protein yield is low there could be four possible causes:

- Poor quality of starting material.
- Too little starting material. In this case, increase the number of sample sections.
- Insufficient deparaffinization step. During deparaffinization, the supernatant must be completely

---

[3] Other quantification methods can be used, such as the Bradford one (Bio-Rad Protein Assay Kit II, cat. No. 500-0002), Bio-Rad DC Protein Assay Kit, cat. No. 500-0111, Pierce Micro BCA Protein Assay Kit, cat. No. 23235 or Quant-iT Protein Assay Kit (Molecular Probes Invitrogen, cat. No. Q33211).

[4] The paraffin blocks should be kept at –20°C or on dry ice in aluminium foil before cutting the sections. Clean the microtome with xylene after each cutting.

[5] The cut sections should be stored at –20°C before performing the deparaffinization step (if not immediately used).

[6] In order to minimise exposure to xylene fumes, it is better to perform this step under a chemical hood.

[7] The volume of buffer depends on the tissue size.

[8] Do not put the samples on ice because the SDS could precipitate.

[9] Due to the presence of detergents, it could be necessary to dilute the lysate with the same volume of distilled water before performing the protein quantification; otherwise, you should have some interference.

[10] The extracted proteins can be stored for up to 1 week at –20°C.

and carefully removed without disturbing the tissue pellet. If samples under processing contain large amounts of paraffin, repeat the xylene treatment once again.

- Loss of the pellet or incomplete resuspension. During the deparaffinization steps, do not pour the supernatant, bur aspire it with a pipette to avoid disturbing the pellet itself.

References

1. Nirmalan NJ, Harnden P, Selby PJ, Banks RE (2009) Development and validation of a novel protein extraction methodology for quantitation of protein expression in formalin-fixed paraffin-embedded tissues using western blotting. J Pathol 217(4):497–506

2. Crockett DK, Lin Z, Vaughn CP, Lim MS, Elenitoba-Johnson KS (2005) Identification of proteins from formalin-fixed paraffin-embedded cells by LC-MS/MS. Lab Investig 85(11): 1405–1415

3. Guo T, Wang W, Rudnick PA, Song T, Li J, Zhuang Z, Weil RJ, DeVoe DL, Lee CS, Balgley BM (2007) Proteome analysis of microdissected formalin-fixed and paraffin-embedded tissue specimens. J Histochem Cytochem 55(7): 763–772

4. Hood BL, Darfler MM, Guiel TG, Furusato B, Lucas DA, Ringeisen BR, Sesterhenn IA, Conrads TP, Veenstra TD, Krizman DB (2005) Proteomic analysis of formalin-fixed prostate cancer tissue. Mol Cell Proteomics 4(11):1741–1753

5. Shi SR, Liu C, Balgley BM, Lee C, Taylor CR (2006) Protein extraction from formalin-fixed, paraffin-embedded tissue sections: quality evaluation by mass spectrometry. J Histochem Cytochem 54(6):739–743

Protein Extraction Protocol from Tissues Treated with Formalin-Free Fixatives

39

Valentina Faoro and Giorgio Stanta

Contents

39.1 Introduction and Purpose

This protocol allows the extraction of proteins from tissues fixed with formalin-free fixatives.

The following protocol is a modified version of two original protocols, the first one published by Stanta et al. [1] and the second one by Bellet et al. [2] that were developed to extract proteins from FineFix and RCL2-fixed tissues, respectively. The protein extracted by following this protocol could also be used for 2D analyses, because the extraction buffer doesn't contain ionic detergents that could modify the protein charge. The use of high temperature improves the extraction of those proteins that might be structurally altered or masked by the action of the fixative.

39.2 Protocol

39.2.1 Reagents[1]

- *Xylene* (Fluka or Sigma-Aldrich)
- *100%, 96%, 70% (v/v) ethanol*
- *10% (v/v) glycerol*
- *Extraction buffer* (50 mM Tris-HCl pH 7.5, 7 M urea, 2 M thiourea, 2% (w/v) CHAPS, 1% (w/v) MEGA, 0.5% (v/v) Triton x-100, 1% (w/v) OGP, and 50 mM DTT)[2]

V. Faoro (✉)
Department of Medical, Surgical and Health Sciences,
University of Trieste, Trieste, Italy

G. Stanta
Department of Medical Sciences,
University of Trieste, Cattinara Hospital,
Strada di Fiume 447, Trieste, Italy

---

[1] If not specified, all the buffers are made in distilled water.

[2] CHAPS, 3-[(3-Cholamidopropyl)dimethylammonio]-1-propane-sulfonate; OGP, n-Octyl β-D-glucopyranoside; MEGA, N-Decanoyl-N-methylglucamine; DTT, Dithiothreitol.

- *Protease inhibitor mixture* (e.g., Amersham Biosciences, cat. No. 80-6501-23 or other similar)
- *Bradford reagent for quantification* (e.g. Bio-Rad Protein Assay Kit II, cat. No. 500-0002)[3]
- *Reagents required for the preparation of the SDS PAGE* (see Chap. 41)

39.2.2 Equipment

- *Adjustable pipettes*, range 2–20 µl, 20–200 µl, 100–1,000 µl
- *1.5 ml tubes*
- *Microcentrifuge* for 1.5 ml tubes
- *Microtome*
- *Vortex*
- *Thermomixer*
- *Equipment required for preparation of the SDS PAGE* (see Chap. 41)

39.2.3 Method

39.2.3.1 Sample Preparation

- Cut about three sections up to 10 µm thickness using a microtome.[4]
- Place the sections in a 1.5 ml collection tube with the help of a scalpel or a little brush.[5,6]

39.2.3.2 Deparaffinization

- Pipette 1 ml of xylene into the tube. Close the tube tightly, vortex vigorously.[7,8]
- Centrifuge the tube in a microcentrifuge at maximum speed for 10 min at room temperature.
- Carefully remove the supernatant using a pipette and discard it. To avoid disturbing the pellet, do not pour the supernatant.
- Repeat washes using 1 ml of fresh xylene.
- Pipette 1 ml of 100% ethanol into the tube containing the pellet, close tightly, vortex vigorously.
- Centrifuge the tube in a microcentrifuge at maximum speed for 10 min at room temperature.
- Carefully remove the supernatant using a pipette and discard it.
- Repeat washes using 1 ml of fresh 96% and 70% ethanol.
- Leave the tube at least for 5 min at room temperature[9] in order to air-dry the pellet.

39.2.3.3 Protein Extraction

- Resuspend the pellet in an appropriate volume of extraction buffer with protease inhibitors diluted according to the manufacturer's instruction.[10]
- Close the tube tightly and vortex vigorously for approximately 10 s.
- Incubate on ice for 15 min, and mix by vortexing.[11]
- Using a thermomixer, incubate the tube at 99°C for 15 min with stirring at 750 rpm.
- Centrifuge at maximum speed for 15 min at 4°C and recover the supernatant.
- The extracted proteins can be stored for up to 1 week at 4°C. For longer-term storage, aliquot the extracted proteins and store at −20°C. Avoid repeated freeze–thaw cycles.

---

[3]Other quantification methods can be used, such as the BCA method (e.g. Pierce Micro BCA Protein Assay Kit, cat. No. 23235), the Lowry method (e.g. Bio-Rad DC Protein Assay Kit, cat. No. 500-0111), the EZQ Protein Quantitation Kit (Molecular Probes Invitrogen, cat. No. R33200), and Quant-iT Protein Assay Kit (Molecular Probes Invitrogen, cat. No. Q33211).

[4]The paraffin blocks should be kept at −20°C or on dry ice in aluminium foil before cutting sections. Clean the microtome with xylene after each cutting.

[5]The cut sections should be stored at −20°C before performing the deparaffinization step (if not immediately used).

[6]If you are interested in performing microdissection of the tissue, it's possible to cut the sections and mount them on a slide on which the area of interest can be excised. Before performing the extraction, the slides should be kept in an oven at 62°C at least for 1 h, or alternatively at 37°C overnight (see Chaps. 4 and 6).

---

[7]In order to minimize exposure to xylene fumes, it's better to perform this step under a chemical hood.

[8]If the extraction is performed directly on slide-mounted sections, you should use staining dishes for the deparaffinization. At the end of these steps you should transfer the excised area on a 1.5 ml tube with the appropriate volume of extraction buffer.

[9]This step is not necessary if you are performing the extraction directly on slide-mounted sections.

[10]The volume of the extraction buffer should be adjusted according to the size of the tissue section.

[11]Be sure that the 1.5 ml tubes are properly sealed.

- Quantify the protein yield[12] by Bradford method (or others) following the manufacturer's instructions.
- Verify the quality of the extraction by loading 1/10 of the lysate into a SDS-PAGE (see Chap. 41).

39.2.4 Troubleshooting

If the protein yield is low there could be several causes:

- Poor quality of starting material.
- Presence of fat on the tissues that could limit the extraction of proteins.
- In case of FineFix-treated tissues, the use of isopropanol instead of JFC solution[13] during the dehydratation-clearing step could limit the extraction of proteins.
- Incorrect fixation procedure that could minimize the extraction. Be careful of the fixation protocol required for each type of fixative.
- Too little starting material. In this case, increase the number of sample sections, even if too much material could limit the deparaffinization step and could decrease the protein yield.

- Insufficient deparaffinization. During deparaffinization, ensure that supernatants are completely and carefully removed without disturbing the tissue pellet. If the samples contain large amounts of paraffin, repeat the xylene treatment once again.
- Loss of the pellet or incomplete precipitation. During the deparaffinization steps, do not pour the supernatant, bur aspire it with a pipette to avoid disturbing the pellet itself.
- Impairment of protein extraction could be due to contamination of the histoprocessor, routinely used in the pathology department, by formalin. If possible, it is suggested to use a dedicated device for fixation of tissues with formalin-free fixatives.

References

1. Stanta G, Mucelli SP, Petrera F, Bonin S, Bussolati G (2006) A novel fixative improves opportunities of nucleic acids and proteomic analysis in human archive's tissues. Diagn Mol Pathol 15(2):115–123
2. Bellet V, Boissiere F, Bibeau F, Desmetz C, Berthe M, Rochaix P, Maudelonde T, Mange A, Solassol J (2007) Proteomic analysis of rcl2 paraffin-embedded tissues. J Cell Mol Med [Epub ahead of print]

---

[12]Due to the presence of detergents, the lysate should be diluted with the same volume of distilled water before performing the protein quantification to avoid interference.

[13]The JFC solution is a mixture of ethanol, isopropanol and long chain hydrocarbons that could be used instead of isopropanaol, during the dehydratation-clearing step, after the microwave fixation with FineFix.

Trichloroacetic Acid (TCA) Precipitation of Proteins

40

Valentina Faoro and Giorgio Stanta

Contents

40.1 Introduction and Purpose

Protein precipitation is an optional step that can be carried out prior to SDS-PAGE analyses in order to separate proteins in the samples from contaminating and interfering species (salts, lipids, nucleic acids…) or to concentrate diluted samples. However, this technique is not fully efficient, because some proteins may not be readily resuspended after precipitation [1–3].

40.2 Protocol

40.2.1 Reagents[1]

- *10% (v/v) TCA*
- *100% ice-cold ethanol*
- *Bromophenol blue stock solution* (1% (w/v) Bromophenol blue, 50 mM Tris-base)
- *Loading Buffer 4X* (50 mM Tris-HCl, pH 6.8, 2% (w/v) SDS, 10% (v/v) glycerol, 1% (v/v) β-mercaptoethanol, 0.02% (v/v) of Bromophenol blue stock solution)[2]
- *Protein samples*

V. Faoro (✉)
Department of Medical, Surgical and Health Sciences,
University of Trieste, Trieste, Italy

G. Stanta
Department of Medical Sciences,
University of Trieste, Cattinara Hospital,
Strada di Fiume 447, Trieste, Italy

---

[1] If not specified, all the buffers are made in distilled water.

[2] It is better to prepare this solution under a chemical hood, because β-Mercaptoethanol is very toxic.

G. Stanta (ed.), *Guidelines for Molecular Analysis in Archive Tissues*,
DOI: 10.1007/978-3-642-17890-0_40, © Springer-Verlag Berlin Heidelberg 2011

40.2.2 Equipment

- *Adjustable pipettes*, range: 2–20 μl, 20–200 μl, 100–1,000 μl
- *Thermo block*
- *Refrigerate microcentrifuge*

40.2.3 Method

- Add 1 volume of 10% TCA[3] to the sample (i.e. add 100 μl of 10% TCA to 100 μl of protein lysate).
- Incubate on ice for at least 30 min.
- Centrifuge for 15 min in a microcentrifuge at 4°C at maximum speed.
- Remove the supernatant, leaving the protein pellet intact.
- Add 1 ml of ice-cold ethanol[4] and centrifuge briefly.

- Remove the ethanol and dry the pellet by placing the tube at 37°C in a thermo block for 5–10 min.
- Resuspend the pellet in 4X Loading Buffer,[5] boil for 5 min at 95°C and then load the sample onto a SDS-PAGE (see Chap. 41).

References

1. Granier F (1988) Extraction of plant proteins for two-dimensional electrophoresis. Electrophoresis 9(11):712–718
2. Mechin V, Damerval C, Zivy M (2007) Total protein extraction with TCA-acetone. In: Plant proteomics: methods and protocols, vol 355, Methods in molecular biology. Humana Press, Clifton, pp 1–8
3. Meyer Y, Grosset J, Chartier Y, Cleyet-Marel JC (1988) Preparation by two-dimensional electrophoresis of proteins for antibody production: antibodies against proteins whose synthesis is reduced by auxin in tobacco mesophyll protoplasts. Electrophoresis 9(11):704–712

---

[3]For a more effective result, the lysate could also be precipitated in 10% TCA in acetone.

[4]Also ice-cold acetone could be used.

[5]The presence of some traces of TCA can give a yellow colour as a consequence of the acidification of the sample buffer; adjust the pH with 1 N NaOH or 1 M TrisHCl pH 8.5 to obtain the normal blue sample buffer colour.

One-Dimensional Sodium-Dodecyl-Sulfate (SDS) Polyacrylamide Gel Electrophoresis

41

Valentina Faoro, Karl-Friedrich Becker, and Giorgio Stanta

Contents

V. Faoro (✉)
Department of Medical, Surgical and Health Sciences,
University of Trieste, Trieste, Italy

G. Stanta
Department of Medical Sciences,
University of Trieste, Cattinara Hospital,
Strada di Fiume 447, Trieste, Italy

K.-F. Becker
Institute of Pathology, Technical University of Munich,
Munich, Germany

41.1 Introduction and Purpose

Protein electrophoresis is the most used method to resolve protein mixtures (also those extracted from fixed tissues) into their individual components in order to visualize, identify, and characterize them. In one-dimensional sodium-dodecyl-sulfate polyacrylamide gel electrophoresis (SDS-PAGE), proteins are separated on the basis of their molecular mass [1–6]. The most common application of SDS-PAGE is the characterization of proteins after protein purification. It is also possible to blot the separated proteins onto a positively charged membrane and to probe them with protein-specific antibodies (such as western blot procedure, see Chap. 43).

41.2 Protocol

41.2.1 Reagents[1]

- *Acrylamide/NN,-methylene-bis-acrylamide 40% (stock solution 29:1)*
- *1.5 M Tris-HCl pH 8.8*
- *0.5 M Tris-HCl pH 6.8*
- *10% (w/v) SDS[2]*
- *TEMED[3]*

---

[1] If not specified, all the buffers are made in distilled water.

[2] SDS (Sodium-dodecyl-sulfate) powder is toxic. You should weigh it under a chemical hood and use a closed box when you transport it out in order to avoid breathing the dust.

[3] *N,N,N′,N′*-Tetramethylethylenediamine; *N,N,N′,N′*-Di-(dimethyl-amino) ethane; *N,N,N′,N′*-Tetramethyl-1-, 2-diaminomethane.

G. Stanta (ed.), *Guidelines for Molecular Analysis in Archive Tissues*,
DOI: 10.1007/978-3-642-17890-0_41, © Springer-Verlag Berlin Heidelberg 2011

- *10% (w/v) APS*[4]
- *10X Running Buffer stock solution* (250 mM Tris, 1.92 M Glycine, 1% (v/v) SDS[5])
- *1% (w/v) agarose sealing solution*[6]
- *Protein molecular weight standard* (e.g., Bench Mark™ Pre-Stained Protein Ladder, Invitrogen, cat. No 10748-010 or Precision Plus Protein Standard Dual Color, BioRad cat. No 161/0374 or PageRuler™ Prestained Protein Ladder, Fermentas cat. No SM1811)
- *Bromophenol blue stock solution* (1% (w/v) Bromophenol blue, 50 mM Tris-base)
- *4X Loading Buffer* (50 mM Tris-HCl, pH 6.8, 2% (w/v) SDS[5], 10% (v/v) glycerol, 1% (v/v) β-Mercaptoethanol, 0.02% (v/v) Bromophenol blue stock solution)[7]
- *n-Butanol*

41.2.2 Equipment

- *Adjustable pipettes*, range: 2–20 µl, 20–200 µl, 100–1,000 µl
- *Gel apparatus and electrophoresis equipment* (i.e., glass plate, combs and spacers,…)
- *Hamilton™ syringe*
- *Test tubes*
- *1.5 ml tubes*
- *Some pieces of filter paper* (e.g., Whatman 3 MM paper)

41.2.3 Method

41.2.3.1 Gel Preparation

- Assemble the gel unit with glasses[8] and spacers according to the manufacturer's instructions.

- Select the gel percentage. The concentration of acrylamide used for the gel depends on the size of proteins that have to be analyzed (see Table 41.1)[9]:
- Prepare the running solution in a test tube mixing the reagents listed in Table 41.2.
- Pipette the gel solution between the assembled glass sandwich a few millimeters below the wells that will be formed by the comb. Overlay with *n*-butanol[10] to the top of the gel unit. A very sharp liquid-gel interface will be visible once the gel has polymerized.[11]

Table 41.1 Recommended acrylamide concentrations for protein separation

Acrylamide percentage in running gel (%)	Separation size range (kDa)
8	24–200
10	14–200
12	14–100[a]
14	14–60[a]

[a]Larger proteins fail to move significantly into the acrylamide gel

Table 41.2 Running solution (final volume 20 ml[a])

Final gel concentration	8%	10%	12%	14%
40% Acrylamide/bis 29:1	4 ml	5 ml	6 ml	7 ml
1.5 M Tris-HCl pH 8.8	5 ml	5 ml	5 ml	5 ml
SDS 10%	200 µl	200 µl	200 µl	200 µl
Distilled water	11 ml	10 ml	9 ml	8 ml
APS[b]	200 µl	200 µl	200 µl	200 µl
TEMED[b]	40 µl	40 µl	40 µl	40 µl

[a]The volumes of the reagents can be adapted to match the volume required by the used electrophoretic apparatus
[b]Add them just before pouring to avoid early polymerization of the gel solution

---

[4]Ammonium persulfate.

[5]A stock solution of 10% (w/v) SDS can also be used.

[6]This solution could be made also in Running buffer 1X.

[7]It is better to prepare this solution under a chemical hood, because β-Mercaptoethanol is very toxic. The solution should be stored in 0.5 ml aliquots at −20°C for up to 6 months.

[8]The glasses should be thoroughly cleaned with detergent and methanol and dried before use.

---

[9]Improved resolution of protein band can be achieved using an acrylamide gradient gel system.

[10]Water can be used instead of butanol.

[11]For the best polymerization you should bring all the refrigerated gel solutions to room temperature prior to use.

Table 41.3 Stacking gel solution (final volume 5 ml[a])

Acrylamide/bis 40% 29:1	0.5 ml
0.5 M Tris-HCl pH 6.8	1.3 ml
SDS 10%	150 μl
Distilled water	3 ml
APS[b]	50 μl
TEMED[b]	10 μl

[a]The volumes of reagents can be adapted to match the volume required by the used electrophoretic apparatus
[b]Added at the end to avoid early polymerization of the gel solution

- After polymerization is complete, pour off the butanol, wash abundantly with water and dry.[12]
- Prepare the stacking solution in a test tube, mixing the reagent according to Table 41.3.
- Pipette the solution down into the glass plate to the top level. Insert a comb, avoiding introduction of air bubbles.[13,14]

41.2.3.2 Sample Preparation

- While the stacking gel is polymerizing, prepare the samples.[15] Add one volume of 4X Loading Buffer to three volumes of protein samples and heat them at 95°C for 5 min.
- Briefly spin down the samples.

41.2.3.3 Gel Loading

- Fill the chamber with 1X Running Buffer.
- Slowly remove the combs from the gels, by raising them up to the comb to avoid disturbing the wells.

---

[12]The remaining water on the plates can be removed using pieces of filter paper.

[13]Oxygen can inhibit polymerization, and bubbles can cause a local distortion in the gel surface at the bottom of the wells. To avoid this drawback the comb can be inserted prior to filling the stacking gel solution.

[14]A very sharp interface will be visible when the gel has polymerized. The gel should be fully polymerized.

[15]For a minigel the amount of proteins loaded into one well should be 10–35 μg in a total of 5–15 μl for Coomassie Blue Staining or Western Blotting, and should be 5–15 μg in 5–10 μl for highly sensitive silver staining. For a standard-sized gel the volume loaded into each well should be less than 40 μl.

- Rinse each well with the 1X Running Buffer with a Hamilton™ syringe.
- Using a Hamilton™ syringe, load the samples slowly in the wells.[16]

41.2.3.4 Gel Running

- Connect the gel to the power supply according the manufacturer's instructions. The gel should run at 80 V and 15 mA/gel until the proteins enter the running gel, then apply 120 V and 30 mA/gel.
- When the tracking dye reaches the bottom of the gel, turn the power supply off.

41.2.3.5 Disassemble the Gel

- Remove the buffer and disassemble the gel unit by gently loosening and sliding away both spacers. Then separate gently the glass plates by acting from one corner with a spacer.
- Remove the stacking gel by cutting it with a spacer and orientate the running gel by removing a corner.
- Carefully lift the gel into a tray of staining solution or fixative, or proceed with the immunoblotting depending on the method.

41.3 Troubleshooting

- If the gel solution leaks out of the sandwich during casting, the plates and the spacers should be realigned and the edges should be resealed with tape or agarose sealing solution.
- If the polymerization failed or is incomplete the solution temperature is probably too low or the catalysts are insufficient or degraded.
- If the run takes longer than usual, the buffers are too concentrated or at wrong pH. The solutions should be reprepared.
- If the stained material concentrates at the top of the running gel, probably there are insoluble precipitates in the sample. Some proteins precipitate upon

---

[16]One well should be loaded with a protein molecular weight standard (e.g., Bench Mark™ Pre-Stained Protein Ladder, Invitrogen, cat. No 10748-010 or PageRuler™ Prestained Protein Ladder, Fermentas cat. No SM1811).

heating at 95°C and lower heating temperatures (60–70°C) should be used for a longer time.

- If the bands are fuzzy or poorly resolved, the sample was probably degraded or the gel acrylamide concentration was not suitable for optimal resolution.
- If the protein bands are distorted, the possible causes could be:
  - Air bubbles trapped under comb that should be removed; or
  - Unreacted acrylamide that continue to polymerize after the comb was removed, so the wells should be rinsed with tank buffer immediately after the comb is removed.
- If the dye front curves up (smiles), the temperature is too high and you should reduce the power setting.
- If the dye front curves down (frowns), probably air bubbles are trapped between the glass plates or the gel next to the spacers was not fully polymerized.

References

1. Andrews AT (1986) Electrophoresis: theory, technique and biochemical and clinical applications, 2nd edn. Clarendon, Oxford
2. Ausubel FM, Brent R, Kingston RE, Moore DM, Smith JA, Struhl K (1995) Current protocols in molecular biology. Wiley-Interscience, New York
3. Cseke JL, Chang SC (2004) Extraction and purification of proteins. In: Cseke JL, Kaufman PB, Podila GK, Tsai CJ (eds) Handbook of molecular and cellular methods in biology and medicine, 2nd edn. CRC Press, Boca Raton, pp 45–65
4. Hames BD (1998) Gel electrophoresis of proteins, 3rd edn. University Press, Oxford
5. Laemmli UK (1970) Cleavage of structural proteins during the assembly of the head of bacteriophage T4. Nature 227(5259):680–685
6. Westermeier R (2004) Electrophoresis in practice, 4th edn. Wiley-VCH Verlag GmbH, Weinheim

Visualization of Proteins

42

Valentina Faoro, Karl-Friedrich Becker, and Giorgio Stanta

Contents

V. Faoro (✉)
Department of Medical, Surgical and Health Sciences,
University of Trieste, Trieste, Italy

G. Stanta
Department of Medical Sciences,
University of Trieste, Cattinara Hospital,
Strada di Fiume 447, Trieste, Italy

K.-F. Becker
Institute of Pathology, Technical University of Munich,
Munich, Germany

42.1 Introduction and Purpose

Once electrophoresis is complete, the protein bands can be visualized by incubating the gel in a staining solution. The two most commonly used methods are Coomassie Blue and Silver staining. Coomassie Blue staining [1] is based on the nonspecific binding of the dye Coomassie Brilliant Blue R250 to virtually all proteins. Although Coomassie Blue staining is less sensitive than silver staining, it is widely used due to its convenience. Coomassie Blue binds to proteins approximately stoichiometrically; for this reason, this staining method is preferable when relative amounts of protein have to be determined by densitometry. The proteins are detected as blue bands on a clear background.

Silver staining is the most sensitive method for permanent visible staining of proteins in polyacrylamide gels. This sensitivity, however, comes at the expense of high susceptibility to interference for a number of factors. Silver staining is a complex, multistep process, and many variables can influence the results. High-purity reagents and precise timing are essential for reproducible, high-quality results. Impurities in the gel and/or the water used for preparing the staining reagents can cause poor staining results.

In silver staining, the gel is impregnated with soluble silver ions and developed by treatment with formaldehyde, which reduces silver ions to form an insoluble brown precipitate of metallic silver. This reduction is promoted by proteins. There are many variations of the silver staining process. The method described here (Vorum Silver Staining Protocol) is based on the method by Mortz and has been selected for overall convenience, sensitivity, reproducibility, speed, and compatibility with mass spectrometry analyses [2].

G. Stanta (ed.), *Guidelines for Molecular Analysis in Archive Tissues*,
DOI: 10.1007/978-3-642-17890-0_42, © Springer-Verlag Berlin Heidelberg 2011

Once the gel is stained, it can be photographed or dried for a record of the position and intensity of each band and then analyzed.

Once the gel run is completed and the power supply is turned off, the gel sandwiches can be dissembled; in order to facilitate the separation of the glass plates, remove slowly one of the spacers and use it to carefully separate the glass starting from a corner. Remove the top plate; the gel should be on the bottom glass plate. To record the gel orientation, make a small cut at the upper left or right corner of the gel using a blade or the spacers itself. The gel is so transferred into a glass or plastic tray for staining.

42.2 Coomassie Brilliant Blue Staining (Detection Limit: 0.1–0.5 μg of Protein)

42.2.1 Reagents[1]

- *Coomassie Blue Staining Solution* (40% (v/v) methanol, 10% (v/v) acetic acid, 0.2% (w/v) Coomassie Brilliant Blue R-250[2])
- *Destaining Solution* (25% (v/v) methanol, 10% (v/v) acetic acid)
- *1% (v/v) acetic acid*
- *1% (v/v) glycerol*

42.2.2 Equipment

- *Shaker*
- *Glass staining trays*

42.2.3 Method

Staining should be performed at room temperature. Covered plastic trays work well and minimize exposure

to methanol and acetic acid vapors. When covers are not used, these procedures should be done in a fume hood. Wear clean gloves to avoid keratin contamination (hairs or skin desquamed cells).

- Submerge the gel in Coomassie Blue Staining Solution for 30 min–2 h with gentle shaking.[3]
- Replace the staining solution with Destaining Solution for 30 min–2 h. Shake slowly until the background signal becomes clear.[4,5]
- Store the gel in a 1% acetic acid solution. To minimize cracking, add 1% glycerol before drying the gel.

42.3 Silver Staining (Detection Limit: 1–5 ng of Protein)

42.3.1 Reagents[6]

- *Fixing Solution* (50% (v/v) methanol, 12% (v/v) acid acetic, 0.05% (v/v) formaldehyde 37% [7])
- *35% (v/v) ethanol*
- *1% (v/v) acetic acid*
- *1% (v/v) glycerol*
- *Double distilled water*
- *Sensitizing Solution* (0.02% (v/v) $Na_2S_2O_3$)[8]
- *Silver Reaction Solution* (0.2% (w/v) silver nitrate, 0.076% (v/v) formaldehyde 37%)
- *Developing Solution* (6% (w/v) Na_2CO_3, 0.05% (v/v) formaldehyde 37%, 0.0004% (v/v) $Na_2S_2O_3$[7])
- *Stop Solution* (50% (v/v) methanol, 12% (v/v) acetic acid)

---

[1]If not specified, all the buffers are made in distilled water.

[2]The solution should be prepared some days before because it needs a long mix. The solution is light sensitive. Do not breathe the dust.

[3]The time required to stain the gel depends in part on the thickness of the gel: a 0.75-mm-thick gel will stain faster than a 1.5 mm gel and is completely stained in an hour.

[4]A folded paper towel placed in the destaining bath soaks up excess stain and allows the reuse of the Destaining Solution.

[5]Use caution because excessive destaining will lead to loss of band intensity.

[6]All the buffers should be done in double-distilled water.

[7]Formaldehyde 37% is the available commercial solution. Formaldehyde is toxic and should not be inhaled. Use covered glass trays.

[8]You can use a stock solution of 10% $Na_2S_2O_3$ (w/v).

42.3.2 Equipment

- *Shaker*
- *Glass staining trays*

42.3.3 Method

All steps can be performed at room temperature in gentle shaking. In order to avoid keratin contamination, touch the gels only on the edges and always with clean gloves,[9] and use only double-distilled water. Glass staining trays are particularly useful because they are easy to clean.[10]

- Put the gel in the Silver Staining Fixing Solution for 2 h to overnight.[11]
- Rinse the gel with 35% (v/v) ethanol for 20 min.
- Remove the solution and add the Sensitizing Solution for 2 min.
- Rinse the gel three times in double-distilled water for 5 min.
- Put the gel for 30 min in the Silver Reaction Solution.
- Rinse briefly with double-distilled water for 30 s.
- Put the gel in the Developing Solution till all the bands appear (approximately 5 min, but the time of development can vary). This step may be visually monitored. The gels should be transferred to the Stop Solution when the spots have reached the desired intensity and before the staining background becomes too dark.
- Add the Stop Solution.
- First wash the gel in the Fixing Solution for about 15 min and then in double-distilled water.
- Store the gel in a solution of 1% (v/v) acetic acid. To minimize cracking, add 1% (v/v) glycerol before drying the gel.[12]

- Gel can be photographed or used for other analyses.
- If mass spectrometry analysis should be carried out, rinse the gel with abundant double-distilled water for a minimum of 30 min.

42.3.4 Troubleshooting

42.3.4.1 Coomassie Stain

- If bands are poorly stained, the staining time should be increased.
- If the background is blue, the destaining time should be increased.

42.3.4.2 Silver Stain

- If the bands develop poorly, the solutions should be prepared again.
- If the background is too dark, the interfering substances may not have been completely washed out, so the fixing time should be increased.

References

1. Merril CR (1990) Gel-staining techniques. Meth Enzymol 182:477–488
2. Mortz E, Krogh TN, Vorum H, Gorg A (2001) Improved silver staining protocols for high sensitivity protein identification using matrix-assisted laser desorption/ionization-time of flight analysis. Proteomics 1(11):1359–1363

---

[9]Any minimal trace of skin proteins on the gel will be stained.

[10]The glass staining trays should be cleaned with ethanol and methanol shaking slowly on a laboratory shaker.

[11]Formaldehyde can be replaced by glutaraldehyde for higher sensitivity, but the method will become incompatible with mass spectrometry analysis.

[12]For gel drying, place the gel between two layers of a plastic film and remove any air bubbles inside the sandwich to prevent any cracks and dry the gel at 70–80°C under vacuum for 1 h. Drying time may vary, depending on gel thickness and strength of vacuum pump. Be careful not to disturb the vacuum until the gel is dried, otherwise it may crack.

Western Blotting

43

Valentina Faoro, Karl-Friedrich Becker, and Giorgio Stanta

Contents

V. Faoro (✉)
Department of Medical, Surgical and Health Sciences,
University of Trieste, Trieste, Italy

G. Stanta
Department of Medical Sciences,
University of Trieste, Cattinara Hospital,
Strada di Fiume 447, Trieste, Italy

K.-F. Becker
Institute of Pathology, Technical University of Munich,
Munich, Germany

43.1 Introduction and Purpose

Western blot (or immune blot) is traditionally used to detect the expression of a protein in complex samples using a specific antibody. In the blotting, protein samples, previously separated on a SDS PAGE gel, are electrophoretically transferred to a positively charged membrane filter that binds the SDS-coated negatively charged proteins [1, 2].

The most frequently used membrane materials in blotting are nitrocellulose and polyvinylidene difluoride (PVDF). The choice depends on the type of analysis and characteristics of the detection system. Nitrocellulose is the most often applicable membrane while PVDF is usually used when the bound protein is ultimately analyzed by automated solid-phase protein sequencing.

Transfer efficiency can be checked by staining proteins directly on the membrane using Ponceau S.

Once transferred to the membrane, the proteins can be probed with the antibodies of interest. Most commonly, a horseradish peroxidase-linked secondary antibody is used in combination with a chemiluminescent agent, and the reaction produces luminescence in proportion to the amount of protein. The horseradish peroxidase (HRP) chemiluminescent reaction is based on the catalyzed oxidation of luminol by peroxide. Oxidized luminol emits light as it decays to its ground state. This technique has the speed and safety of chromogenic detection methods (a system usually used for immunohistochemistry), at higher sensitivity levels.

G. Stanta (ed.), *Guidelines for Molecular Analysis in Archive Tissues*,
DOI: 10.1007/978-3-642-17890-0_43, © Springer-Verlag Berlin Heidelberg 2011

43.2 Protocol

43.2.1 Reagents[1]

- *100% methanol*
- *10X Transfer buffer stock solution* (250 mM Tris, 1.92 M Glycine)
- *1% (w/v)Ponceau S Staining Solution (e.g. Sigma-Aldrich cat. no P3504-10G)[2] in 1% (v/v) Acetic Acid*
- *Double-distilled water*
- *Washing buffer*: either PBS or TBS with 0.1% Tween®-20 (v/v)
- *Blocking buffer*: 5% (w/v) blocking agent (e.g., BSA or nonfat dried milk) in washing buffer
- *Primary antibody* specific for the protein of interest
- *HRP-conjugated secondary antibody*, specific for primary antibody
- *Chemiluminescent reagents for immunodetection* (e.g., Amersham Biosciences ECL Western Blotting Detection Reagents or Millipore Immobilon™ Western Chemiluminescent Substrates)
- *Special chemical solutions for development and fixing of the X-ray film* (e.g., Kodak)

43.2.2 Equipment

- *Adjustable pipettes*, range: 2–20 µl, 20–200 µl, 100–1,000 µl
- *Transfer electroblotting apparatus* (i.e., electroblotting cassettes, sponges …)
- *Orbital shacker*
- *PVDF or nitrocellulose membrane* (e.g., Amersham Bioscience Hybond-P PVDF Membrane, cat. No. RPN303F or Millipore Immobilon-P Transfer membranes, cat. No. IPVH00010)
- *Sheets of filter paper* cut to the dimension of the gel
- *Shallow trays*, large enough to hold the blot
- *Plastic wrap*

- *X-ray film* (e.g., Kodak Biomax XAR film) *and cassette*
- *Dark room or chemiluminescence-compatible imaging systems*

43.2.3 Method

43.2.3.1 Electroblotting

- Cut four pieces of filter papers and a piece of membrane (PVDF or nitrocellulose membrane) of the same size as the gel on which proteins are previously resolved.[3]
- If working with a PVDF membrane:
  – Wet the membrane in 100% methanol for about 10 s, or until the membrane appearance uniformly changes from opaque to semitransparent;
  – Wash it in distilled water for 5 min.[4]
- Alternatively, if working with a nitrocellulose membrane, proceed directly to the following step, because nitrocellulose membranes do not require prewetting.
- Equilibrate the membrane, the sponges, and filter papers in 1X Transfer Buffer for at least 10 min.
- Avoiding air bubbles, assemble the electroblotting sandwich, placing the sponge on the cathode (negative, usually black), two sheets of filter papers, the gel, the membrane, and the other sponge to the anode (positive, usually red).[5]
- Fill the chamber with 1X Transfer Buffer added with 20% of methanol[6] (i.e., 800 ml of 1X Transfer Buffer + 200 ml of methanol).
- Carry out the protein transfer. For power, voltage, and transfer time consult the manufacturer's instruction

---

[1]If not specified, all the buffers are made in distilled water.

[2]Ponceau S solution is a reversible staining solution designed for rapid (1 min) staining of protein bands on nitrocellulose or PVDF membranes. Ponceau S stain is easily reversed with water washes, facilitating subsequent immunological detection.

[3]Avoid touching and folding the membrane; to avoid contamination, always handle the filter papers and the membrane with gloves and blunt end forceps.

[4]The membrane must be kept wet all the time. Should it dry out, rewet in methanol and water as previously described.

[5]To ensure a complete transfer, remove air bubbles by carefully rolling a clean Pasteur pipette over each layer in the sandwich. Avoid excessive pressure that could damage the gel and the membrane.

[6]Methanol is necessary to achieve efficient binding to nitrocellulose and PVDF membranes.

of the used apparatus (usually, we perform the transfer in a cold room for 1 h at 350 mA or overnight at 10 mA).[7]

43.2.3.2 Immunoblotting

- Once the transfer is complete, its efficiency could be monitored directly on the membrane by staining it with Ponceau S Staining Solution for 2 min and then destaining it in distilled water until bands are visible.
- If the transfer is good, proceed with blocking non-specific binding sites with the Blocking Buffer for 1 h at room temperature on an orbital shaker.[8]
- Incubate the membrane with the diluted primary antibody for 1 h at room temperature on an orbital shaker. Ensure that the solution moves freely across the entire surface of the membrane.
- Rinse the membrane for five times 5 min each with fresh wash buffer on an orbital shaker.
- Incubate the membrane with the HRP-conjugated secondary antibody[9] for 1 h at room temperature on an orbital shaker.[10] Ensure that the solution moves freely across the entire surface of the membrane.
- Wash the membrane for 5 min with five fresh changes of wash buffer.[11]
- Prepare the working chemiluminescent reagents as HRP substrates for immunodetection according to the supplier's guidelines.
- Place the blotted protein side up in a clean container, add the HRP substrate, and incubate the blot following the usage guidelines.

- Adsorb the excess of substrate bumping into a paper by holding the membrane gently with forceps.
- Place the membrane, with the blotted protein side up, in the X-ray film cassette. Cover it with a clean plastic wrap or sheet protector and remove any air bubble. Ensure that the surface of the plastic wrap or sheet protector is dry and unwrinkled.
- In the dark room, place a sheet of X-ray film on the top of the membrane. Close the cassette and expose the blot to a suitable X-ray film for an appropriate time.[12]
- In the dark room, remove the film and place it in the developing solution until the bands compare.
- Rinse the film with water and place it in fixing solution for some minutes.
- Rinse the film again with water and leave it to air dry.
- Scan the film and save it as a TIFF image.

43.2.4 Troubleshooting

- High background: it could be caused by many problems, such as insufficient washes, poor quality of reagents, cross-reactivity between blocking reagent and antibody, membrane drying during incubation processes, poor-quality antibodies, protein-protein interactions or insufficient blocking. If high background is present several solutions are possible:
  - Increase wash buffer volumes and wash cycle repetitions.
  - Use high-grade reagents and double-distilled water.
  - Use sufficient volumes of solutions to cover the entire surface of the membrane during incubations.
  - Use high-quality affinity-purified antibodies.
  - Decrease the antibody concentration.
  - Reduce x-ray exposure time.
  - Increase the concentration or volume of the blocking agent.

---

[7]The time of transfer depends on the size of the proteins (larger proteins take longer to transfer), on the percentage of acrylamide, and on the gel thickness.

[8]Alternatively, membranes may be left in the blocking solution overnight in gently agitation at 2–8°C. The membrane at this stage can be dried by wrapping with plastic film and stored at 4°C until use.

[9]The antibody should be diluted according to the supplier's datasheet.

[10]Alternatively, membranes may be left in a cold room over night by shaking.

[11]Additional or longer washes may further reduce the background.

---

[12]The first exposure time should be 1 min. On the basis of its appearance, estimate how long the second film should be exposed. Because of the light sensitivity of the HRP substrate, a shorter exposure time may be required.

- Weak or absent signal: protein transfer can be optimized. The antibody concentration could be too low, the antibody reaction time could be insufficient, or the antibody could be inactive (due to multiple freeze-thaws). If the problem persists more sample can be loaded on the gel or several different times of exposure can be tried.
- Presence of nonspecific bands: the primary or secondary antibody dilution can be increased because the antibody concentration is probably too high. If the problem persists, the amount of loaded protein can be decreased.

References

1. Knudsen KA (1985) Proteins transferred to nitrocellulose for use as immunogens. Anal Biochem 147(2):285–288
2. Towbin H, Staehelin T, Gordon J (1992) Electrophoretic transfer of proteins from polyacrylamide gels to nitrocellulose sheets: procedure and some applications. 1979. Biotechnology 24:145–149

Dot Blot

44

Valentina Faoro and Giorgio Stanta

Contents

V. Faoro (✉)
Department of Medical, Surgical and Health Sciences,
University of Trieste, Trieste, Italy

G. Stanta
Department of Medical Sciences,
University of Trieste, Cattinara Hospital,
Strada di Fiume 447, Trieste, Italy

44.1 Introduction and Purpose

This technique is similar to the western blot technique. However, in the dot blot procedure proteins are not separated electrophoretically but are spotted directly onto the membrane by applying a vacuum. Proteins bind to the membrane while the other sample components pass through. The proteins on the membrane are then available for analysis [1]. This technique can be used either as a qualitative method for rapid screening of a large number of samples or as a quantitative technique. It is especially useful for testing the suitability of experimental design parameters (such as the optimization of primary antibody concentration).

44.2 Protocol

44.2.1 Reagents[1]

- *100% methanol*
- *Double-distilled water*
- *PBS or TBS*
- *Wash buffer*: either PBS or TBS with 0.1% Tween®-20 (v/v)
- *Blocking buffer*: 5% (w/v) blocking agent (e.g., BSA or nonfat dried milk) in wash buffer
- *Primary antibody* specific for the protein of interest
- *HRP-conjugated secondary antibody* specific for primary antibody
- *Chemiluminescent reagents for immunodetection* (e.g., Amersham Biosciences ECL Western Blotting

---

[1]If not specified, all the buffers are made in distilled water.

G. Stanta (ed.), *Guidelines for Molecular Analysis in Archive Tissues*,
DOI: 10.1007/978-3-642-17890-0_44, © Springer-Verlag Berlin Heidelberg 2011

Detection Reagents, or Millipore Immobilon™ Western Chemiluminescent Substrates)
- *Special chemical solutions for development and fixing of the X-ray film*

44.2.2 Equipment

- *Adjustable pipettes*, range: 2–20 μl, 20–200 μl, 100–1,000 μl
- *Orbital shaker*
- *Dot blot apparatus*
- *PVDF or nitrocellulose membrane* (e.g., Amersham Bioscience Hybond-P PVDF Membrane, cat. No. RPN303F or Millipore Immobilon-P Transfer membranes, cat. No. IPVH00010)
- *Shallow trays* large enough to hold the blot
- *X-ray film* (e.g., Kodak Biomax XAR film) and *cassette*
- *Dark room or chemiluminescence-compatible imaging systems*

44.2.3 Method

- Prepare serial dilutions of the protein samples in either PBS or TBS (e.g., undiluted (1/10 of the total lysate), 1:2, 1:4, 1:8, 1:16)[2].
- Prepare the membranes[3]. If different antibody conditions should be tested, label each membrane with a pencil according to the dilutions that will be screened.

---

[2]The volume for spotting should be at least 10 μl in each well. Using lower volume may result in irregular spotting.

[3]See also the chapter relative to the western blot method.

- Carefully place the membrane in wash buffer and let it equilibrate for at least 5 min.
- Assemble the blot unit according to the manufacturer's instructions
- Once the membrane is placed on the blot unit, connect the unit to the vacuum line.
- Carefully pipette the samples into the wells. Minimize the area penetrated by the solution by applying it slowly.
- When all of the samples have filtered through the membrane, turn off the vacuum.
- Remove the blot and let it dry, until no visible moisture remains.
- For immunodetection, follow the immunoblotting protocol described in Chap. 43.

Reference

1. Oprandy JJ, Olson JG, Scott TW (1988) A rapid dot immunoassay for the detection of serum antibodies to eastern equine encephalomyelitis and St. Louis encephalitis viruses in sentinel chickens. Am J Trop Med Hyg 38(1):181–186

Membrane Stripping

45

Valentina Faoro and Giorgio Stanta

Contents

45.1 Introduction and Purpose

The stripping process is applied after western blot or dot blot analyses. This process disrupts the antigen-antibody interaction and allows probing sequentially a single blot on a membrane with a different primary antibody. The method described hereafter [1] is useful especially for process optimization or when the sample amount is limited.

45.2 Protocol

45.2.1 Reagents[1]

- *Stripping solution (100 mM β-mercaptoethanol, 2% (v/v) SDS,[2] 62.5 mM Tris–HCl, pH 6.7)[3]*
- *Double-distilled water*

45.2.2 Equipment

- *Orbital shaker*
- *Shallow trays, large enough to hold the blot*
- *Oven*

V. Faoro (✉)
Department of Medical, Surgical and Health Sciences,
University of Trieste, Trieste, Italy

G. Stanta
Department of Medical Sciences,
University of Trieste, Cattinara Hospital,
Strada di Fiume 447, Trieste, Italy

[1]If not specified, all the buffers are made in distilled water.

[2]You can use a stock solution of 10% SDS (w/v). SDS powder is toxic. You should weigh it under a chemical hood and use a close box when you transport it out in order to avoid breathing the dust.

[3]It is better to prepare this solution under a chemical hood, because β-Mercaptoethanol is very toxic. The waste has to be collected with hazardous chemical waste.

G. Stanta (ed.), *Guidelines for Molecular Analysis in Archive Tissues*,
DOI: 10.1007/978-3-642-17890-0_45, © Springer-Verlag Berlin Heidelberg 2011

45.2.3 *Method*

- Place the membrane for 30 min in Stripping solution and incubate with agitation for 30 min at 50°C.[4]
- Wash the membrane with double-distilled water two to three times.
- Test if the enzyme conjugate and the primary antibody are completely removed by incubating the membrane, with substrate or with the secondary antibody and substrate and by imaging the blot respectively.[5]
- Perform the immunodetection as described in the western blot protocol from incubation with blocking solution.[6]

---

[4]It is convenient to cover the plastic trays well in order to minimize exposure to fumes. When covers are not used, these procedures should be done in a fume hood. The waste has to be collected with hazardous chemical waste.

[5]If no signal is detected after 5 min exposure, we know that the enzyme conjugate and the primary antibody have been successfully removed from the membrane. If the signal is still detected, repeat the incubation of the membrane in Stripping solution for additional time or at higher temperature.

[6]Alternatively, the membrane at this stage can be dried by wrapping it with plastic film and stored at 4°C until use.

Reference

1. Western blotting protocols – Stripping by heat and detergent. http://www.millipore.com/immunodetection/id3/western blottingprotocols

Reverse-Phase Protein Microarrays

46

Christina Schott and Karl-Friedrich Becker

Contents

C. Schott and K.-F. Becker (✉)
Institute of Pathology, Technical University of Munich,
Munich, Germany

46.1 Introduction and Purpose

Protein microarrays are an emerging class of nano-technology assays for tracking many different proteins simultaneously. They may be subdivided as follows: (1) arrays for protein profiling and (2) arrays for functional studies [1–3]. Arrays for protein profiling can be further divided into (a) forward and (b) reverse-phase protein microarrays, depending on the way the sample is applied. In the case of the forward array setting, the protein lysate is analyzed on a single microarray containing up to several thousand different capture molecules, e.g. antibodies or aptamers. Here, many proteins of interest can be analyzed in one sample during the same experiment. In contrast, in the so-called reverse-phase protein microarrays (or protein lysate microarrays), hundreds or thousands of different protein lysates are immobilized onto one array. Using an appropriate ligand or antibody, one protein can be assayed in a large number of samples [4–7]. In these guidelines, only reverse-phase protein microarrays are described.

46.2 Protein Spotting

46.2.1 Reagents

- *Qproteome FFPE Tissue Kit buffer* (QIAGEN, cat No. 37623)
- *Double -distilled water (autoclaved)*
- *Desiccant* (e.g., from Roth GmbH, Germany)
- *Ice*

G. Stanta (ed.), *Guidelines for Molecular Analysis in Archive Tissues*,
DOI: 10.1007/978-3-642-17890-0_46, © Springer-Verlag Berlin Heidelberg 2011

46.2.2 Equipment

- *Pipette*, range: (2–20 µl)
- *1.5 ml tubes*
- *Ice box*
- *Microtiter plates*
- *Sealing foils*[1]
- *Centrifuge* (with plate-adapters)
- *Vortex*
- *Spotting device* (e.g., BioOdyssey Calligrapher MiniArrayer (BioRad,) including waterbath with cooling possibility[2]
- *Pins*[3] (e.g., ANOPOLI, size PTS 200)
- *FastSlides* (e.g., Whatman/Schleicher & Schuell, cat No. 10484182)[4]

46.2.3 Method

46.2.3.1 Preparing Lysates

1. Thaw protein extracts and Qproteome FFPE Tissue Kit buffer on ice and mix them by vortexing.
2. Dilute the lysates with the extraction buffer (undiluted, 1:2, 1:4, 1:8, 1:16).[5]
3. Place a microtiter plate (64-well or 384-well) on ice and arrange the lysates into the well, starting on the left corner on top of the plate; put one case with all the dilutions after another, ending after each dilution-row is with one well of buffer (as negative control).
4. Seal the plate with a foil.
5. Centrifuge the plate at 4°C and about 1,000 rcf (g) for 1–2 min to get rid of air bubbles.
6. Keep the plate on ice until the protein spotter is ready to use.

---

[1]Sealing foils are preferred which allow easy peal-off while avoiding evaporation.

[2]Alternatively, a manual hand-held spotter, e.g. MicroCaster™ Hand-Held Microarrayer System, Schleicher & Schuell, order No. 10485047, may be used.

[3]Solid pins are preferred.

[4]These glass slides are coated with nitrocellulose which has high protein binding capacity.

[5]The volume for spotting should be at least 10 µl in each well. Using less volume may result in irregular spotting.

46.2.3.2 Spotting

Prepare spotter, e.g. BioOdyssey Calligrapher Mini-Arrayer, as follows:

1. Set the humidity value between 55% and 60% and the temperature on 8°C.
2. Prepare waste and supply bottles for washing steps.
3. Load the pins into the print head.
4. Place the slides (max. 16) into the spotter.
5. Create your run (e.g., define number of spots, distance between spots, printing directions, number of reprints, and sample picking direction).
6. Place the source plate into the spotter.
7. Start the run.

After the run you can keep the spotted slides dry at 4°C until protein detection.

46.3 Protein Detection on Reverse-Phase Protein Microarrays

46.3.1 Reagents

- *Wash buffer*: *TBST Buffer* (20 mM Tris-HCl, 150 mM NaCl, pH 7.6 Tween20 0.1%) or PBST Buffer (PBS + Tween20 0.1%)
- *Peroxidase blocking reagent* (e.g., from DAKO cat. No S2023)
- *Milk powder*
- *BSA*
- *Casein*
- Primary antibody specific for the protein of interest, diluted in blocking or wash buffer
- *HRP-conjugated secondary antibody, specific for primary antibody*, diluted in blocking buffer
- *Chemiluminescent reagents for immunodetection* (e.g. Amersham Biosciences ECL Western Blotting Detection Reagents, or ECL plus or ECL Advanced or Millipore Immobilon™ Western Chemiluminescent Substrates)
- *Special chemicals for development and fixing of the X-ray film* (e.g., Kodak)
- *Sypro Ruby* (SYPRO® Ruby Protein Gel Stain, Molecular Probes *cat. No* S12000)

Table 46.1 List of primary antibodies and conditions

Antibody against	Blocking	Dilution primary Ab	Dilution solution
E-cad, Clone 36, BD Biosciences, San Diego, USA	5% MP/TBST	1:5,000	5% MP/TBST
EGFR, #2,232, New England Biolabs	5% MP/TBST	1:2,000	5% BSA/TBST
β-Actin, Sigma Aldrich Chemie GmbH, Steinheim, Germany	5% MP/TBST	1:10,000	5% MP/TBST
Her2, Code A0485, DakoCytomation, Hamburg, Germany	5% MP/TBST	1:500	TBST
ERα, 578–595, Sigma	0.5% Casein/TBS	1:3,000	0.5% Casein/TBST

46.3.2 Equipments

- *Shaker*
- *Slide incubation chambers*
- *Pipette*, range: (0.5–10; 2–20; 20–100 µl)
- *Cold storage room* (4°C) with shaker
- *Glass-plate*
- *Sheet protector foils* (clear and as thin as possible)
- *Autoradiography cassettes*
- *Dark room with red-light*
- *X-ray films*
- *Special chemicals for development and fixing of the X-ray film* (e.g., Kodak)

46.3.3 Method

46.3.3.1 Incubating Slides

1. Take slides from 4°C storage and prewet them shortly in TBST buffer by gently shaking (in an incubation chamber) on a shaker or rocker.[6]
2. Discard the TBST buffer and pour peroxidase blocking reagent on the slides and let them gently shake at room temperature for 1 h.
3. Discard the blocking reagent and wash three times for 2 min each with TBST
4. Incubate the slides with the blocking solution which is suitable for the desired antibody (in general 5% milk powder in TBST will do, but it can also be 5% BSA in TBST or a mixture of both or casein-based).

5. Gently shake at room temperature for at least 1 h.
6. Incubate the slides with the primary antibody at 4°C over night (16 h)[7], while gently shaking (the dilution and also the kind of diluent depends on the suggestion of the antibody company or your own experience).

Possible conditions for using the following primary antibodies (examples) are reported in Table 46.1.

46.3.3.2 Detection of Slides

The next day, after discarding antibody solution, wash three times for 10 min each with TBST

1. Incubate the slides with the secondary antibody for 1 h at room temperature (22°C) or 2 h at 4°C with shaking.
2. Discard the secondary antibody and wash again in TBST several times.[8]
3. Let TBST rinse a little from the slides.[9]
4. Place the slides on a glass plate and pour the detection reagent directly on top of the slides (appr. 500 µl per slide) and leave them there for 5 min (switch off room lights).[10]
5. Let the detection reagent rinse a little and put the slides in a sheet protector, heat seal, put in an autoradiography cassette.

---

[6]For example Heraeus Quadriperm incubation chambers can be used; here, only 3 ml of solution is needed.

[7]Shorter incubation times at room temperature may also be possible, depending on the workflow.

[8]For example: 2×5 min, 1×10 min with two or three changes, 1×5 min.

[9]IMPORTANT: do not let them dry out!

[10]Tip: for even distribution of detection reagent place a blown film for the last 2 min over the slides.

Table 46.2 Conditions for the secondary antibodies

Secondary Ab	Dilution secondary Ab	Diluent	Detection
Antimouse	1:10,000	5% MP/TBST	ECL plus
Antirabbit	1:2,000	5% MP/TBST	ECL plus

6. In the dark room expose films for different times.
7. Develop films (similar to western blot) and scan them individually on a scanner (e.g. Scanjet 3,770, Hewlett-Packard, Hamburg, Germany) with 1,200 dpi.
8. Save as tiff files.
9. Analyze with appropriate software, e.g. ScionImage (Scion Corporation, Frederick, MA, USA) or MicroVigene™ (VigeneTech, Carlisle, MA 01741, USA)
10. Normalize to total protein (Sypro Ruby detection, see below).

Possible conditions for secondary antibodies are reported in Table 46.2.

46.4 Estimation of Total Protein with Sypro Ruby Staining

One of the spotted slides is stained with SYPRO Ruby protein stain in order to normalize the signal intensity to total protein content:

1. Prewet the slide briefly in TBST.
2. After discarding TBST, incubate the slide in 7% acetic acid and 10% methanol for 15 min in a staining dish by gentle agitation.
3. Wash four times in deionized water for 5 min each.
4. Incubate the slide in SYPRO Ruby protein stain for 15 min with gentle shaking.

5. Wash the slide four to six times for 1 min each in deionized water.
6. Visualize the staining on an imaging system, e.g. Eagle Eye (Stratagene, La Jolla, CA).
7. Analyze with appropriate software, e.g. ScionImage (Scion Corporation, Frederick, MA, USA) or MicroVigene™ (VigeneTech, Carlisle, MA 01741, USA).

References

1. Liotta LA, Espina V, Mehta AI, Calvert V, Rosenblatt K, Geho D, Munson PJ, Young L, Wulfkuhle J, Petricoin EF 3rd (2003) Protein microarrays: meeting analytical challenges for clinical applications. Cancer Cell 3(4):317–325
2. MacBeath G (2002) Protein microarrays and proteomics. Nat Genet 32(Suppl):526–532
3. Poetz O, Schwenk JM, Kramer S, Stoll D, Templin MF, Joos TO (2005) Protein microarrays: catching the proteome. Mech Ageing Dev 126(1):161–170
4. Becker KF, Metzger V, Hipp S, Hofler H (2006) Clinical proteomics: new trends for protein microarrays. Curr Med Chem 13(15):1831–1837
5. Berg D, Hipp S, Malinowsky K, Bollner C, Becker KF (2010) Molecular profiling of signalling pathways in formalin-fixed and paraffin-embedded cancer tissues. Eur J Cancer 46(1):47–55
6. Paweletz CP, Charboneau L, Bichsel VE, Simone NL, Chen T, Gillespie JW, Emmert-Buck MR, Roth MJ, Petricoin IE, Liotta LA (2001) Reverse phase protein microarrays which capture disease progression show activation of pro-survival pathways at the cancer invasion front. Oncogene 20(16): 1981–1989
7. Malinowsky K, Wolff C, Gündisch S, Berg D, Becker KF (2011) Targeted therapies in cancer - challenges and chances offered by newly developed techniques for protein analysis in clinical tissues. J Cancer 2:26–35

Introduction

Valentina Faoro, Giorgio Stanta, and Alessandro Vindigni

Content

V. Faoro (✉)
Department of Medical, Surgical and Health Sciences,
University of Trieste, Trieste, Italy

G. Stanta
Department of Medical Sciences,
University of Trieste, Cattinara Hospital, Strada di Fiume 447,
Trieste, Italy

A. Vindigni
Genome Stability Group, ICGEB (International Centre for
Genetic Engineering and Biotechnology), Trieste, Italy

Mass Spectrometry (MS) is the most powerful technology currently used to identify proteins. Mass spectrometry is a technique that measures the mass-to-charge ratio (m/z) of ions in the gas phase. MS instrumentation has been greatly modified and improved in the recent decades. Highly sensitive, robust instruments have been developed to consistently analyze biomolecules, proteins, and peptides in particular. Mass spectrometers can be automated and can achieve sensitivity down to the femtomole level.

In a typical proteomics experiment, proteins are first separated by two-dimensional electrophoresis (or other separation methods); spots containing the proteins of interest are then excised from the gels and degraded enzymatically to peptides, usually using trypsin. The peptides are then ionized; the most frequent techniques used to volatize and ionize peptides are electrospray ionization (ESI) and matrix-assisted laser desorption/ionization (MALDI). In ESI, a liquid sample is introduced through a needle into the orifice of the mass spectrometer, where a potential difference between the capillary and the inlet to the mass spectrometer results in the generation of a fine mist of charged droplets; as the solvent evaporates, the sizes of the droplets decrease, resulting in the formation of desolvated ions [1]. In MALDI, the sample is incorporated into matrix molecules and then subjected to irradiation by a laser. The laser promotes the formation of molecular ions [2]. The ionized peptides enter the mass spectrometer, which detects the mass of each peptide. Currently, four types of mass analyzers are available for proteomics research:

- The ion trap
- The time-of-flight (TOF)
- The quadrupole
- Fourier transform ion cyclotron (FT-MS) analyzers.

G. Stanta (ed.), *Guidelines for Molecular Analysis in Archive Tissues*,
DOI: 10.1007/978-3-642-17890-0_47, © Springer-Verlag Berlin Heidelberg 2011

These analyzers can be alone or, in some cases, put together in tandem to take advantage of each [3]. The resulting mass spectrum is converted to a list of peptide masses that is searched against specific protein databases, and compared with *in silico* trypsin-digested proteins in order to identify the protein of interest.

In the so-called "shotgun approaches," mass spectrometry can be used to determine relative levels of expression without the need for prior gel separation. Isotope-coded affinity tags (ICAT) is a high-throughput MS-based technique that allows both sequence identity and quantitative analysis of complex protein mixtures. The samples to be compared (e.g., cancer versus normal sample) are each labeled with an ICAT reagent, which exists in an isotopically heavy and light form; this reagent couples to cysteine residues of the proteins. Then the two samples are combined and are enzymatically cleaved to generate peptide fragments, some of which are tagged. The tagged (cysteine-containing) peptides are isolated by avidin affinity chromatography and, finally, analyzed by MS [4].

Other labeling techniques are also available [5]. Proteins can be isotopically tagged by means of enzyme-catalyzed incorporation of heavy water (^{18}O water) during proteolysis; each peptide is labeled at the carboxy terminal [6, 7]. In the global internal standard technology (GIST), the control and experimental samples are tagged with deuterated and nondeuterated form of an acylating agent [8]. Recently, a new generation of tagging reagent, iTRAQ, has been introduced; this technology makes use of amine-specific, stable isotope reagents that can label all peptides of up to four different biological samples simultaneously. However, in this approach, the labeling process occurs after proteolytic digestion [9, 10]. Proteins can also be metabolically labeled by growing two distinct cellular populations in labeled amino acid (e.g., ^{12}C and ^{13}C, ^{14}N or ^{15}N) media. This method is called "stable isotope labeling with amino acids in cell culture" (SILAC). However, this method is valid only for cell culture and for samples capable of undergoing protein synthesis in vitro [11, 12].

Currently surface-enhanced laser desorption and ionization (SELDI) technology appears to be the prevalent approach to proteomics [13]. This technology, arising from the MALDI technique, refers to the process of affinity capture on special chemical surfaces, followed by precise mass analysis using laser desorption/ionization-based detection. The proteins are directly captured on a specific protein array from the original source material, without previous sample preparation. The arrays could have a chemically treated surface (such as cationic, anionic, metal affinity, hydrophobic, or hydrophilic) or a biological surface, coupled with appropriate molecules (such as antibodies or receptors) for specific interaction with proteins of interest. Once washing away the unbound proteins, the array represents a sort of protein map of a specific tissue or disease state in which the bound proteins are detected quantitatively by the instrument reader. The spectra of complex protein mixtures represent the mass-to-charge ratio (m/z) of proteins and their binding affinity to the chip surface. The obtained data give information about differentially expressed proteins in the same tissue; moreover, it is also possible to compare the peak intensities obtained from samples representing different physiological or pathological states [14, 15]. High sensitivity, throughput, and resolving power make this technique suitable for clinical applications in biomarker discovery [16–25].

References

1. Fenn JB, Mann M, Meng CK, Wong SF, Whitehouse CM (1989) Electrospray ionization for mass spectrometry of large biomolecules. Science 246(4926):64–71
2. Karas M, Hillenkamp F (1988) Laser desorption ionization of proteins with molecular masses exceeding 10,000 Daltons. Anal Chem 60(20):2299–2301
3. Aebersold R, Mann M (2003) Mass spectrometry-based proteomics. Nature 422(6928):198–207
4. Gygi SP, Rist B, Gerber SA, Turecek F, Gelb MH, Aebersold R (1999) Quantitative analysis of complex protein mixtures using isotope-coded affinity tags. Nat Biotechnol 17(10): 994–999
5. Schneider LV, Hall MP (2005) Stable isotope methods for high-precision proteomics. Drug Discov Today 10(5): 353–363
6. Mirgorodskaya OA, Kozmin YP, Titov MI, Korner R, Sonksen CP, Roepstorff P (2000) Quantitation of peptides and proteins by matrix-assisted laser desorption/ionization mass spectrometry using (18)o-labeled internal standards. Rapid Commun Mass Spectrom 14(14):1226–1232
7. Rose K, Simona MG, Offord RE, Prior CP, Otto B, Thatcher DR (1983) A new mass-spectrometric c-terminal sequencing technique finds a similarity between gamma-interferon and alpha 2-interferon and identifies a proteolytically clipped gamma-interferon that retains full antiviral activity. Biochem J 215(2):273–277
8. Chakraborty A, Regnier FE (2002) Global internal standard technology for comparative proteomics. J Chromatogr A 949(1–2):173–184

9. Aggarwal K, Choe LH, Lee KH (2006) Shotgun proteomics using the iTRAQ isobaric tags. Brief Funct Genomics Proteomics 5(2):112–120

10. Zieske LR (2006) A perspective on the use of itraq reagent technology for protein complex and profiling studies. J Exp Bot 57(7):1501–1508

11. Ong SE, Foster LJ, Mann M (2003) Mass spectrometric-based approaches in quantitative proteomics. Methods 29(2): 124–130

12. Ong SE, Kratchmarova I, Mann M (2003) Properties of 13c-substituted arginine in stable isotope labeling by amino acids in cell culture (SILAC). J Proteome Res 2(2):173–181

13. Merchant M, Weinberger SR (2000) Recent advancements in surface-enhanced laser desorption/ionization-time of flight-mass spectrometry. Electrophoresis 21(6):1164–1177

14. Petricoin EF, Liotta LA (2004) SELDI-TOF-based serum proteomic pattern diagnostics for early detection of cancer. Curr Opin Biotechnol 15(1):24–30

15. Seibert V, Wiesner A, Buschmann T, Meuer J (2004) Surface-enhanced laser desorption ionization time-of-flight mass spectrometry (SELDI TOF-MS) and proteinchip technology in proteomics research. Pathol Res Pract 200(2): 83–94

16. Belluco C, Petricoin EF, Mammano E, Facchiano F, Ross-Rucker S, Nitti D, Di Maggio C, Liu C, Lise M, Liotta LA, Whiteley G (2007) Serum proteomic analysis identifies a highly sensitive and specific discriminatory pattern in stage 1 breast cancer. Ann Surg Oncol 14(9):2470–2476

17. de Bont JM, den Boer ML, Reddingius RE, Jansen J, Passier M, van Schaik RH, Kros JM, Sillevis Smitt PA, Luider TH, Pieters R (2006) Identification of apolipoprotein A-II in cerebrospinal fluid of pediatric brain tumor patients by protein expression profiling. Clin Chem 52(8):1501–1509

18. Li J, Zhao J, Yu X, Lange J, Kuerer H, Krishnamurthy S, Schilling E, Khan SA, Sukumar S, Chan DW (2005) Identification of biomarkers for breast cancer in nipple aspiration and ductal lavage fluid. Clin Cancer Res 11(23): 8312–8320

19. Mazzocca A, Coppari R, De Franco R, Cho JY, Libermann TA, Pinzani M, Toker A (2005) A secreted form of ADAM9 promotes carcinoma invasion through tumor-stromal interactions. Cancer Res 65(11):4728–4738

20. Menard C, Johann D, Lowenthal M, Muanza T, Sproull M, Ross S, Gulley J, Petricoin E, Coleman CN, Whiteley G, Liotta L, Camphausen K (2006) Discovering clinical biomarkers of ionizing radiation exposure with serum proteomic analysis. Cancer Res 66(3):1844–1850

21. Paradis V, Degos F, Dargere D, Pham N, Belghiti J, Degott C, Janeau JL, Bezeaud A, Delforge D, Cubizolles M, Laurendeau I, Bedossa P (2005) Identification of a new marker of hepatocellular carcinoma by serum protein profiling of patients with chronic liver diseases. Hepatology 41(1):40–47

22. Seibert V, Ebert MP, Buschmann T (2005) Advances in clinical cancer proteomics: SELDI-TOF-mass spectrometry and biomarker discovery. Brief Funct Genomics Proteomics 4(1):16–26

23. Smith FM, Gallagher WM, Fox E, Stephens RB, Rexhepaj E, Petricoin EF 3rd, Liotta L, Kennedy MJ, Reynolds JV (2007) Combination of seldi-tof-ms and data mining provides early-stage response prediction for rectal tumors undergoing multimodal neoadjuvant therapy. Ann Surg 245(2):259–266

24. Srinivasan R, Daniels J, Fusaro V, Lundqvist A, Killian JK, Geho D, Quezado M, Kleiner D, Rucker S, Espina V, Whiteley G, Liotta L, Petricoin E, Pittaluga S, Hitt B, Barrett AJ, Rosenblatt K, Childs RW (2006) Accurate diagnosis of acute graft-versus-host disease using serum proteomic pattern analysis. Exp Hematol 34(6):796–801

25. Ward DG, Cheng Y, N'Kontchou G, Thar TT, Barget N, Wei W, Billingham LJ, Martin A, Beaugrand M, Johnson PJ (2006) Changes in the serum proteome associated with the development of hepatocellular carcinoma in hepatitis c-related cirrhosis. Br J Cancer 94(2):287–292

Sample Preparation and In Gel Tryptic Digestion for Mass Spectrometry Experiments

48

Valentina Faoro and Giorgio Stanta

Contents

48.1 Introduction and Purpose

Once the protein lysates are separated on polyacrylamide gels (SDS-PAGE), the interested proteins can be excised from gels, digested with trypsin, and identified by Mass Spectrometry (MS). The protocol described here applies to one- or two-dimensional gels stained with Coomassie brilliant blue R250 or G250 or with silver [1, 2]. The method is compatible with downstream MS experiments; by including $H_2^{18}O$ into the digestion buffer [3–5] or by mixing SILAC-labeled protein mixtures before separation, it enables quantification of the digestion products [6, 7].

Careful attention should be paid in casting the gels and handling the excised spots or bands of interest in order to avoid the risk of contaminating samples with keratin and enhanced chemical noise in analyzed samples. (Measures to minimize this contamination are described in paragraph "Protein Handling" in Chap. 36).

48.1.1 Reagents[1]

- *Double-distilled water*
- *30 mM Potassium ferricyanide*
- *100 mM Sodium thiosulfate*
- *100 mM Ammonium bicarbonate*
- *100 mM Ammonium bicarbonate/Acetonitrile (50:50, v/v)*
- *10 mM DTT in 100 mM Ammonium bicarbonate*
- *50 mM Iodoacetamide in 100 mM Ammonium bicarbonate*

V. Faoro (✉)
Department of Medical, Surgical and Health Sciences,
University of Trieste, Trieste, Italy

G. Stanta
Department of Medical Sciences,
University of Trieste, Cattinara Hospital,
Strada di Fiume 447, Trieste, Italy

---

[1]If not specified, all the buffers are made in distilled water.

G. Stanta (ed.), *Guidelines for Molecular Analysis in Archive Tissues*,
DOI: 10.1007/978-3-642-17890-0_48, © Springer-Verlag Berlin Heidelberg 2011

- *20 mM Ammonium bicarbonate/Acetonitrile (50:50, v/v)*
- *Trypsin* (e.g. Promega)
- *60% Acetonitrile, 1% Trifluoroacetic acid (TFA)*

48.1.2 Equipment

- *Adjustable pipettes, range*: 2–20 µl, 20–200 µl, 100–1,000 µl
- *Scalpel*
- *1.5 ml tubes*
- *Teflon stick*
- *Speed vacuum centrifuge*
- *Light box*

48.1.3 Method

- Rinse the entire slab of a one- or two-dimensional gel with water for few hours.
- Put a plastic tray with gel onto a light box.
- Excise manually with a scalpel the selected spots for further MS analysis from the preparative gel[2].
- Place the individual gel samples in a 1.5 ml tube.
- Add 300 µl of double-distilled water[3].
- To remove silver staining, place the samples in a 1:1 mix of 30 mM potassium ferricyanide and 100 mM sodium thiosulfate until the gel pieces turn clear.
- Discard this solution and wash each gel sample with 300 µl of double-distilled water for 15 min. If identification of Coomassie-stained bands (spots) is intended, skip these two steps and proceed with the next one.
- Add 300 µl of 100 mM ammonium bicarbonate into the solution, and incubate the gel pieces for additional 15 min.
- Discard the solution and wash the bands with 100 mM of ammonium bicarbonate/acetonitrile for 15 min.

- Remove this solution and crush the gel samples with a Teflon stick.
- Add 100 µl of acetonitrile for 5 min.
- Remove the acetonitrile and dry the bands in a Speed vacuum centrifuge for 5 min.
- Resuspend the sample in 50 µl of 10 mM DTT in 100 mM ammonium bicarbonate and incubate for 1 h at 56°C.
- Remove the solution and place the samples in 50 µl of 50 mM iodoacetamide in 100 mM ammonium bicarbonate at room temperature for 30 min in the dark.
- Remove the solution and wash the gel samples with 300 µl of 100 mM ammonium bicarbonate for 15 min.
- Place the samples in 300 µl of 20 mM ammonium bicarbonate/acetonitrile (50:50, v/v) for 15 min.
- Replace the previous solution with 100 µl of acetonitrile and leave at room temperature for 5 min.
- Dry the samples in a Speed vacuum centrifuge for 10 min.
- Resuspend with 5 µl of trypsin[4] (0.1 µg/µl) in 100 mM ammonium bicarbonate pH 8.0 for 10 min.
- Add 100 µl of 100 mM ammonium bicarbonate and carry out the digestion overnight at 37°C.
- Collect the supernatant in a second microcentrifuge tube.
- Wash the gel pieces once with 100 µl of double-distilled water and twice with 100 µl of 60% acetonitrile/1% TFA.
- Pool all the washes and add to the previously collected supernatant. Reduce the volume of the solution to 5–10 µl in a Speed vacuum centrifuge and store the sample at 4°C until analyzed with the mass spectrometer.

---

[2]Only the darkest zone of the spot must be cut-off.

[3]The cut spots can be stored for many weeks at 4°C before performing MS experiments by the specialist group.

[4]Trypsin specifically hydrolyzes peptide bonds at the carboxylic sides of lysine and arginine residues. Unmodified trypsin is subject to autolysis, generating fragments that can interfere with mass spectrometry analysis of the peptides. In addition, autolysis can result in the generation of pseudotrypsin, which has been shown to exhibit an additional chymotrypsin-like specificity. Promega trypsin has been modified by reductive methylation, rendering it extremely resistant to autolytic digestion. It is very important to maintain an adequate enzyme/substrate ratio to maximize the digestion. This ratio is much higher than the one needed for *in-solution* digestion.

References

1. Shevchenko A, Tomas H, Havlis J, Olsen JV, Mann M (2006) In-gel digestion for mass spectrometric characterization of proteins and proteomes. Nat Protoc 1(6):2856–2860
2. Shevchenko A, Wilm M, Vorm O, Mann M (1996) Mass spectrometric sequencing of proteins silver-stained polyacrylamide gels. Anal Chem 68(5):850–858
3. Mason CJ, Therneau TM, Eckel-Passow JE, Johnson KL, Oberg AL, Olson JE, Nair KS, Muddiman DC, Bergen HR 3rd (2007) A method for automatically interpreting mass spectra of 18O-labeled isotopic clusters. Mol Cell Proteomics 6(2):305–318
4. Shevchenko A, Shevchenko A (2001) Evaluation of the efficiency of in-gel digestion of proteins by peptide isotopic labeling and maldi mass spectrometry. Anal Biochem 296(2):279–283
5. Yao X, Freas A, Ramirez J, Demirev PA, Fenselau C (2001) Proteolytic 18O labeling for comparative proteomics: Model studies with two serotypes of adenovirus. Anal Chem 73(13):2836–2842
6. Ong SE, Blagoev B, Kratchmarova I, Kristensen DB, Steen H, Pandey A, Mann M (2002) Stable isotope labeling by amino acids in cell culture, silac, as a simple and accurate approach to expression proteomics. Mol Cell Proteomics 1(5):376–386
7. Ong SE, Mann M (2006) A practical recipe for stable isotope labeling by amino acids in cell culture (silac). Nat Protoc 1(6):2650–2660

MALDI Imaging Mass Spectrometry on Formalin-Fixed Paraffin-Embedded Tissues

49

Axel Walch, Sandra Rauser, and Heinz Höfler

Contents

A. Walch (✉) and S. Rauser
Institute of Pathology, Helmholtz Zentrum München,
Neuherberg, Germany

H. Höfler
Institute of Pathology, Technische Universität München,
Munich, Germany and
Institute of Pathology, Helmholtz Zentrum München,
Neuherberg, Germany

49.1 Introduction and Purpose

Most of the tissue samples in pathology are formalin-fixed and paraffin-embedded for preservation. This huge reservoir cannot be accessed by standard MALDI Imaging Mass Spectrometry since the proteins in the FFPE samples are cross-linked [1–5]. Therefore the proteins have to be made accessible by digestion into peptides before Imaging Mass Spectrometry.

The required time for the sample preparation is one working day (6–8 h) and the time for the sample measurement is between 1 and 12 h, depending on its size.

49.2 Protocol

49.2.1 Reagents[1]

- *Acetic acid* (HPLC grade, Sigma-Aldrich)
- *Acetonitrile (ACN)* (HPLC grade, Sigma-Aldrich)
- *Ammonium bicarbonate* (reagent grade, Sigma-Aldrich)
- *α-Cyano-4-hydroxycinnamic acid (CHCA)*[2] (MALDI TOF-MS grade, Sigma-Aldrich)
- *EDTA* (reagent grade, Sigma-Aldrich)
- *Ethanol* (HPLC grade, Sigma-Aldrich)
- *Trifluoroacetic acid (TFA)*[2] (HPLC grade, Sigma-Aldrich)
- *Tris base* (reagent grade, Sigma-Aldrich)
- *Trypsin Gold*[2] (Promega)

---

[1]If not differently stated, the reagents can be stored at room temperature.

[2]Store at 4°C.

G. Stanta (ed.), *Guidelines for Molecular Analysis in Archive Tissues*,
DOI: 10.1007/978-3-642-17890-0_49, © Springer-Verlag Berlin Heidelberg 2011

- *Water* (HPLC grade, Sigma-Aldrich)
- *Xylene* (HPLC grade, Sigma-Aldrich)
- Ammonium bicarbonate solution 100 mM in ddH$_2$O
- CHCA matrix solution[3] (7 g/L in ACN:H$_2$O 1:1 with 0.2% TFA)
- Trypsin stock solution (0.5 µg/µL trypsin in 50 mM acetic acid)
- Trypsin solution[3] (dilute 50 µL trypsin stock solution with 500 µL 100 mM ammonium bicarbonate solution)
- Tris-EDTA pH 9.0 solution (10 mM Tris Base, 1 mM EDTA in ddH$_2$O)

49.2.2 Equipment

- *Cuvettes* (100 mL)
- *Hot plate* (40°C)
- *ImagePrep spray roboter* (Bruker Daltonics)
- *Incubator* (65°C)
- *Ultraflex III MALDI-TOF/TOF mass spectrometer* (Bruker Daltonics)
- *Flex Control 3.0, Flex Imaging 2.1, and ClinProTools 2.2 software* (Bruker Daltonics)
- *Indium-tin-oxide (ITO)-coated slides* (Bruker Daltonics)
- *Portrait 630 reagent multispotter* (Labcyte)
- *Microtome*
- *Microwave*

49.2.3 Method

49.2.3.1 Tissue Preparation

- Cool the FFPE tissue block at −20°C before cutting.
- Using a clean and sharp microtome blade make a 5 µm section of FFPE sample and mount it onto an ITO-coated slide.
- Incubate for 10 min on a hot plate (60°C).
- Remove the paraffin by washing twice with xylene in a 100 mL cuvette for 20 min.

- Rehydrate the sample by washing with 100%, 95%, 80%, and 70% ethanol for 5 min each (in 100 mL cuvettes).
- Warm up 1 L of Tris-EDTA pH 9.0 buffer to boiling temperature in the microwave.
- Transfer the sample slide on a rack into the pre-heated buffer and boil it for 20 min in the microwave (350 W). [4]
- Wash with double-distilled water for 5 min in 100 mL cuvette.
- Dry for 10 min on a hot plate (40°C).

49.2.3.2 On-Tissue Digestion

- Spot trypsin solution onto the sample using a multi-spotter. The spots have a centre to centre distance of 250 µm and each spot is 175 µm in diameter. On each spot 30 cycles of trypsin solution are added. In each cycle 160 pl are added.
- After each cycle, the sample is incubated for 5 min at room temperature.
- After digestion continue directly with the matrix coating.

49.2.3.3 Matrix Coating

- The sample is spray-coated with CHCA matrix solution using the ImagePrep spray roboter. For this, the ImagePrep is set up according to the manufacturer's protocol and the standard programme for CHCA is used.

49.2.3.4 MALDI Imaging Mass Spectrometry and Data Analysis

Analysis is performed using an Ultraflex III MALDI-TOF/TOF mass spectrometer equipped with a smart beam N$_2$ laser. The spectrometer is operated using the Flex Control 3.0 software in positive polarity in reflectron mode. The acquisition range is 700–5,000 *m/z*. 200 spots per spot position are collected.

For data analysis the Flex Imaging 2.1 and ClinProTools 2.2 software are used.

---

[3]Store the solution for less than 1 week.

[4]In order to prevent evaporation of large amount of water, loosely place a lid onto the containment.

References

1. Groseclose MR, Massion PP, Chaurand P, Caprioli RM (2008) High-throughput proteomic analysis of formalin-fixed paraffin-embedded tissue microarrays using MALDI imaging mass spectrometry. Proteomics 8(18):3715–3724

2. Lemaire R, Desmons A, Tabet JC, Day R, Salzet M, Fournier I (2007) Direct analysis and MALDI imaging of formalin-fixed, paraffin-embedded tissue sections. J Proteome Res 6(4):1295–1305

3. Ronci M, Bonanno E, Colantoni A, Pieroni L, Di Ilio C, Spagnoli LG, Federici G, Urbani A (2008) Protein unlocking procedures of formalin-fixed paraffin-embedded tissues: application to MALDI-TOF imaging MS investigations. Proteomics 8(18):3702–3714

4. Stauber J, Lemaire R, Franck J, Bonnel D, Croix D, Day R, Wisztorski M, Fournier I, Salzet M (2008) MALDI imaging of formalin-fixed paraffin-embedded tissues: application to model animals of Parkinson disease for biomarker hunting. J Proteome Res 7(3):969–978

5. Walch A, Rauser S, Deininger SO, Hofler H (2008) MALDI imaging mass spectrometry for direct tissue analysis: a new frontier for molecular histology. Histochem Cell Biol 130(3): 421–434

Part XV

Elements of Good Laboratory Practice

Introduction

50

Giorgio Stanta, Ermanno Nardon, and Renzo Barbazza

Content

In recent years, both basic and clinical research have joined their efforts to translate scientific discoveries into practice, from bench to bedside applications. Therefore a rapidly increasing number of laboratories are now establishing the molecular technologies to be used both in a clinical and a translational research setting. This results in an obvious need for standardization and harmonization of the developed test systems and of the laboratory procedures among different institutions.

Good Laboratory Practice (GLP) generally refers to a quality system concerned with the organizational process and the conditions under which studies are planned, performed, monitored, recorded, archived, and reported. The main goal of this system is to ensure the consistency and reliability of the results and therefore their mutual acceptance among all countries [1]. To this regard, many regulatory and guidance materials have been developed over time by governmental authorities and by accrediting or nonaccrediting organizations.

A first set of regulatory standards was developed in the late 1970s by the Organization for Economic Cooperation and Development (OECD) and by the Food and Drug Administration (FDA) [2] for assessing chemicals and testing safety of chemical products, including pharmaceuticals. The OECD standards, reviewed in 1997, dealt with a set of core elements including organization, quality assurance, facilities, test systems, controls, study design, and reporting of results relevant to nonclinical health and environmental safety studies (Table 50.1) [1]. Almost in the same time frame, the International Organization for Standardization (ISO) extended the applicability of GLP to laboratories performing clinical analysis (ISO/DIS 15189:1998, Quality Management in the

G. Stanta (✉)
Department of Medical Sciences,
University of Trieste, Cattinara Hospital,
Strada di Fiume 447, Trieste, Italy

E. Nardon and R. Barbazza
Department of Medical, Surgical and Health Sciences,
University of Trieste, Trieste, Italy

G. Stanta (ed.), *Guidelines for Molecular Analysis in Archive Tissues*,
DOI: 10.1007/978-3-642-17890-0_50, © Springer-Verlag Berlin Heidelberg 2011

Table 50.1 GLP and GCLP core elements

GLP[a]	GCLP[b]
Organization and personnel.	Organization and personnel.
Defines management-, sponsor-, study director-, principle investigator- and personnel- responsibilities	Defines responsibilities, job description, training, competency assessments, performance evaluation, and education programme for all the personnel.
Quality assurance programme.	Quality control programme.
Responsibilities of the quality assurance personnel related to study conduct, facilities adequacy, and process compliance to GLP.	Procedures for monitoring: test standards and controls, reagents, specimens, review of QC data, QC logs, and inventory.
c	Verification of performance specifications.
	Standards for assay validation in clinical trials.
Facilities.	Physical facilities.
Definition of facilities for testing, for sample and reference items storage, for waste disposal, for archives.	Requirements for equipment placement, workplace environment, work areas.
Apparatus, materials, and reagents.	Equipment, materials, and reagents.
Standards for equipment maintenance and inspection, reagents labeling and storage.	Standards for equipment maintenance and inspection. Requirements for reagents and reference material verification, labeling and storage.
Test systems; test and reference items. Standards for test systems and controls handling, verification, labeling, and storage.	See "Quality control programme" and "Equipment, material, and reagents" core elements.
Standard operating procedures.	Standard operating procedures.
Guidelines for SOP writing and applicability.	Guidelines for SOP writing and applicability.
Performance of study.	Planning and conduct.
Definition of the study plan and conduct of study.	Definition of the study plan and conduct of study.
c	Specimen transport and management.
	Required activities for collection, transportation, and receipt.
c	Personnel safety.
	Required activities, equipment, and training.
Reporting. Retention of records.	Records and reports.
Requirements for reporting of results and for retention of records.	Requirements for reporting of results and for archiving reports. Standards for laboratory information system.

Elements described especially according to [a]OECD 1997 guidelines [1] and FDA's 21 CFR part.58 [10];
[b]ISO 15189 [4], CLIA [11], CAP [19], and BARQA [12].
[c]The topic has not been adequately described

Medical Laboratory). This standard introduced guidelines on specimen management, personnel safety, and verification of performance (mainly based on the ISO 17025:1999, the generic standard for testing and calibration laboratories), as core elements of the GLP. Currently, the ISO 15189 and 17025 guidelines are still relevant to all general clinical chemistry laboratories, and compliance to their standards is required for laboratory accreditation [3–5]. However, the ever-growing demand for molecular genetic testing still awaits coverage from ISO standards [6]. A draft of the best practice guidelines addressing testing for inherited disorders was developed by the European Molecular Genetics Quality Network (EMQN) in 2001 [7] and the draft for quality assurance in genetic testing was finally issued by OECD in 2006 [6]. Importantly, this latter draft recognized that research laboratories play a pivotal role in providing tests for genetic disorders, and supplied guidance standards for the validation and translation into clinical practice of new tests.

Particular surveillance and standardization are needed by activities of laboratories that perform the analysis or evaluation of samples collected as part of a clinical trial. In 1997, the European Medicine Agency (EMEA) introduced a set of regulatory elements concerning ethical and scientific quality standards for designing, conducting, recording, and reporting trials that involve the participation of human subjects (Good Clinical Practice, GCP) [8]. More recently, GCP was recognized as too vague with respect to sample analysis to ensure practical implementation in clinical trials [9]. Consequently, GCP was implemented with applicable portions of GLP and ISO 15189 standards [4, 10] by the joined effort of the British Association of Research Quality Assurance (BARQA) and USA governmental authorities [11, 12]. This resulted in the creation of the hybrid Good Clinical Laboratory Practice (GCLP) guidelines, which allowed connecting activities of basic research and clinical research laboratories (Table 50.1) [13–15].

To date, many relevant fields are still waiting for dedicated protocols from GLP standards. For example, only peculiar branches of basic and translational molecular research on human archive tissues have been addressed so far. Among these, requirements for biospecimen resource centers were formalized by OECD and NCI in 2007 [16, 17]. Although these guidelines recognize that the reliability of data from molecular assays is heavily dependent on the quality and consistency of the analyzed biospecimens, they do not specify in detail the technical requirements for preanalytical managing of tissues [18]. Many other relevant issues of translational research, such as RNA testing and proteomics, do not benefit from dedicated items in current international standards. This is even more important because, in our experience, standardization of the methods by commercial kits is not sufficient to guarantee reproducibility; good laboratory practice is absolutely necessary to obtain reliable results [20].

References

1. Organisation for Economic Co-operation and Development. OECD principles on good laboratory practice (as revised in 1997). Paris: OECD; 1998. OECD series on principles of GLP and compliance monitoring, no. 1, env/mc/chem(98)17. Available from: http://www.Olis.Oecd.Org/olis/1998doc. Nsf/linkto/env-mc-chem(98)17. Accessed 30 Mar 2010
2. U.S. Food and Drug Administration, non-clinical laboratory studies, good laboratory practice regulations, U.S. Federal Register. Vol. 41, no. 225, 19 November 1976, pp. 51206–51226 (proposed regulations) and Vol. 43, no. 247, 22 December 1978, pp. 59986–60020 (final rule)
3. Burnett D (2002) A practical guide to accreditation in laboratory medicine. London: ACB Venture Publications. www.Acb.Org.Uk
4. ISO 15189 (2003) Medical laboratories, particular requirements for quality and competence, international organization for standardization. http://www.Iso.Org/iso/en/cataloguede-tailpage.Cataloguedetail?Csnumber=26301. Accessed 30 Mar 2010
5. ISO/IEC 17025 (2005) General requirements for the competence of testing and calibration laboratories. International Organization for Standardization, Geneva
6. Organisation for Economic Co-operation and Development. Working party on biotechnology on draft guidelines for quality assurance in molecular genetic testing dsti/stp/bio (2006)60rev1. http://www.Oecd.Org/dataoecd/43/26/37103271.Pdf. Accessed 30 Mar 2010
7. European Molecular Genetics Quality Network. Best Practice Guidelines. http://www.Emqn.Org/emqn/bestpractice.Html. Accessed 30 Mar 2010
8. EMEA (2002) Guideline for good clinical practice, cpmp/ich/135/95: http://www.Emea.Europa.Eu/pdfs/human/ich/013595en.Pdf. Accessed 30 Mar 2010
9. Stevens W (2003) Good clinical laboratory practice (GCLP): the need for a hybrid of good laboratory practice and good clinical practice guidelines/standards for medical testing laboratories conducting clinical trials in developing countries. Qual Assur 10(2):83–89
10. Code of Federal Regulations (2005). 21 CFR part 58. Good laboratory practice for nonclinical laboratory studies. Available: http://www.Access.Gpo.Gov/nara/cfr/waisidx_01/21cfr58_01.Html. Accessed 31 Mar 2010
11. Code of Federal Regulations (2005) 42 cfr part 493. Laboratory requirements. Available: http://www.Access.Gpo.Gov/nara/cfr/waisidx_05/42cfr493_05.Html. Accessed 31 Mar 2010
12. Stiles T, Grant V, Mawby N (2003) Good clinical laboratory practice (GCLP): a quality system for laboratories that undertake the analyses of samples from clinical trials, Barqa. http://www.Barqa.Com/. Accessed 30 Mar 2010
13. DAIDS Guidelines for Good Clinical Laboratory Practice Standards-Training (2007). http://www3.Niaid.Nih.Gov/research/resources/daidsclinrsrch/pdf/labs/gclp.Pdf. Accessed 31 Mar 2010
14. Sarzotti-Kelsoe M, Cox J, Cleland N, Denny T, Hural J, Needham L, Ozaki D, Rodriguez-Chavez IR, Stevens G, Stiles T, Tarragona-Fiol T, Simkins A (2009) Evaluation and

recommendations on good clinical laboratory practice guidelines for phase i-iii clinical trials. PLoS Med 6(5): e1000067

15. Ezzelle J, Rodriguez-Chavez IR, Darden JM, Stirewalt M, Kunwar N, Hitchcock R, Walter T, D'Souza MP (2008) Guidelines on good clinical laboratory practice: bridging operations between research and clinical research laboratories. J Pharm Biomed Anal 46(1):18–29

16. Office of Biorepositories and Biospecimen Research, NCI. NCI Best Practices for Biospecimen Resources. 2007.

17. Organisation for Economic Co-operation and Development. OECD Best Practice Guidelines For Biological Resource Centers. Paris, France 2007 pp 1-116

18. Lippi G, Guidi GC, Mattiuzzi C, Plebani M (2006) Preanalytical variability: the dark side of the moon in laboratory testing. Clin Chem Lab Med 44(4):358–365

19. College of American Pathologists (2010) Surveys and anatomic pathology education programs. http://www.cap.org/apps/cap.portal?_nfpb=true&_pageLabel=education Accessed 30 Mar 2010

20. Bonin S, Hlubek F, Benhattar J, Denkert C, Dietel M, Fernandez PL, Hoefler G, Kothmaier H, Kruslin B, Mazzanti CM, Perren A, Popper H, Scarpa A, Soares P, Stanta G, Groenen PJTA (2010) Multicentre Validation Study of Nucleic Acids Extraction from FFPE Tissues. Virchows Archive. 457:309–317

Internal Quality Control (IQC) Organization

51

Giorgio Stanta, Ermanno Nardon, and Renzo Barbazza

Contents

G. Stanta (✉)
Department of Medical Sciences,
University of Trieste, Cattinara Hospital,
Strada di Fiume 447, Trieste, Italy

E. Nardon and R. Barbazza
Department of Medical, Surgical and Health Sciences,
University of Trieste, Trieste, Italy

51.1 Introduction and Purpose

There is some contiguity between translational research in human tissues and clinical application of the same type of analysis. For this reason, the authors deemed it pertinent to report some suggestions on Internal Quality Control (IQC) in this book. Most of these considerations can be useful also for a correct internal research procedure and for multicentric studies to obtain more reproducible results.

The World Health Organization defines the internal quality control as the set of procedures undertaken by the staff of a laboratory for continuously assessing laboratory work and the emergent results. This definition proposed in 1981 can still be used in molecular medicine clinical analyses.

The following IQC procedures are derived from the direct experience of the IMPACTS group (www.impactsnetwork.eu) and compared with guidelines reported from EMQN (European Molecular Genetics Quality Network – http://www.emqn.org) [1, 2], SGMG guidelines (Schweizerische Gesellshaft für Medizinische Genetik – http://www.ssgm.ch), College of American Pathologists (http://www.cap.org), and CLIA (Clinical Laboratory Improvement Amendments - http://www.cms.gov/clia/) [3].

The following IQC procedures are especially dedicated to archive tissue molecular analysis and are meant to represent a short, basic report also for further discussion and implementation.

51.1.1 IQC Procedures

It is possible to divide the IQC procedures into two main parts, those related to the examined samples and those referred to the entire laboratory.

G. Stanta (ed.), *Guidelines for Molecular Analysis in Archive Tissues*,
DOI: 10.1007/978-3-642-17890-0_51, © Springer-Verlag Berlin Heidelberg 2011

- Sample-related procedures:
  - Sample reception
  - Sample storage
  - Sample handling
  - Controls
  - Interpretation of the results
  - Reporting

- Laboratory-related procedures:
  - Documentation
  - Biological and chemical hazards
  - Audit
  - Staff training
  - Test validation procedures

51.2 Sample-Related Procedures

51.2.1 Sample Reception

- Record of the sample reception and request form should be done on a computer data base system with specific characteristics of security, such as one having a restricted access with frequently changed personal passwords and with a continuous system of back-up.
- Patient identification should come with at least two different types of information (for example, the printed label with name and date of birth plus hospital or Health System code, etc.). In the cases in which this is impossible, such as in cases of outside patients or foreign people, a specific partial data procedure should be present in the reception system.
- The technician who accepts the sample should check that the specimen corresponds to the characteristics reported in the request form (number of containers and of samples, sample size, etc.). In case of incorrect samples reception, this should be reported in the record system.
- In case the sample has already been studied during histopathological examination in the Pathology Department, the reference information should be conveyed to the pathology reception, but in a specific separate file and computerized system.

51.2.2 Sample Storage

- The tissue samples or the extracted macromolecules should be stored in well-defined conditions of temperature, humidity and, if possible, in aliquots with a safe system of storage, with a reliable procedure of labels and with a safe system of recovery[1]. The availability of aliquots should be reported in the computerized system to repeat the molecular examination or to add new types of molecular analysis. The residual samples or the aliquots should be stored for at least 1 year.
- Internal or consult paraffin blocks should be stored as usual in the Pathology Departments (for archive tissues, country-specific rules have been adopted). It is better to cut sections from the paraffin block just before analysis.
- Cores of tissues punched from donor paraffin blocks can be transferred by means of a tissue arrayer apparatus into a recipient block. In this way, most of the original tissue is preserved for further examination (see Chap. 5).
- If possible the preanalytical conditions of any sample should be recorded and, when available, standardized procedures should be used[2].
- Special rules should be followed when (handling) storing RNA (see Sect. 12.2) and also when (handling) storing proteins (see paragraph "Protein handling" in chapter 36).

51.2.3 Sample Handling

- Dedicated areas in which the tissue samples can be handled should be well defined, with safety cabinets, dedicated instruments, pipettes etc., especially for the molecular amplification procedures (separate areas for pre- and postamplification – see Sect. 12.2).
- Mix-up risk of samples should be minimized, with minimum tube to tube transfer; for each laboratory

---

[1]For cryopreservation, see http://www.cryobiosystem-imv.com/CBS/Cryobiology/cons_cbs.asp

[2]See Chap. 2.

method, a procedure sheet should be kept ready to check all the steps of the protocol.

- A record of all the laboratory solutions and batches of reagents (with preparation dates and name of the operator also on the bottle) should be kept and the reference should also be reported in any procedure sheet.

- Procedure automation, and commercial reagents and solutions should improve the standardization of the methods, but correct and validated laboratory procedures are always necessary to obtain reproducible results [4].

- Special rules should be followed when handling RNA (see Sect. 12.2). These rules can be very useful, for example, in laboratories that already have well-developed facilities for DNA analysis and where an underevaluation of the importance of different precautions in RNA treatment could lead to detrimental consequences.

- Special rules should be followed when handling proteins (see paragraph "Protein handling" in chapter).

51.2.4 Controls

The controls can be method and sample related.

- The *method-related controls* are:
  - Positive control as sample processing control (positive tissue, specific cell line, or specific molecular construct); this control should be performed for each analysis battery;
  - Negative control as contamination control (for FFPE tissues, a good control is the processing of sections cut from a paraffin block without tissue included); this control should be performed for each analysis battery;
  - Quantitative analysis and sensitivity control. This can be performed once, when the method is settled or each time the method is modified (determination of the limits of the sensitivity and of the quantitative dynamic range); but whenever a quantitative (e.g., real-time PCR) response is expected, the use of control materials at more than one level, such as a "high" and "low" control could be advisable in each run [3];

- Proper molecular markers should be used, if necessary, for the interpretation of results (molecular weight ladder, etc.);
  - Position control: this control is a reference point to correctly establish the relative positions of the samples in an analysis battery (for example, in a dot-blot, this can be a specific sign put in the upper right part of the blotted membrane, etc.…);
  - Inhibition control: this must be done only once when setting the method. For example, in PCR procedures, add to the tested extraction solution a sequence different from the target, absent in the sample, and check amplification efficiency in comparison with amplification of the same sequence in optimally performed PCR solution. (Examples of methods to establish the presence of inhibitors have already been reported in Chaps. 10 and 25).

- The *sample-related controls* are (they must be performed for each sample):
  - Control of representativeness of tissues: it is mandatory to control the representativeness of tissues by histological examination and, if necessary, a mechanical or laser microdissection must be performed. The final sample should contain at least 70% of cells representing the analyzed lesion, but a higher percentage allows better-defined analysis results;
  - Target accessibility control, to establish that macromolecules are amenable, and molecular degradation control, to establish the level of degradation of nucleic acids, are suggested in most of the cases. The use of house-keeping genes to standardize the results is necessary, especially in AT. The target accessibility and the molecular degradation controls can be the same house-keeping genes analysis[3].
  - Biological variation controls: these must be performed only once to assess for each new type of tissue or test if there is any variation range related e.g. to age or gender that could be confused with pathological alterations.

_____

[3]See Chaps. 17 and 21, and Sect. 25.5.

51.2.5 Results Interpretation

An internal grading of the results should be applied, dividing the results obtained into at least 3 levels: excellent (clear result), less than excellent but reportable, not reportable. The rules and limits of this grading should be reported for any procedure result interpretation on the laboratory manual (see Sect. 51.3).

51.2.6 Reporting

- Define local procedures of reporting with an established agreement about the content of the clinical report (by telephone, by e-mail, by fax, by internal mail, etc.…).
- All the reports should be signed by at least two staff components (the technician who performed the examination, the lab leader, and, if necessary, the secretary who typed the report – the initials of the names could be sufficient for identification).
- All the raw data (gels picture, graphics, etc.) should be stored for at least 5 years (this can be done also on magnetic support).

51.3 Laboratory-Related Procedures

51.3.1 Documentation

- Collection of all the laboratory procedures and protocols should be kept in a laboratory manual as constant reference text (Standard Operating Procedures – SOP) that should be accessible to all laboratory personnel. All the procedures to prepare solutions, aliquots, and the places and temperature of storage should also be reported, along with the procedures for instrumentation set up and correct use, and for equipment cleaning and sterilization (see Chap. 52).
- Periodic review of the protocols should be performed and the old protocols should be recorded for future interpretation of old data.

51.3.2 Biological and Chemical Hazards

The precautions for chemical hazards are already reported here in the guidelines related to the specific methods.

As for the biological hazards, all the fresh human tissues should be considered as potentially infected and infectious [5], and the precautions should be the same as those taken in the pathology departments for the treatment of human tissues.

The necessary precautions should be reported for each protocol on the laboratory manual.

51.3.3 Audit

- A periodical revision of procedures, such as traceability of analysis results and patient information, security of data files, reproducibility of results, etc. should be performed.
- Periodical revision of the results and related statistics should be recorded.

51.3.4 Staff Training

- The technical staff must be specifically trained for molecular analysis in tissues. A basic requirement is that technicians must be specifically trained and technicians with other professional skills cannot be directly involved in this type of analysis if not properly prepared.
- A continuous training with internal and external seminars and courses is necessary.

51.3.5 Test Validation Procedures[4]

- Validation of home-made protocols adopting already validated technical platforms:
  - Method validation as reproducibility, accuracy, specificity, and sensitivity.

---

[4]These procedures represent suggestions that can be modified for specific tests.

- Sample validation as quality, quantity, and controls.
- Preparation of internal quality control-specific programme and of Standard Operating Procedures (SOP) that should be reported in the laboratory manual.
- Validation of already diffused commercial kits:
  - Minimum requirements: the kit must give an added value in comparison with already existing technical--clinical procedures.
  - Validation of test components should meet acceptable performance standards (intra- and interlaboratory validation).
  - Well-defined internal quality control procedures and common SOPs among laboratories are absolutely necessary to obtain comparable results [3, 4].
- For new research and industry proposals:
  - Technological platform validation: the platforms can be already validated or, in case of new platforms, a new analytical validation procedure is necessary.
  - Specificity test validation must be performed with the use of at least one different method that gives comparable results.
  - Internal quality control and SOPs should be prepared.
  - Interlaboratory evaluation is strongly recommended.
  - A clinical performance evaluation should be performed also during the possible future application of the test. This should be done to evaluate different clinical situations that laboratories may encounter in patient testing.

It is of paramount importance that the laboratory joins an organized External Quality Assessment (EQA) available to obtain a sufficient level of proficiency. In case specific EQAs are not available, a possible alternative could be the regular performance of the analysis in different laboratories, as reference laboratories, on the same sample and comparison of the results obtained.

Laboratory accreditation for AT molecular analysis should be considered specifically different from that for genetic or general type of hospital laboratories because of its peculiarities, starting from the material that needs special competence and cautions to be profitably analyzed. IQC procedures, clearly defined SOPs, and participation in EQA programmes should be prerequisites for laboratory accreditation.

References

1. http://www.Emqn.Org/emqn/bestpractice.Html. Accessed 30 Mar 2010
2. Dequeker E, Ramsden S, Grody WW, Stenzel TT, Barton DE (2001) Quality control in molecular genetic testing. Nat Rev Genet 2(9):717–723. doi:10.1038/35088588 [pii]
3. Code of Federal Regulations (2005) 42 CFR part 493. Subpart k (493.1256), standard: Control procedures. Available: http://www.Access.Gpo.Gov/nara/cfr/waisidx_05/42cfr493_05.Html. Accessed 31 Mar 2010
4. Bonin S, Hlubeck F, Benhattar J, Denkert C, Dietel M, Fernandez PL, Höfler G, Kothmaier H, Kruslin B, Mazzanti CM, Perren A, Popper H, Scarpa A, Soares P, Stanta G, Groenen P (2010) Multicentre validation study of nucleic acids extraction from FFPE tissues. Virchows Arch 457(3): 309–317, iv Epub
5. Grizzle WE, Fredenburgh J (2001) Avoiding biohazards in medical, veterinary and research laboratories. Biotech Histochem 76(4):183–206

Standards for Equipment and Reagents

52

Ermanno Nardon and Renzo Barbazza

Contents

E. Nardon (✉) and R. Barbazza
Department of Medical, Surgical and Health Sciences,
University of Trieste, Trieste, Italy

52.1 Introduction and Purpose

Herein follows a proposed collection of Good Laboratory Practice (GLP)-compliant suggestions for optimal management of basic laboratory equipment and reagents. These elements were chosen because of their (often neglected) relevance to an assay performance. Most often researchers are faced with unexpected assay results or unexplainable test failures: the way to discover the underlying cause is often cumbersome and time-wasteful. Laboratory compliance to GLP principles is indubitably demanding but allows us to get rid of many confounder elements. The following proposal is based partly on applicable parts of existing guidance elements (in particular, the Laboratory Requirements as described by 42 CFR part 493 of Clinical Laboratory Improvement Amendments) [1] and partly on sound lab practices followed in the IMPACTS institutions.

52.2 Equipment

The laboratory should be endowed with the equipment necessary to perform all the analysis within the scope of the laboratory. All equipment should be easily accessible to the personnel. Instrumentation must be kept clean, avoiding any build up of dust, dirt, and spills that may adversely affect performance. Regular inspections and preventive maintenance operations should be scheduled on weekly, monthly, half-yearly, or annual basis, depending on the type of instrument. These same operations are also aimed to verify instrumentation performance. Instrumentation suppliers usually indicate upper and lower ranges around the expected instrumental output of a given function (e.g., the temperature for a heating block or a refrigerator), typically set ±2 std. dev. of the intended mean output monitored over time. The laboratory must establish tolerance limits for instrumentation performance that are consistent with manufacturers' specifications and/or with a given procedural activity. All instrumentations whose performance is outside established limits should be returned for service to the manufacturer or to a specialized facility. It is advisable to keep documentation of all scheduled maintenance operations or inspections performed, as well as of calibrations, performance tests, and services. All these activities must be performed according to Standard Operating Procedures that must be written and accessible to the personnel.

Importantly, instruments must be protected from fluctuations and interruptions in electric power that could adversely affect test results, reagents, and specimen preservation. To this regard, laboratory's electric power supply must be backed with an emergency power system (EPS) to provide power resources whenever regular systems fail. Specific instruments, current fluctuation-sensitive (e.g., a real-time thermal cycler), should also be protected with an uninterruptible power source (UPS) [2].

Herein are shown some examples of expected maintenance and sound GLP operations for molecular pathology laboratory equipment.

52.2.1 Automatic Pipettors and Pipetting Aids

All automatic pipettors should be autoclavable. If they are used for nucleic acids processing, UV irradiation (at 254 nm) is recommended after use, to prevent carryover.

Volumetric accuracy and reproducibility of adjustable and fixed-volume automatic pipettors should be checked before they are placed in service and at regular time intervals, at least twice a year. For this latter operation and for calibration intervention it is preferable to avail of an external, specialized contracting service.

52.2.2 Refrigerators and Freezers

Set temperature tolerance limits in accordance with the needs of all reagents, supplies, and specimens stored

within them. Place low-temperatures (e.g., −80°C) freezers in facilities that are well ventilated and far from heat sources. Actual temperatures should be monitored by means of external thermometers and daily recorded. Low-temperature freezers and nitrogen freezers are usually equipped with a 7-day battery-operated chart recorder. Change the paper weekly.

Perform defrosting operations at the time intervals suggested by the manufacturer. Time interval, however, may vary according to the number of times the door is opened. The air condenser of low-temperature freezers should be cleaned every 2 months, with soft brushes or by blowing compressed air on it.

52.2.3 Incubators, Ovens, Thermoblocks, and Water Baths

Establish tolerance limits for temperatures, carbon dioxide level, and humidity, as applicable. Maintain daily record of temperatures. For cell culture incubators, a tolerance limit of ±1°C between set temperature and measured temperature is usually adopted. Regular cleaning and disinfection of internal surfaces must be scheduled.

52.2.4 Centrifuges

Operating speeds should be measured at least annually, by means of a tachometer. To this purpose, it is recommended to avail an external, specialized contracting service. Performance of centrifuge timers by comparing to a known standard should also be checked. Cleaning of rotor and rotor chamber and lubrication of rotor and swinging baskets should be performed in accordance to the workload and manufacturer specifications.

52.2.5 Autoclaves

Autoclave performance check and temperature calibration must be performed once a year, by a specialized service. Calibration ensures that correlation between set temperature and saturated steam pressure value is compliant with that expected (e.g., to a set temperature of 121°C should correspond a saturated steam pressure of 1.08 bar, correlation tables are available). Compliance should be periodically monitored by laboratory personnel along with a check of mechanical timing device. Keep records of these operations. Effective processing of material must be verified with each autoclave run, by means of heat-sensitive tape. Efficiency of sterilization can also be monitored, using an appropriate biological indicator (e.g., commercially available biological spores). Sterilization chamber should be regularly cleaned and sanitized. Autoclave feeding with demineralized water is recommended. If this is not possible, possible limestone concretions can be removed with a suitable detergent.

52.2.6 Laminar Air Flow Hoods/ Biosafety Cabinets

Both chemical fume hoods and biosafety cabinets must be daily inspected to verify that filters and intake grills are not obstructed. If the hood is equipped with an inflow velocity meter, effectiveness of protective function should also be daily verified, checking that inflow velocity is consistent with manufacturer's specifications. The filter should be replaced in accordance with their operative run life, by qualified personnel only. Laminar air flow hoods need to be serviced at least once per year, by a certified maintenance company. Servicing should include airflow control and its calibration, mechanical (front panel and sliding window) and electrical system control, and UV lamp (if present) [3]. Cabinet's work surfaces should be cleaned after each use with 70% ethanol or other disinfectant as recommended by the manufacturer. It is recommended that the UV bulbs be cleaned weekly with 70% ethanol to optimize the light output and enhance germicidal effectiveness.

52.2.7 Analytical Balances

Set up the balance on a stable, even surface. Balance must be protected from vibrations, air currents, heat sources, chemical vapors, or moisture. Always ensure that the instrument is leveled before use. Keep the balance clean, especially the weighing pan. The balance should be checked for accuracy using standard weights of the appropriate class at a scheduled frequency (based on manufacturer suggestions) or at least once a year. Availing of a certified, external service for calibration is recommended.

52.2.8 Thermal Cyclers

Although usually designed for research purposes and rarely intended for clinical use, thermal cyclers have become standard equipment in the molecular pathology lab. These instruments need particular surveillance since their correct performance can affect the results of many clinical tests. Maintenance procedures can be performed in accordance to manufacturer's guidelines. Regular controls should include verification of thermal accuracy across the entire sample block and monitoring of cycle time reproducibility and of heating and cooling rates. These controls are usually performed by the instrument's software. Whenever these instrument functions are not available, monitoring should be performed using an external calibrated device.

52.2.9 Glassware, Plasticware, Magnetic Stirrers, and Spatulas

We deal with these items because of their potential impact on home-made solution preparation and on sample quality and suitability. Glass (bottles, beckers, cylinders, trays, etc.) is sometimes preferred to plastic because it is relatively inert, transparent, and more heat resistant. All the glassware and equipment intended for nucleic acids or proteins preparation must be soil- or dirt-free and sterile and nuclease- or protease-free. Wash equipment glassware after each use, using an appropriate detergent and rinse with demineralized water. Then, all items must be autoclaved at 120°C for 20′or at 134°C for 15′. This treatment is not sufficient if material is intended for RNA handling, due to the heat-resistant nature of RNases, which can be overcome by overnight baking of the glassware at 180°C. Alternatively, glassware and plasticware can be cleaned by soaking them for at least 2 h in a 10% solution of hydrogen peroxide or 1% SDS followed by thorough washing with DEPC-treated water (see Chap. 12). Note, however, that metal and some plastics may be attacked by these reagents. Commercially available solutions for RNase removal (e.g., Ambion RNaseZap #AM9780) may provide an alternative solution.

52.3 Materials and Reagents

Materials include consumable supplies such as cleansing liquid, disposable gloves, disposable caps, pipettes, pipette tips, test tubes, PCR microtubes, disposable weighing paper, and bench paper; reagents include water, chemicals, in-house prepared solutions, enzymes, test kits, and control and calibration material.

The laboratory must have an established documented inventory control system for all supplies. This system should include the recording of lot numbers and expiration dates of all relevant reagents, control materials, and calibrators; the date of receipt in the laboratory; and the date the material is placed in service. In order to avoid any interruption of laboratory ability to perform testing, the inventory control system should be structured to support the need to place supplies order, whenever necessary [2].

Importantly, hazardous reagents should be handled according to established safety rules. Waste disposal must be in keeping with local legislation.

52.3.1 Consumable Plasticware

Materials such as single-use pipettes, pipette tips, test tubes, and PCR microtubes can affect sample quality and/or suitability; therefore, all previous recommendations concerning "glassware" apply also to plasticware. However, since sterilization operations may adversely affect plasticware properties, purchasing of sterile, DNase/RNase-free-certified plasticware is the best solution for laboratories performing molecular analysis. Specific rules about the handling of disposable plasticware should be defined every time a new package is opened, especially if intended for RNA procedures.

52.3.2 Water

Laboratory water is classified as Type 1, 2, or 3 and each type has different specifications for maximum microbial content, resistivity, maximum silicate contents, and particulate matter (Table 52.1). Type 1 is the purest grade and it is suited for molecular biology applications. Type 3 is acceptable for glassware rinsing, heating baths and filling autoclaves, or to feed

Table 52.1 Laboratory water specifications

	Type 1	Type 2	Type 3
Max. spec. conductance (μmhos/cm)	0.056	1	4
Min. spec. resistance (MΩ • cm @ 25°C)	18.2	1	0.25
Max. silicate (μg/l)	3	3	500
Max. total organic carbon (μg/l)	10	50	200
Max. sodium (μg/l)	1	5	10
Max. chlorides (μg/l)	1	5	10
Max. bacterial growth (cfu/ml)	1	100	1,000
Max. endotoxin (EU/ml)	0.03	n.a.	n.a.
pH	n.m.	n.m.	n.m.

n.m. not measurable, *n.a.* no data available

Type 1 lab water systems [4]. Notice that the largest part of Type 1 water-producing devices allow for monitoring only water resistivity or Total Organic Carbon (TOC). In order to ensure that water is sterile and free of RNase contamination, it should be treated with diethylpyrocarbonate (DEPC) and then autoclaved.

52.3.3 Chemicals

Chemicals storage areas must be compliant with local institutional rules for safety. Storage conditions (temperature, humidity, exposure to sunlight) must be in keeping with requirements specified on the label of chemical product. Labels also provide information about chemical product quality class, which is based on the purity of the basic substance of the product. In the "pure chemicals" class, the content of the basic substance is at least 98% and the content of individual impurities is of an order of magnitude of $10^{-2}\%$; in the "chemicals for analysis," the content of the basic substance usually ranges from 99.0% to 99.8% and the content of individual impurities is only of an order of magnitude of $10^{-3}\%$. In addition, labels can indicate if the product is suited for a given use (e.g., "for synthesis," "for analysis," "for molecular biology") and/or meets the standard requirements of American Chemical Society (ACS) or of International Standardization Organization (ISO). An ACS reagent is usually suited for nucleic acids preparations and less expensive than the same chemical labeled "for molecular biology."

52.3.4 In-House Prepared Solutions

Buffers and solutions intended for nucleic acids and proteins extractions should be prepared according to well-defined SOPs, using the highest grade of reagents available and type 1 water. All solutions should be sterilized by autoclaving or filtration through a 0.22 μm filter if solution components do not withstand autoclaving (such as Magnesium, Calcium, and Ammonium salts, Dithiothreitol, ß-Mercaptoethanol). In addition, solutions can be made RNase-free by DEPC-treatment (see Chap. 12).

All buffer and solutions must be labeled to indicate the following: identity; titer, strength, or concentration as applicable; storage conditions; preparation date or opening date; unopened and opened expiration date if pertinent to the performance of the reagent; the identity of the preparer [2].

52.3.5 Enzymes and Test Kits

These items are expected to come from accredited suppliers who are ready to provide certificates of analysis and, where appropriate, evidence of compliance with the national pharmacopeia and other compendia or formularies [5]. Enzymes and kits must be stored according to manufacturer's instructions. According to laboratory needs, aliquoting could be advisable for enzymes (e.g., Taq Polymerases). Reagents must not be used when they have exceeded the manufacturer's

stated expiration date, have deteriorated, or are of substandard quality. Components should not be interchanged between different kits, or between different lots of the same kit, unless otherwise specified by the manufacturer or verified by the laboratory.

52.3.6 Controls and Calibrating Material

For the intended use of controls and calibrating material refer to Section 51.2.4 and to section "Equipment" of this chapter. For purchasing and storing, the same abovementioned standards for enzymes and kits generally apply. In addition, an expiration date must be assigned to quality control (QC) materials that do not have a manufacturer-provided expiration date. The manufacturer should be consulted in case this situation arises.

References

1. Code of federal regulations (2005) 42 cfr part 493. Laboratory requirements. Available: Http://www.Access.Gpo.Gov/nara/cfr/waisidx_05/42cfr493_05.Html. Accessed 31 Mar 2010
2. Code of federal regulations (2005) 42 cfr part 493. Subpart k (493.1252), standard: Test systems, equipment, instruments, reagents, materials, and supplies. Available: Http://www.Access.Gpo.Gov/nara/cfr/waisidx_05/42cfr493_05.Html. Accessed 31 Mar 2010
3. BS EN 12469:2000 (standard). Biotechnology. Performance criteria for microbiological safety cabinets. 15/07/00. Pp.1–48
4. Clinical and Laboratory Standards Institute. Preparation and testing of reagent water in the clinical laboratory-third edition. NCCLS document c3-a3. Clinical and Laboratory Standards Institute, Wayne, 1998
5. Organisation for economic co-operation and development. OECD series on principles of GLP and compliance monitoring number 5 (revised). Consensus document. Compliance of laboratory suppliers with glp principles. Env/jm/mono(99)21. Paris, 2000

Appendices

Appendix A: Agarose Gel Electrophoresis

Agarose gel electrophoresis is used to separate and identify DNA fragments according to their size. Nucleic acids are negatively charged at pH 8 and are moved through an agarose matrix by an electric field (electrophoresis). Smaller fragments migrate faster than larger ones and the distance migrated on the gel is inversely proportional to the logarithm of the molecular weight [1].

Gel electrophoresis analysis might be used for the following purposes:

- To assess the effectiveness of PCR amplification
- To quantify or to isolate a particular PCR product
- To estimate the size of DNA molecules following restriction enzyme digestion
- To assess the integrity of a nucleic acid

The most common dye used to visualize DNA or RNA bands in agarose gel electrophoresis is ethidium bromide (EtBr).[1] This dye binds strongly to the double-stranded DNA by intercalating between the bases. In this state, it absorbs the UV light and transmits the energy as visible fluorescence, thus allowing nucleic acids to be viewed.

Most agarose gels are made between 0.7% and 2.5%. A 0.7% gel is preferable for separation of large DNA fragments, while a 2% gel will allow a good separation of small fragments (Table A.1) [2].

Table A.1 Agarose concentrations and size range of DNA to separate [2]

Agarose concentrations (%)	Size range of DNA fragments (kb)
0.9	7–0.5
1.2	6–0.4
1.5	4–0.2
2.0	3–0.1

Since most PCR amplificates obtainable from FFPE material do not exceed this latter size range, here the description of 2% agarose gel is reported.

Reagents

- *Agarose* (Sigma-Aldrich)
- *1× Tris-borate-EDTA (TBE) buffer, pH 8.3* (made by diluting a 5X concentrated TBE stock solution. Composition of TBE 5X per liter: 54 g Tris base, 27.5 g boric acid, 20 ml 0.5 M EDTA pH 8.0).
- *Ethidium bromide stock solution 10 mg/ml.* Stock solution of EtBr should be stored at 4°C in a dark bottle.
- *100 bp DNA Step Ladder* (Promega)
- *6× Loading Dye Blue/Orange buffer* (Promega)

Equipment

- Clean[2] adjustable pipettes, range: 2–20 μl
- Horizontal gel electrophoresis apparatus with spacers and combs

---

[1]EtBr is a potentially carcinogenic compound. Always wear gloves. Used EtBr solutions must be collected in containers for chemical waste and discharged according to the local hazardous chemical disposal procedures.

[2]Clean the pipette with DNase Away™ to avoid DNase and DNA contamination. Alternatively, it is possible to clean the pipette with alcohol or another disinfectant and leave them under the UV lamp for almost 10 min.

G. Stanta (ed.), *Guidelines for Molecular Analysis in Archive Tissues*,
DOI: 10.1007/978-3-642-17890-0, © Springer-Verlag Berlin Heidelberg 2011

- Power supply
- Microwave oven
- UV-transparent plastic gel formers
- UV transilluminator.[3]

Method

1. Make 2% agarose gel by adding 2.0 g of agarose to 100 ml of 1× TBE buffer. Agarose is solubilized by heating the solution in a microwave oven. Avoid boiling.
2. Add 10 μl of EtBr to the gel and allow the agarose to cool to 60°C before pouring. Mix thoroughly, avoid making bubbles.
3. Pour the agarose solution into the gel former[4] mold.
4. Put the well-forming comb near the upper edge of the gel. Clip the comb to bridge above the gel and ensure that the fingers of the comb are slightly above the plate without touching it.
5. Let the gel harden until it becomes opaque (approximately 30 minutes), then gently remove the comb from the gel and place it horizontally in an electrophoresis tank.
6. Pour the right volume of 1x TBE buffer into the electrophoresis tank: the gel should be totally submerged in the buffer, but not covered by more than 1 cm. Remove bubbles from the wells with a pipette.
7. Add 2 μl of loading buffer to each 10 μl sample and for the size marker sample mix 2 μl of the DNA ladder, 8 μl of H_2O and 2 μl of loading buffer. Carefully add sample mixtures to individual wells.
8. Turn the power on at 80–100 mV. Let it run for about 40 minutes. The DNA will move from minus to plus pole.
9. Stop the run when the first dye (Bromophenol Blue) front has covered approximately 50% of the gel length.
10. Turn off the power. Transfer the gel on the transilluminator surface. Check the ethidium bromide stained bands in the gel using a UV protection shield.

---

[3]A UV source between 310 and 330 nm is recommended: lower wavelengths could damage nucleic acids.

[4]Commercially available gel formers with two walls positioned to hold well-forming combs are now commonly used.

Appendix B: Polyacrylamide Gel Electrophoresis

Polyacrylamide gel electrophoresis is used to analyze small DNA fragments (e.g., 50–2,000 bp) and to determine their size by comparing the distance migrated during the run to the distance migrated by fragments of known length. Polyacrylamide gel is formed by the polymerization of the monomer molecule-acrylamide cross-linked by N,N'-methylene bisacrylamide, in the presence of free radicals generated by ammonium persulfate (APS) and a catalyst as TEMED (N,N,N',N'-tetramethylethylenediamine). Polyacrylamide solutions are a mixture of acrylamide and N,N'-methylene bisacrylamide in water, and acrylamide/bisacrylamide ratio can also affect migration of DNA fragments.

The polyacrylamide solution is poured between two glass plates that are separated by spacers. In this disposition, most of the acrylamide solution is protected from exposure to the air and thus unaffected by inhibition of polymerization by oxygen. Polyacrylamide gels are run in vertical position [3].

The polyacrylamide percentage should be modified according to the size of the fragments to be separated (see Table A.2) [3]. The length of the gel may be varied according to the desired resolving power, e.g., a gel length of at least 30 cm is required to resolve DNA fragments differing by 1 basepair. This is better achieved using a 19/1 (w/w) acrylamide/bisacrylamide ratio.

According to the length of PCR amplicates usually obtainable from FFPE tissues (60–400 bp), the preparation of an 8% polyacrylamide gel is described as follows.

Table A.2 Polyacrylamide concentrations for effective separation of DNA fragments [3]

Polyacrylamide concentrations (%)	Size range of DNA fragments (bp)
3.5	100–1,000
5.0	80–500
8.0	60–400
12.0	40–200
20.0	10–100

Reagents

- *40% acrylamide,[5] 29/1 (w/w) acrylamide/bisacryl-amide* stock solution.
- *1× Tris-borate-EDTA (TBE) buffer, pH 8.3* (made by diluting a 5X concentrated TBE stock solution. Composition of TBE 5X per liter: 54 g Tris base, 27.5 g boric acid, 20 ml 0.5 M EDTA pH 8.0).
- *double distilled water*
- *10% Ammonium Persulfate (APS) solution.* Store in aliquots at −20°C.
- *Temed (N,N,N′,N′-tetramethylethylenediamine).* Store at 4°C.
- *Ethidium bromide stock solution 10 mg/ml.[6]* Stock solution of EtBr should be stored at 4°C in a dark bottle.
- *100 bp DNA Step Ladder* (Promega)
- *6× Loading Dye Blue/Orange buffer* (Promega)

Equipment

- Electrophoresis power supply
- Gel plates and spacers (0.8 mm thick) for vertical gel
- Vertical Gel apparatus
- Well formers (0.8 mm thick).
- UV transilluminator

Method

1. First assemble plates and spacers for casting the gel. Spacers may be sealed to a watertight fit by the use of a melted 2% agarose solution. Spacers are clamped to maintain their position. Calculate the volume of chamber between glass plates and spacers, to determine the amount of polyacrylamide solution to prepare.

2. For a final volume of 15 ml, mix in the following order: 9 ml of dd H_2O, 3 ml of 5× TBE, 3 ml of 40% acrylamide solution. Keep this mixture in ice.

3. Prepare the glass plates to pour the gel.

4. Immediately before adding the gel mix into the chamber, add 150 µl of 10% APS and 15 µl of TEMED. Mix gently and dispense the mixture into the chamber. Fill the space entirely. Avoid bleb formation.

5. Insert the comb into the top of the chamber and clamp tightly in place.

6. Let the gel polymerize (15 min to 1 h).

7. After the gel has polymerized, remove the well-forming comb, and place the gel in the electrophoresis apparatus.

8. Fill the upper and lower tanks with 1× TBE buffer, submerging wells.

9. Determine the volume of the sample to be loaded. If the desired volume is 10 µl, add 2 µl of loading buffer to each sample. For the marker, add 2 µl of DNA Step Ladder to 8 µl H_2O mq + 2 µl of loading buffer.

10. Load samples into the gel well, with a Hamilton syringe or a flat-tip micropipette.

11. Run the gel at a constant voltage of 200 V until the Bromophenol Blue dye has migrated to the bottom of the gel.[7]

12. Turn off the power supply and remove the gel from the apparatus. Separate the plates. The gel should stick to one of them.

13. The DNA can be detected by EtBr staining: add 5 µl of EtBr stock solution 10 mg/ml in 50 ml of 1× TBE buffer to a final concentration of EtBr 1 µg/ml. Remove the gel from the glass plate and soak it into the EtBr solution. Allow the ethidium bromide solution to diffuse into the gel and stain DNA for 5 min.

14. Gently rinse the excess stain from the gel surface with H_2O, then place the gel over the UV transilluminator to visualize the DNA bands.

---

[5]Always wear gloves and do not breathe or touch unpolymerized polyacrylamide. This chemical is a neurotoxin and carcinogen. The use of an already prepared commercial stock of 40% acrylamide is highly recommended to avoid manipulation of polyacrylamide powder.

[6]EtBr is a potentially carcinogenic compound. Always wear gloves. The used EtBr solutions must be collected in containers for chemical waste and discharged according to the local hazardous chemical disposal procedures.

---

[7]This dye is expected to comigrate with a DNA fragment of approximately 45 bp in an 8% polyacrylamide gel.

References

1. Davis LG, Dibner MD, Battey JF (eds) (1986) Agarose gel electrophoresis. Basic methods in molecular biology. Elsevier, New York, pp 58–61
2. Maniatis T, Fritsch EF, Sambrook J (eds) (1982) Agarose gel electrophoresis. Molecular cloning (a laboratory manual). Cold Spring Harbor, New York, pp 150–162
3. Maniatis T, Fritsch EF, Sambrook J (eds) (1982) Polyacrylamide gel electrophoresis. Molecular cloning (a laboratory manual). Cold Spring Harbor, New York, pp 173–177
4. Davis LG, Dibner MD, Battey JF (eds) (1986) Polyacrylamide gel electrophoresis of DNA restriction fragments. Basic methods in molecular biology. Elsevier, New York, pp 115–118

Index

Printing and Binding: Stürtz GmbH, Würzburg